S0-BKU-080

CHABOT COLLEGE-HAYWARD
2 555 000 025327 -

# Light and Sound for Engineers

R. C. Stanley Lecturer in Applied Physics Brighton College of Technology

HART PUBLISHING COMPANY, INC.
NEW YORK CITY

FIRST AMERICAN PUBLICATION, 1968
PRINTED IN THE UNITED STATES OF AMERICA

# Preface to the American Edition

There has been a steady growth in recent years in the number of engineering students requiring light and sound to be presented in a manner satisfying their especial requirements. These two topics are commonly covered in a physics department, leaving those topics of more immediate concern to engineers to be covered by the respective engineering departments themselves.

Available textbooks are mainly directed towards school physics courses, and so are of too low a standard for engineering degree students; or towards a physics degree, and so are too specialized. This volume presents light and sound in a way most useful to engineers.

The treatment assumes only a minimal knowledge of physics, and therefore theoretical derivations are given in full. All mathematical steps are shown, with explanation where necessary. Consequently the mathematics, including calculus, should present no difficulties to students at this level. Wherever possible the practical applications discussed are those of particular interest to engineers. To suit different syllabus requirements theoretical sections can be omitted without detriment to the understanding of the discussions on practical aspects.

The units used throughout are mks, conforming to the Système International d'Unités, but alternatives are given where these are in common use.

# Contents

## PART I—LIGHT

### *Chapter 1 Reflection and refraction at plane surfaces*



### *Chapter 2 Reflection and refraction at spherical surfaces*



### *Chapter 3 Thick lenses*



### *Chapter 4 Aberrations of mirrors and lenses*

## *Chapter 9 Diffraction*



## *Chapter 10 Polarization*



# PART II—SOUND

## *Chapter 11 Waves and vibrations*



## *Chapter 12 Velocity of sound*



## *Chapter 13 Transverse vibrations*

# 1 Reflection and refraction at plane surfaces

## 1-1 Introduction

When light from a large distant source, such as the sun, passes through a hole in a screen, it forms a shaft of light with a well-defined boundary. This is a *beam*; the term is normally restricted to mean a shaft of light whose sides are parallel. If the cross-section of this beam is reduced by making the hole in the screen smaller, it will be seen that the light is travelling in a straight line, and so it should theoretically be possible to obtain a beam whose cross-section is infinitesimally small just by reducing the size of the hole sufficiently. In this case the beam of light would become a *ray*; this has neither thickness nor breadth and is therefore the equivalent of a mathematical line. But a 'ray' is a theoretical rather than a practical concept, since if the hole diameter is made smaller than a few tenths of a millimetre, the sides of the emergent beam will not remain parallel but will diverge. This is due to diffraction, caused by the wave nature of the light, which is discussed in Chapter 9.

Our study of light may thus be conveniently split into two parts: *geometric optics*, which concerns the macroscopic aspects of light and is studied by beams and rays of light, which are assumed always to travel in straight lines; and *physical optics*, which concerns the microscopic aspects arising from the wave nature of light in systems in which distances occur comparable to the wavelength of light.

The study of geometric optics is based on three laws, the first of which states that light travels in straight lines, and is simply demonstrated using a series of screens with pinholes in them. Only when all the pinholes are in the same straight line can light pass through all the screens. The other two laws deal with reflection and refraction.

## 1-2 Laws of reflection and refraction

These two laws, originally based only on experimental observations, may be derived by a consideration of principles to be discussed in the following sections.

When a ray of light falls on the interface between two transparent media in which the velocity of light is different, the incoming ray generally divides into two: one ray does not enter the second medium but is reflected, whilst the other is bent or refracted into the second medium. The angle at which the incoming ray meets the interface is designated, in optics, by the angle that the ray makes with the normal to the interface. The incoming ray and the normal together define a plane, the *plane of incidence*. Similarly, the angles of reflection and refraction are the angles between the rays and the normal. For the reflected ray, the law may be stated as:

*The reflected ray lies in the plane of incidence, with the angle of reflection equal to the angle of incidence.*

Similarly, for a refracted ray the law, usually called Snell's law, is:

*The refracted ray lies in the plane of incidence, with the ratio of the sine of the angle of incidence to the sine of the angle of refraction a constant for any two particular media and for light of a particular wavelength.*

Thus in Fig. 1-1

$$\theta = \theta''$$

and

$$\frac{\sin \theta}{\sin \theta'} = \text{const.}$$

If in Fig. 1-1 the upper medium is vacuum, then this constant is the *refractive index*, $n$, of the lower medium. Generally, for refraction at the

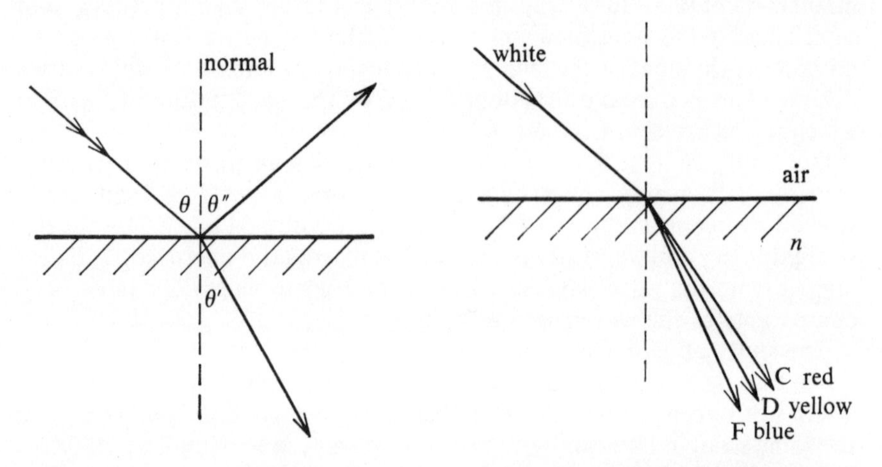


Fig. 1-1 Refraction and reflection

Fig. 1-2 Refraction of white light

boundary between two transparent media whose refractive indices are $n$ and $n'$, Snell's law may be written as

$$n \sin \theta = n' \sin \theta' \tag{1-1}$$

where $\theta'$ is the angle the ray makes with the normal in the medium of refractive index $n'$.

In practice, since the refractive index of air is about 1·000293 for yellow light, little error occurs in measuring refractive indices relative to air rather than to vacuum and, of course, they are considerably easier to measure experimentally.

Also, when rays pass from optically less dense to optically denser media they are bent towards the normal, and conversely. When light enters an optically denser medium its velocity is reduced. Thus the refractive index of a medium may also be defined as the ratio of the velocity of light in vacuum (or in practice, air) to its velocity in the medium.

In some circumstances these laws of reflection and refraction at a boundary are not true. For example, the law of reflection only applies to smooth boundaries—if the surface is rough, the incident light may be scattered in all directions. Snell's law for refraction applies only to isotropic materials; for anisotropic crystals, which have a different atomic configuration in different directions, the light velocity varies with direction and Snell's law does not hold.

Taking the velocity of light in vacuum (or air) as the maximum, the refractive indices of other media vary from unity upwards, although most substances have a refractive index between 1 and 2. Since the refractive index of a medium depends on the colour (wavelength) of the light, we must state the wavelength when quoting a refractive index. Normally we use only certain wavelengths—those corresponding to certain dark lines

Table 1-1

| *Designation* | *Wavelength* Å | *Origin* | *Colour* |
|---|---|---|---|
| A′ | 7,682 | K | red |
| C | 6,562·8 | H | red |
| $D_1$ | 5,895·9 | Na | yellow |
| $D_2$ | 5,889·9 | Na | yellow |
| d | 5,876 | He | yellow |
| e | 5,460·7 | Hg | green |
| F | 4,861·3 | H | blue |
| g | 4,358·3 | Hg | blue |
| G′ | 4,340 | H | blue |
| h | 4,047 | Hg | violet |

occurring in the sun's spectrum—for this purpose. These lines are termed Fraunhofer lines, and have been designated by the letters A, B, . . ., starting at the extreme red end of the spectrum. Thus, C refers to red light, F to blue, and D to yellow, which is about the middle of the visible spectrum (Fig. 1-2). Table 1-1 gives the wavelengths in angstroms (1 Å = $10^{-10}$ m) of the more important spectral lines. Thus, $n_d$ is the refractive index of a medium as measured using yellow light of wavelength 5,876 Å. The angular separation of the blue and red light, shown in Fig. 1-2, is a measure of the

Table 1-2 Some Chance-Pilkington optical glasses

| *Glass type* | $n_d$ | $n_F$ | $n_C$ | $\nu$ |
|---|---|---|---|---|
| Borosilicate crown | 1·50970 | 1·51518 | 1·50727 | 64·4 |
| Borosilicate crown | 1·51757 | 1·52316 | 1·51509 | 64·1 |
| Hard crown | 1·52400 | 1·53019 | 1·52129 | 58·9 |
| Light barium crown | 1·54065 | 1·54696 | 1·53788 | 59·5 |
| Medium barium crown | 1·56938 | 1·57650 | 1·56629 | 55·8 |
| Dense barium crown | 1·60557 | 1·61258 | 1·60249 | 60·0 |
| Dense barium crown | 1·60982 | 1·61781 | 1·60637 | 53·3 |
| Extra light flint | 1·54769 | 1·55612 | 1·54411 | 45·6 |
| Light flint | 1·57860 | 1·58851 | 1·57444 | 41·1 |
| Barium flint | 1·60483 | 1·61453 | 1·60073 | 43·8 |
| Dense flint | 1·62576 | 1·63813 | 1·62062 | 35·7 |
| Extra dense flint | 1·70035 | 1·71677 | 1·69364 | 30·3 |
| Double extra dense flint | 1·80120 | 1·82359 | 1·79217 | 25·5 |

*dispersion*, while the angle through which the yellow ray is bent gives the average *deviation* of the spectrum. The dispersion, $\theta_F - \theta_C$, depends on the difference in refractive indices for these colours, $n_F - n_C$, and the deviation for the d-ray depends on $n_d - 1$. These values may be combined to give the *dispersive power*, defined as

$$\frac{1}{\nu} = \frac{n_F - n_C}{n_d - 1} \tag{1-2}$$

This is normally a small number and, for convenience, its reciprocal, $\nu$, is usually quoted. This varies between 30 and 60 for most common optical glasses. Table 1-2 gives the refractive indices and $\nu$-values (or $V$-values) for some optical glasses.

## 1-3 Critical angle

Snell's law also shows that there is a maximum value of the angle of refraction for a ray travelling from an optically denser to an optically less dense medium (Fig. 1-3). By Snell's law, $n' \sin \theta' = n \sin \theta$, but $\sin \theta$ can have a

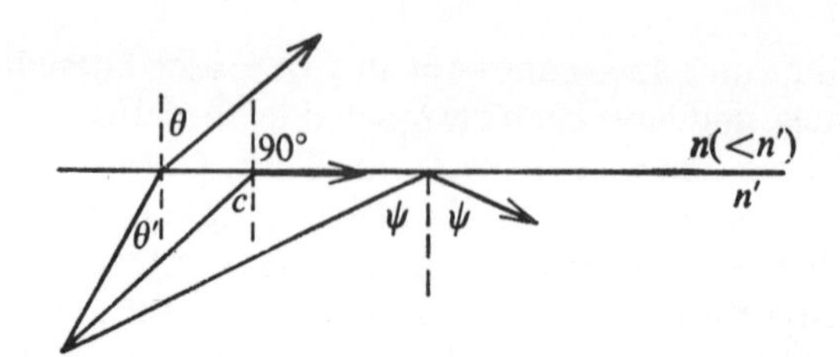


Fig. 1-3 Critical angle and total internal reflection

maximum value, unity, when $\theta = 90°$ and the refracted ray travels along the boundary. Angle $\theta'$ is then termed the *critical angle*, and $\theta' = c$; thus

$$\sin c = n/n' \tag{1-3}$$

If $\theta'$ is greater than the critical angle, there is no refracted ray and *total internal reflection* occurs. The incident and reflected rays then obey the simple law of reflection, with equal angles of incidence and reflection.

## REFLECTION AT PLANE SURFACES

### 1-4 Image formation by mirrors

It follows from the law of reflection and by consideration of the symmetry of Fig. 1-4 that the image formed by reflection at a plane mirror is:

1. Virtual, that is, the rays only *appear* to pass through the image but do not actually do so, as they would if the image were real.
2. Situated as far behind the mirror as the object is in front of it.
3. Laterally reversed.
4. The same size as the object.

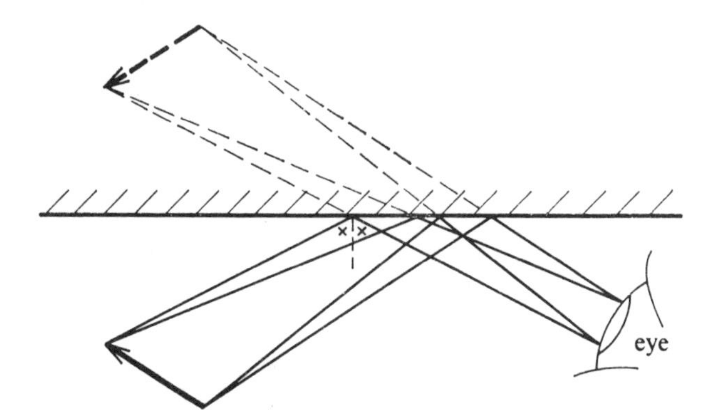


Fig. 1-4 Formation of an image by reflection

### 1-5 Parallel mirrors

When an object is situated between two parallel mirrors, an image is formed by each mirror. Each image then acts as an object and in turn is imaged by the opposite mirror. This process continuously repeats itself, and in theory an infinite number of images are formed. In practice, however, the number is limited by the loss of intensity at each reflection, causing successive images to become progressively fainter.

### 1-6 Inclined mirrors

When an object is placed between two mirrors that are inclined to each other, then the number of images formed by multiple reflection depends

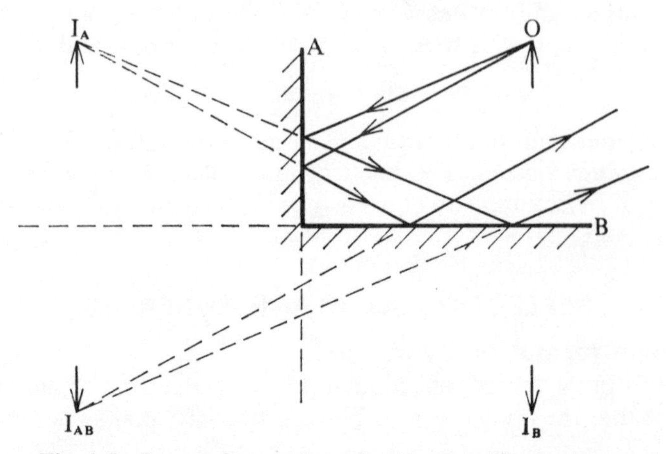


Fig. 1-5 Images formed by reflection at inclined mirrors

on the angle $\alpha$ between the two mirrors. For example, Fig. 1-5 shows the position of images of an object placed between two mirrors A and B which are at right angles to each other. The two mirrors produce two images $I_A$ and $I_B$ of the object O, $I_A$ by reflection in mirror A, and $I_B$ by reflection in B. A third image, $I_{AB}$, which may be imagined as the image of $I_A$ in the mirror B (or the image of $I_B$ in mirror A), is also formed. Since the eye is only concerned with the direction in which the light is travelling when it reaches the eye, and is not aware that the light ray has been bent, the images appear to be in the positions indicated. Similarly, Fig. 1-6 shows the positions of images formed when the mirrors are inclined at 60° to each other.

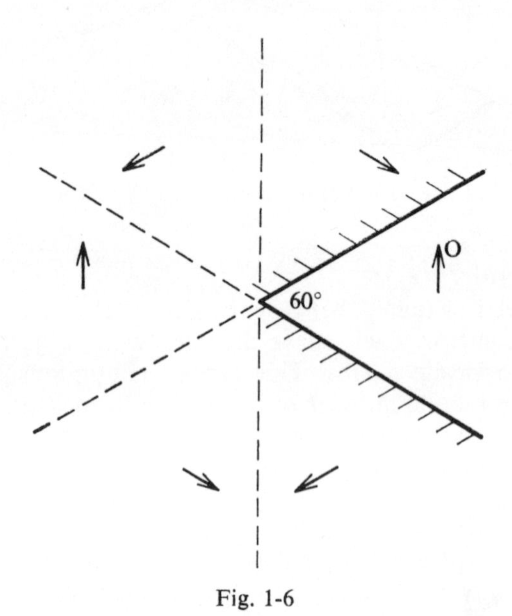


Fig. 1-6

We can readily deduce that, if the mirrors are inclined at $\alpha°$ to each other, $360/\alpha$ sectors are formed, of which one is the real sector between the mirrors and contains the object, and each of the others contains an image. That is, $(360/\alpha - 1)$ images are formed—for example, for mirrors inclined at 10° to each other, 35 images are formed. It may not, however, be possible to see all the images simultaneously from one viewing direction, and the direction may, therefore, have to be changed to bring different groups of images into view.

## 1-7 Rotating mirrors

If a ray is reflected by a mirror that is being rotated, the reflected ray is turned through twice the angle turned through by the mirror. In Fig. 1-7, a ray AO is reflected along the path OB, making angle $\theta$ with the normal ON. If the mirror is now rotated through an angle $\alpha$, the incident ray AO then makes an angle $(\theta + \alpha)$ with the normal ON′ and is reflected along OC. Thus

$$\angle AOC = 2(\theta + \alpha)$$

also

$$\begin{aligned} \angle BOC &= \angle AOC - \angle AOB \\ &= 2(\theta + \alpha) - 2\theta \\ &= 2\alpha \end{aligned}$$

The reflected ray is thus turned through an angle $2\alpha$ as the mirror is turned through an angle $\alpha$.

This principle is used as an optical lever to amplify small movements, as in the mirror galvanometer and also in the sextant.

Figure 1-8 shows the basic arrangement of a sextant. A fixed mirror, A, the *horizon glass*, is silvered on only half of its surface. Thus an observer looking through the telescope can see the horizon through the clear area and also receive rays reflected from a movable mirror B (the *index glass*) via the silvered part of A. When the mirrors are parallel to each other, rays from a distant object, received through the clear area of A and by reflection at B and A, coincide at the telescope. The position of the mirror B as indicated on the angle scale CDE should then be at zero. To find the

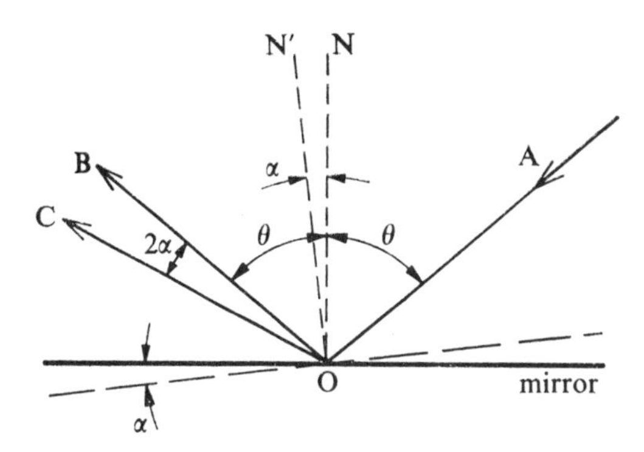


Fig. 1-7 Reflection at a rotating mirror

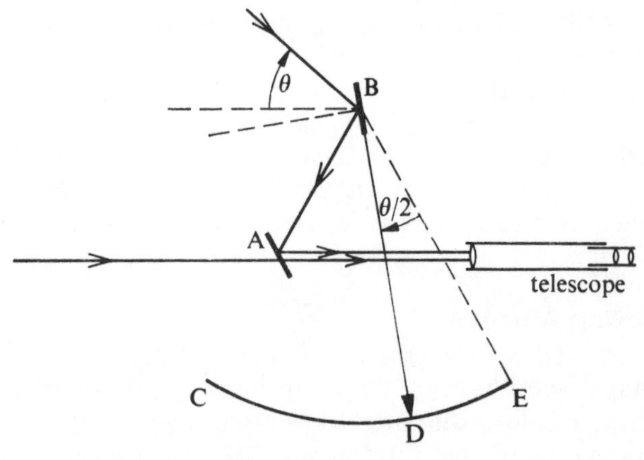


Fig. 1-8 Sextant

elevation of a star or the sun, the instrument is held so that the horizon can be viewed directly by the telescope through the clear part of mirror A, and mirror B is rotated so that the star or lower edge of the sun can be seen at the same time in the telescope and appears to coincide with the horizon. The altitude is then read direct from the position of the mirror B as indicated on the angle scale CDE. Each degree of rotation of mirror B is usually numbered as two degrees so that the scale CDE—usually 60°, hence the name 'sextant'—is numbered from 0° to 120°, thus reading altitude directly. For sighting on the sun, filters are necessary to reduce the light intensity.

## 1-8 Deviation by successive reflections

The deviation of a ray on reflection can be deduced by reference to Fig. 1-9. Ray AB is reflected along the path BC such that $\angle ABC = 2\theta$; that is, the ray travelling in the original direction AB is turned through an angle $(\pi - 2\theta)$. Let this reflected ray be in turn reflected by a second mirror, inclined to the first at an angle $\alpha$. If the angle of incidence of the ray BC with the second mirror is $\theta'$, then ray BC is turned through an

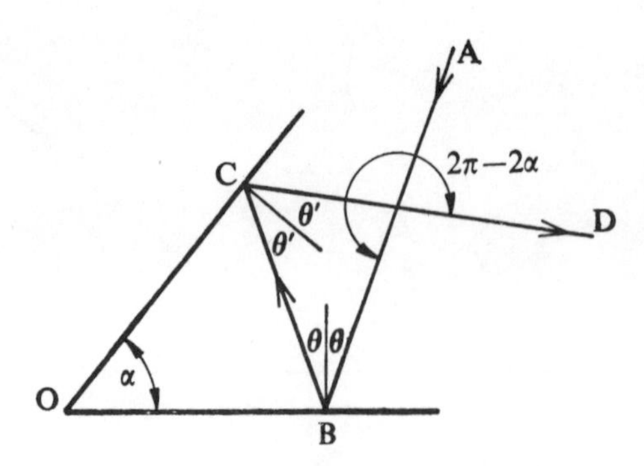


Fig. 1-9

angle $(\pi - 2\theta')$. Thus, the original ray AB has suffered two deviations, totalling

$$(\pi - 2\theta) + (\pi - 2\theta') = 2\pi - 2(\theta + \theta')$$

before emerging in the direction CD. But, in Fig. 1-9,

$$\angle BOC = 180^\circ - \angle OBC - \angle OCB$$

i.e.,

$$\alpha = 180^\circ - (90^\circ - \theta) - (90^\circ - \theta')$$
$$= \theta + \theta'$$

Thus, for two reflections,

$$\text{deviation} = 2\pi - 2\alpha = \text{const.} \tag{1-4}$$

That is, the deviation depends only on the angle between the mirrors and not on the angle made by the incident ray.

## 1-9 Reflecting prisms

Prisms, as well as simple mirrors, are used for reflection. This normally involves the property of total internal reflection. Here no energy is lost at the total internal reflection surface, but some small losses will occur due to absorption in the material of the prism and by reflection at the surface

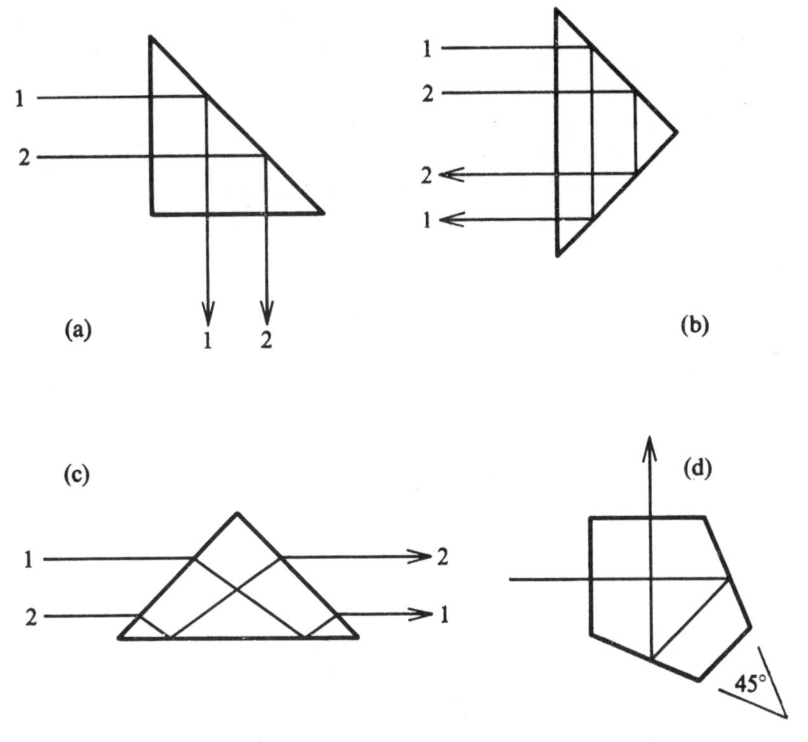


Fig. 1-10 Reflecting prisms: (a) total reflecting; (b) Porro; (c) Dove or inverting; (d) constant deviation

where the light enters and leaves the prism. Besides high reflectivity, prisms have the advantage of permanence: there is no silvering to deteriorate and the angle between the reflecting faces cannot change.

Most reflecting prisms have 45° reflecting angles and air as the surrounding medium; the refractive index, $n$, of the prism must be such that

$$n \geqslant 1/\sin 45°$$

i.e.,

$$n \geqslant 1{\cdot}414$$

Thus glass is a satisfactory material for the prisms.

Figure 1-10 shows some common prisms. The total reflecting prism (a) and the Porro prism (b) merely change the direction of the rays and at the same time laterally invert them, while the Dove prism (c) inverts them without changing their direction. The constant deviation prism (d) has the advantage that small rotations of the prism do not affect the deviation produced. As has been shown, the deviation is equal to $(2\pi - 2\alpha)$ where $\alpha$ is the angle between the reflecting faces. In this case the deviation is 270°, which gives 45° as the angle between the reflecting faces. If the prism is rotated by a small amount, the angles of incidence and emergence are no longer 90° and refraction occurs, but since the refractions at incidence and emergence are equal and in the opposite sense, they cancel out so that the total deviation remains unchanged.

## REFRACTION AT PLANE SURFACES

### 1-10 Refraction by a plate

When a ray is refracted at a surface, it obeys Snell's law. If the ray is passing through a parallel-sided slab of transparent material, say glass, it is deviated at the second surface by an amount equal, but in the opposite sense, to that produced by the refraction at the first surface. Thus, as in Fig. 1-11, the emergent ray is parallel to the incident ray, but is laterally

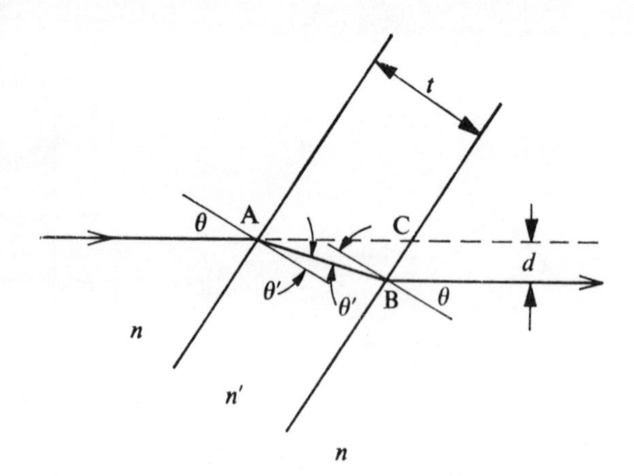


Fig. 1-11 Refraction by a plate

shifted. If $t$ is the thickness of the plate, the lateral shift, $d$, of the ray is given by

$$\begin{aligned} d &= \text{AB} \sin \angle\text{BAC} \\ &= \frac{t}{\cos \theta'} \sin (\theta - \theta') \\ &= t\left(\sin \theta - \sin \theta' \frac{\cos \theta}{\cos \theta'}\right) \\ &= t \sin \theta\left(1 - \frac{n \cos \theta}{n' \cos \theta'}\right) \end{aligned} \tag{1-5}$$

## 1-11 Refraction of paraxial rays

All rays diverging from a point object appeared to diverge from a single point image after reflection at a plane surface. But this is not the case for refracted rays in which there is a large angle of divergence. For example, in Fig. 1-12 rays are diverging from a point A; then, considering the lower ray drawn,

$$h = u \tan \theta = v \tan \theta'$$

i.e.,

$$v = u \frac{\tan \theta}{\tan \theta'} = u \frac{\sin \theta \cos \theta'}{\cos \theta \sin \theta'}$$

But

$$\sin \theta / \sin \theta' = n'/n = \text{const.}$$

$$\therefore\ v = u \frac{n'}{n} \frac{\cos \theta'}{\cos \theta}$$

But the ratio of the cosines is not constant, and hence rays of different angles of incidence $\theta$ appear to diverge from points whose distance $v$ from the screen varies with the angle. That is, the rays no longer appear to diverge from one point.

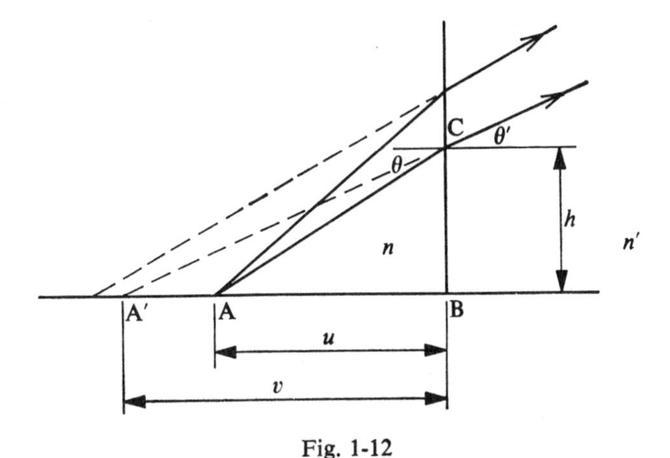


Fig. 1-12

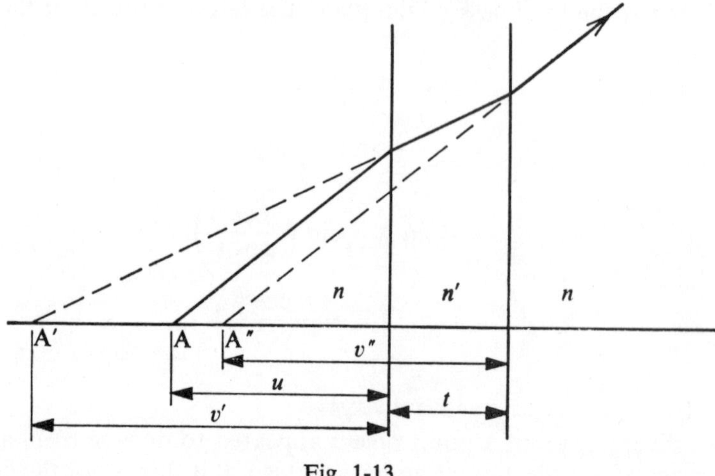


Fig. 1-13

If the rays involved are *paraxial*—that is, make angles of incidence and refraction small enough for the sines to be equal to the angles, and the cosines to equal unity—then

$$v = u \frac{n'}{n} \tag{1-6}$$

In this case the rays do appear to diverge from a virtual image that is a point.

If we have a parallel-sided slab of material instead of a single surface, we may calculate the displacement of the image for paraxial rays. In Fig. 1-13, paraxial rays are diverging from a point A. After refraction at the first surface they then appear to diverge from a point A′, such that

$$v' = u \frac{n'}{n}$$

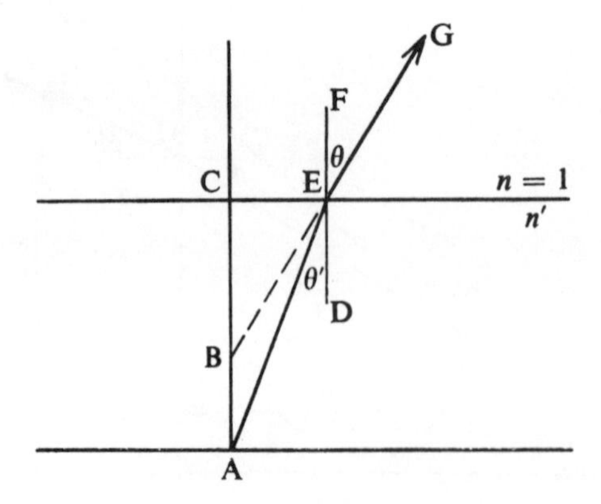


Fig. 1-14 Real and apparent depth

A′ then acts as an object for the second surface. After refraction at the second surface the rays appear to diverge from a point A″, such that

$$v'' = (v' + t)\frac{n}{n'} = \left(u\frac{n'}{n} + t\right)\frac{n}{n'} = u + \frac{n}{n'}t$$

Therefore, the image appears to be displaced from the object by a distance

$$AA'' = (u + t) - \left(u + \frac{n}{n'}t\right)$$

$$= t\left(1 - \frac{n}{n'}\right) \qquad (1\text{-}7)$$

Thus, on looking straight through a glass slab—say in air, where $n = 1$—an object appears to be an amount $t(1 - 1/n')$ closer than it really is.

## 1-12 Real and apparent depth

The fact that an object appears closer than it really is when viewed through a refracting medium may be used to measure the refractive index of the medium. In Fig. 1-14, a point A on the bottom surface of a parallel-sided slab appears to be at a point B when viewed from above, provided only paraxial rays are used. Then

$$n' = \frac{\sin\theta}{\sin\theta'} = \frac{\sin\angle FEG}{\sin\angle DEA}$$

$$= \frac{\sin\angle CBE}{\sin\angle CAE}$$

$$= \frac{CE/BE}{CE/AE}$$

$$= \frac{AE}{BE}$$

But since only paraxial rays are being considered, AE/BE = AC/BC, i.e.,

$$\text{refractive index, } n' = \frac{\text{real depth}}{\text{apparent depth}} \qquad (1\text{-}8)$$

This is, of course, in agreement with Eqn (1-7), where the distance $AB = AC(1 - 1/n')$.

The refractive index may thus be determined by using a microscope fitted with a vertical scale, and focusing first on the stage without the slab—that is, focusing on A—and then, with the slab in position, on the image of A at B, and again on the surface at C.

## 1-13 Refraction through layers

Consider, as in Fig. 1-15, a number of parallel layers with different refractive indices $n_1, n_2, n_3, \ldots, n_m$, each measured with respect to air.

If the angles of incidence and refraction are as indicated, then for the

$$\begin{array}{lll} \text{1st refraction} & n_1 \sin \theta_1 & = n_2 \sin \theta_2 \\ \text{2nd refraction} & n_2 \sin \theta_2 & = n_3 \sin \theta_3 \\ \vdots & \vdots & \vdots \\ (m-1)\text{th refraction} & n_{m-1} \sin \theta_{m-1} & = n_m \sin \theta_m \end{array}$$

Multiplying these equations together gives

$$n_1 \sin \theta_1 . n_2 \sin \theta_2 . \ldots . n_{m-1} \sin \theta_{m-1}$$
$$= n_2 \sin \theta_2 . n_3 \sin \theta_3 . \ldots . n_m \sin \theta_m$$

i.e.,

$$n_1 \sin \theta_1 = n_m \sin \theta_m \tag{1-9}$$

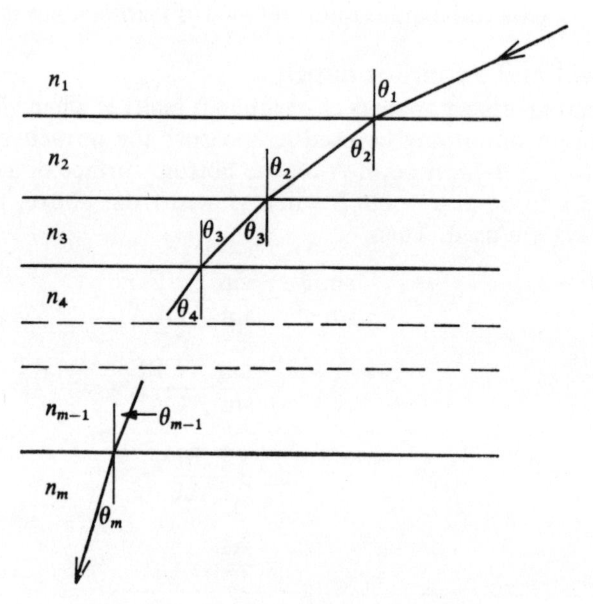


Fig. 1-15 Refraction through layers

Thus, the deviation of the ray depends only on the difference in refractive index between the outer layers, the inner layers causing only a lateral shift of the emergent ray.

Also, if ${}_1n_2$ means the refractive index of the medium of the second layer measured, not with respect to air, but with respect to the first medium, and similarly ${}_2n_3$ is the refractive index of the third medium measured with respect to the second, etc., then for the

$$\begin{array}{lll} \text{1st refraction} & {}_1n_2 & = \sin \theta_1 / \sin \theta_2 \\ \text{2nd refraction} & {}_2n_3 & = \sin \theta_2 / \sin \theta_3 \\ \vdots & \vdots & \vdots \\ (m-1)\text{th refraction} & {}_{m-1}n_m & = \sin \theta_{m-1} / \sin \theta_m \end{array}$$

Multiplying the equations together, as before, gives

$$
\begin{aligned}
{}_1n_2 \cdot {}_2n_3 \cdot \;\ldots\; \cdot {}_{m-1}n_m \\
&= \frac{\sin\theta_1}{\sin\theta_2}\cdot\frac{\sin\theta_2}{\sin\theta_3}\cdot \;\ldots\; \cdot\frac{\sin\theta_{m-1}}{\sin\theta_m} \\
&= \frac{\sin\theta_1}{\sin\theta_m}
\end{aligned}
$$

i.e.,

$$ {}_1n_2 \cdot {}_2n_3 \cdot \;\ldots\; \cdot {}_{m-1}n_m = {}_1n_m \qquad (1\text{-}10) $$

since, by Eqn (1-9), the deviation is independent of the inner layers.

## 1-14 Refractometers

Most instruments designed for the accurate measurement of refractive indices of solids and liquids make use of the critical angle.

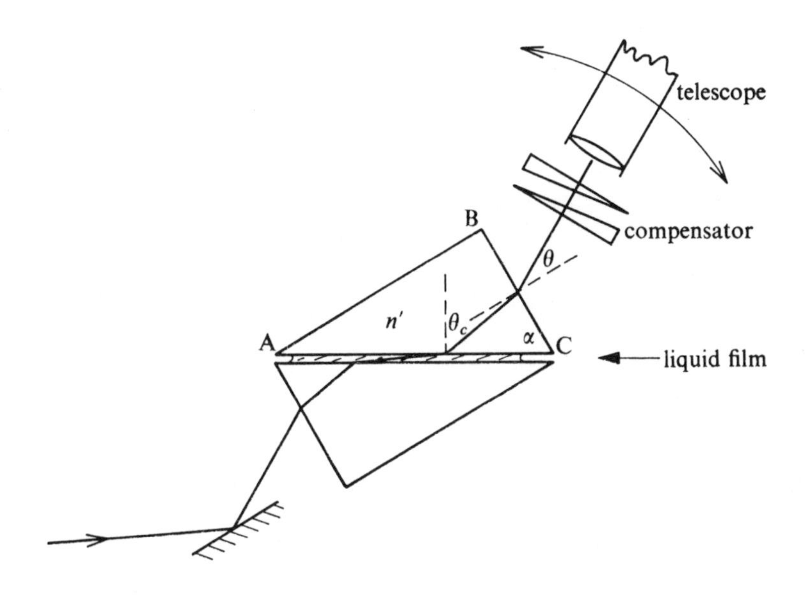


Fig. 1-16 Abbe refractometer

In the *Abbe refractometer*, the liquid under test is sandwiched as a thin parallel film between two prisms (Fig. 1-16). The prisms are of dense flint glass with a refractive index $n_d$ of about 1·75, that is, higher than any refractive index to be measured. The hypotenuse face of the upper prism is polished while that of the lower prism is left matt to prevent the formation of false images. The function of the lower prism, which can be hinged down to allow a drop of the test liquid to be placed on it, and subsequently raised to form the film, is to direct light at grazing incidence into the liquid film. Light is directed into the lower prism by a mirror. All rays of a particular angle of incidence are equally deviated and emerge as parallel rays, when they are then focused by the telescope to form a single bright line in its focal plane. Other rays incident at different angles similarly cause bright lines. Thus the total effect of incident convergent light rays is to form a bright area over part of the telescope's focal plane and to leave the remaining area dark. The boundary between the light and dark areas is the limit set by the critical angle of refraction. The position of the telescope is adjusted to make this boundary coincide with the eyepiece cross-hairs, and the axis of the telescope then makes an angle $\theta$ with the normal to the final face of the prism.

In Fig. 1-16, the critical angle, $\theta_c$, is given by

$$n' \sin \theta_c = n \qquad (1\text{-}11)$$

where $n'$ is the refractive index of the prism, and is greater than the refractive index to be measured, $n$. Also, if the prism angle, $\angle$ACB, is $\alpha$, and the angle between the emergent ray and the normal to the emergent face is $\theta$, then for refraction at the emergent face,

$$n' \sin (\alpha - \theta_c) = \sin \theta \qquad (1\text{-}12)$$

Expanding Eqn (1-12), substituting Eqn (1-11), and rearranging gives

$$\sin \alpha (n'^2 - n^2)^{1/2} = \sin \theta + n \cos \alpha$$

Squaring and rearranging again,

$$n^2 + 2n \cos \alpha \sin \theta + (\sin^2 \theta - n'^2 \sin^2 \alpha) = 0$$

Then, taking the positive square root, since the refractive index cannot be negative,

$$n = \sin \alpha (n'^2 - \sin^2 \theta)^{1/2} - \cos \alpha \sin \theta \qquad (1\text{-}13)$$

Thus the refractive index of the sample liquid can be determined as a function of the emergent angle $\theta$, as measured by the position of the telescope. In practice, the instrument is calibrated to read the refractive index directly.

The instrument can be used with daylight by employing a compensator in front of the telescope, to compensate for the colour dispersion introduced. This consists of two direct-vision prisms that may be rotated in opposite directions at equal rates to form a system of variable dispersion. The position of the compensator prisms is indicated on a scale. The dispersion—that is, the difference in the refractive indices as measured for blue light and for red light, $n_F - n_C$—is obtained from an expression of the form

$$n_F - n_C = A + Bs + Cs^2 \qquad (1\text{-}14)$$

where $s$ is the compensator scale reading and $A$, $B$, and $C$ are constants depending on the particular instrument, and vary with the mean refractive index $n_d$.

In use, the positions of the compensator and telescope are adjusted until the boundary between the light and dark areas as seen in the telescope is free from colour and coincides with the eyepiece cross-hairs. The instrument, calibrated by the makers to read correctly at 20°C, then gives directly the refractive index, $n_d$, over the range 1·3 to 1·7 to an accuracy of about 0·0002, and enables the dispersion to be calculated, from Eqn (1-14) and the maker's table of constants. It is thus very suitable for the quick determination of dispersive power. The prisms are normally mounted in a water jacket so that the temperature may be controlled, though the makers usually give a correcting factor to allow the refractive index of the sample to be determined at temperatures other than 20°C. A correction of +0·0001 in the refractive index is necessary for a temperature rise of about 14°C in the Abbe prisms.

The refractive index of a solid may also be determined on this instrument. The solid must have one of its faces flat and polished and one end, which need not necessarily be polished, should be approximately perpendicular to this face. The solid is mounted on the instrument by removing the lower Abbe prism, inverting the instrument so that the telescope points directly at the light source, and placing the solid centrally on the prism face (Fig. 1-17). Optical contact between the solid and prism face is

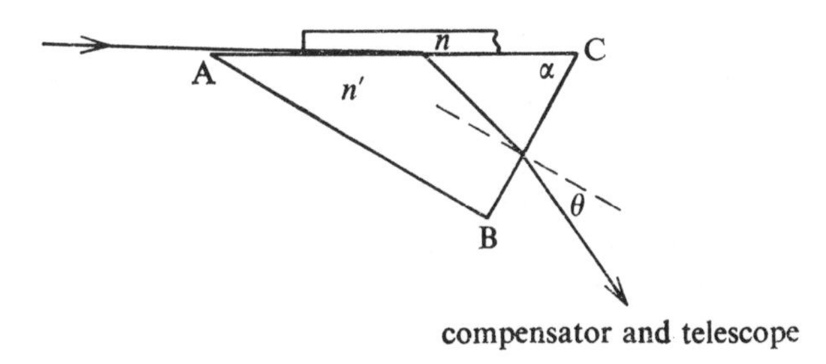


Fig. 1-17 Abbe refractometer for measurement of refractive index of a solid

made by a drop of high refractive index liquid, such an monobromonaphthalene ($n_d = 1{\cdot}66$). The sample must be pressed well down to exclude surplus liquid, care being taken to see that no liquid lies in front of the sample, otherwise a false reading may be given.

The *Pulfrich refractometer* works on similar principles to those of the Abbe refractometer. The Pulfrich prism is made of high refractive index material and consists of two polished surfaces, one vertical and one horizontal (Fig. 1-18). The sample is placed on the horizontal surface. Solids should be in the same form as that required for the Abbe refractometer, optical contact being ensured by the use of a high refractive index liquid as before. A glass cell cemented to the prism is used for holding liquids. Light is directed into the sample in a direction almost parallel to the horizontal surface of the prism so as to make the critical angle $\theta_c$, as shown.

If the refractive index of the sample is $n$, and of the prism, $n'$,

$$n' \sin \theta_c = n \tag{1-15}$$

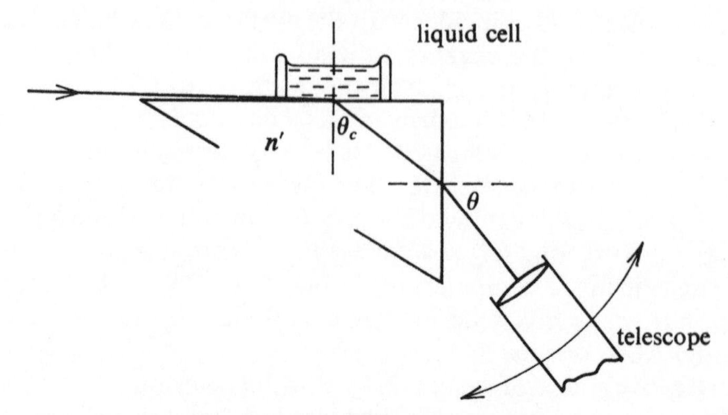


Fig. 1-18 Pulfrich refractometer

Also, if the emergent ray makes an angle $\theta$ with the normal to the prism face, so that

$$n' \sin (90° - \theta_c) = \sin \theta$$

i.e.,
$$n' \cos \theta_c = \sin \theta \tag{1-16}$$

Squaring and adding Eqns (1-15) and (1-16),

$$n'^2 = n^2 + \sin^2 \theta$$

i.e.,
$$n = (n'^2 - \sin^2 \theta)^{1/2} \tag{1-17}$$

As with the Abbe refractometer, an area of light field and of dark field forms at the focal plane of the viewing telescope, with the boundary between them determined by the critical angle, $\theta_c$. When this boundary is made coincident with the centre of the eyepiece cross-hairs, by moving the telescope, the instrument measures the angle $\theta$. Pulfrich refractometers are usually calibrated by the makers to give the refractive index, $n$, directly. Since there is no compensator for dispersion, it is necessary to use monochromatic light. The refractive index is thus determined at this wavelength.

As with the Abbe refractometer, the samples to be measured must have a refractive index less than that of the prism material.

## PROBLEMS

**1-1** If a luminous point object is located between two plane mirrors inclined at an angle of 32°, show that the number of images formed is either 10 or 11, depending on whether the angular separation of the point from the nearer mirror, measured from the intersection of the mirrors, is less or greater than 8°.

**1-2** An object is placed 3 cm in front of a plane mirror. If a second mirror is now placed parallel to and facing the first, find their separation if the image due to the fourth reflection is located 18 cm behind the second mirror. [5 cm]

**1-3** A ray incident at 58° onto a flat glass surface is refracted in a direction at right angles to the direction of the reflected ray. Calculate the refractive index of the glass. [1·60]

**1-4** A circular disc is floating on the surface of a liquid of refractive index 1·4. Find its smallest diameter such that it could prevent a point object located 10 cm below the surface from being seen from above. [20·4 cm]

**1-5** A travelling microscope is focused on a scratch on the top surface of a sheet of glass 5 mm thick and of refractive index 1·52. Find the distance the microscope has to be moved down in order to focus on the image of the scratch formed by one reflection at the lower surface. [6·58 mm]

**1-6** Find the total apparent depth, when viewed vertically from above, of 10 cm of water of refractive index 1·33 lying over a 10 cm thickness of glass of refractive index 1·55. [13·97 cm]

**1-7** Calculate the maximum angle of incidence of a ray onto the first face of a prism of refractive index 1·6 and refracting angle 60°, if the ray is just to undergo total reflection at the second face. [35° 35′]

**1-8** A ray is incident normally onto one surface of a glass wedge of refractive index 1·533 and whose surfaces are partially silvered. If after three internal reflections the ray emerges at an angle of 25° with the incident ray, calculate the angle of the wedge. What is the largest number of internal reflections that can occur if the ray is still to emerge? [4°, 9]

**1-9** A ray is incident at grazing incidence onto one side of a glass cube. Show that if $\theta$ is the angle made with the normal by the ray emerging from the adjacent face of the cube, then $\sin \theta = \cot \theta_c$, where $\theta_c$ is the critical angle.

**1-10** An Abbe refractometer with a prism of refracting angle 60° and refractive index 1·75 is used to measure the refractive index of a liquid. If the angle between the emergent ray and the normal to the emergent face is 9° 7′, what is the refractive index of the liquid? [1·43]

**1-11** What is the lowest refractive index that can be measured with a Pulfrich refractometer, whose prism is made from glass of refractive index 1·71? [1·39]

**1-12** A ray of light is incident normally onto the first face of a 10° refracting angle glass prism of refractive index 1·68. The ray then passes directly into a second prism, of refractive index 1·52. Calculate the refracting angle of the second prism if the final emergent ray is to be parallel to the primary incident ray. [13° 7′]

# 2 Reflection and refraction at spherical surfaces

## 2-1 Reflection at curved surfaces

When light falls onto a curved mirror, it is reflected according to the simple laws of reflection discussed in the previous chapter. Figure 2-1 illustrates the reflection of light from a source at O. The curved surface may be considered as composed of an infinite number of infinitely small tangent planes. A normal to the surface, which is in fact a radius of the curve, can be drawn to each of these tangent planes. Each ray that strikes a particular tangent plane is reflected so that the angle of incidence is equal to the angle of reflection. In Fig. 2-1, O is a point on the *principal axis*, that is, a line passing through the *centre of curvature*, C, and the *pole*, P, or centre, of the

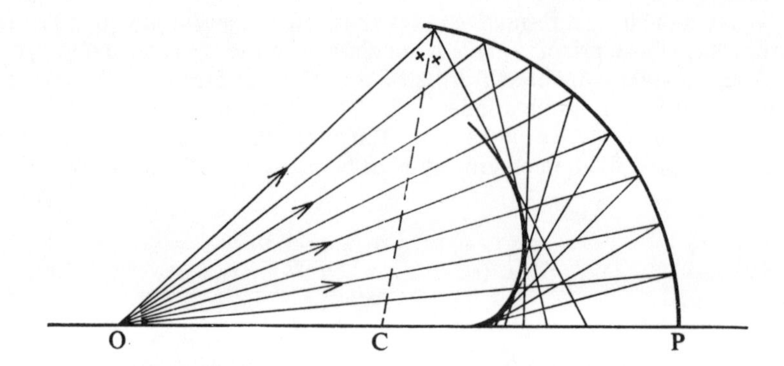


Fig. 2-1 Formation of a caustic by reflection

mirror. The reflected rays from the point object at O do not intersect at one point to form a point image, but form a vague image in the shape of a cusped curve, termed a *caustic curve*. In practice, in order to form a distinct image, it is necessary to restrict the *aperture* of the mirror—that is, its diameter—so that the incoming rays make only a small angle with the principal axis. The smaller this angle is made, the sharper is the image.

Within this small aperture condition, we can derive useful equations connecting the size and position of the image with that of the object and the curvature of the mirror. First, however, let us decide the nature of the image by graphical constructions.

## 2-2 Graphical construction for mirrors

In Fig. 2-2(a), a ray parallel to the axis is reflected to meet the axis at F. By definition, F is the *principal focus* of the mirror. It is assumed that the mirror is of sufficiently small aperture that all such parallel rays come to this same focus. Parallel rays inclined at an angle to the principal axis are similarly brought to a focus at a point on the *focal plane* through F [Fig. 2-2(b)].

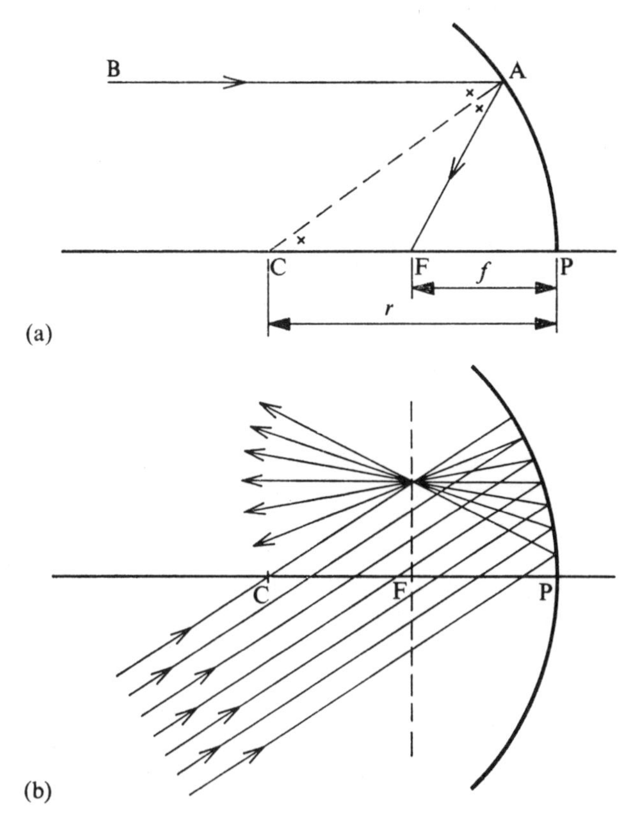


Fig. 2-2 Focusing of a parallel beam by a concave mirror

In Fig. 2-2(a), by the laws of reflection $\angle BAC = \angle FAC$. Also, since BA is parallel to CP, $\angle BAC = \angle FCA$. Thus, AFC is an isosceles triangle, with AF = CF. Since the mirror is of small aperture, A is in fact close to P, so that AF = FP to a first approximation. Hence CF = FP. Therefore, if the radius of curvature, CP, is $r$ and the *focal length*, FP, of the mirror is $f$, then

$$r = 2f \tag{2-1}$$

The mirror in Fig. 2-2 is *concave*; Eqn (2-1) is equally applicable to the *convex* mirror of Fig. 2-3, in which a ray parallel to the axis is reflected and appears to come from the principal focus, F.

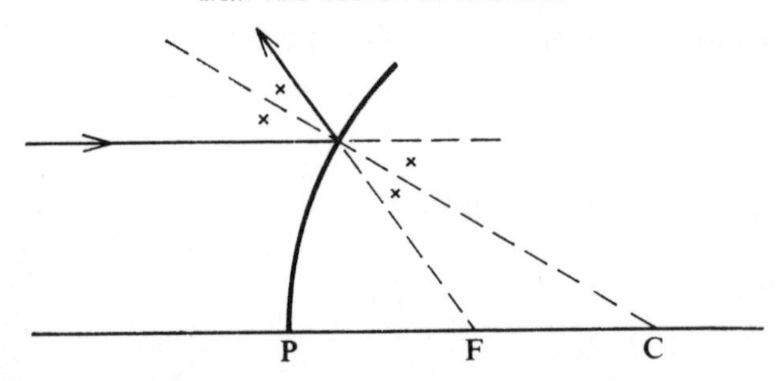


Fig. 2-3 Reflection by a convex mirror

In Fig. 2-4, light from an off-axis object at O′ is brought to a focus at I′. Similarly, light from any point on the object OO′ passes through a corresponding point on II′, in other words, the image of the object OO′ is at II′. Of the infinite number of possible rays between corresponding object and image points, certain rays lend themselves to a graphical construction of the image position. For example, by definition a ray from O′ parallel to the axis is reflected to pass through the principal focus F, and a ray from O′ passing through the centre of curvature C of the mirror strikes the mirror normally and is reflected back along its own path. These two rays alone are sufficient to define the position of I′ and hence II′. In Fig. 2-5 these two rays are used to determine the size and position of the corresponding image for various positions of the object, with both concave and convex mirrors. As the mirrors are of small aperture—that is, their diameters are small in comparison with all other distances—they may be drawn as straight vertical lines, shaded on one side to indicate the back.

If an object is at the centre of curvature of a concave mirror, the image is also at the centre of curvature and the same size, but inverted. If the object is at the focus of the mirror, then the image is at infinity. In Fig. 2-5(a) and (b), if the object is a bright one, the image may be focused onto a screen: this is a *real* image. The objects and images are also said to be *conjugate*, since their positions are interchangeable. In Fig. 2-5(c) and (d), the images may be seen only by looking into the mirror and only appear

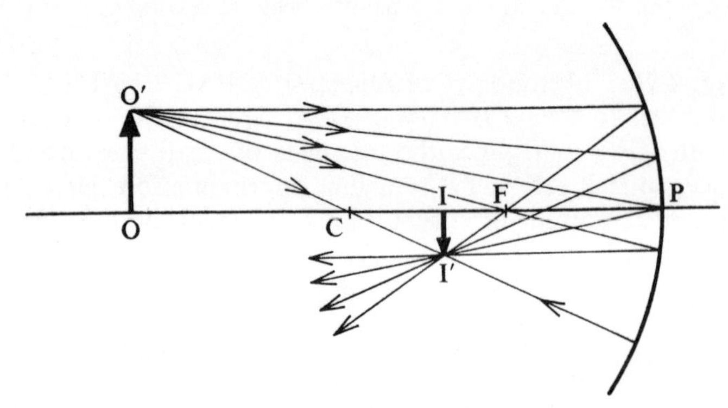


Fig. 2-4 Formation of a real image by a concave mirror

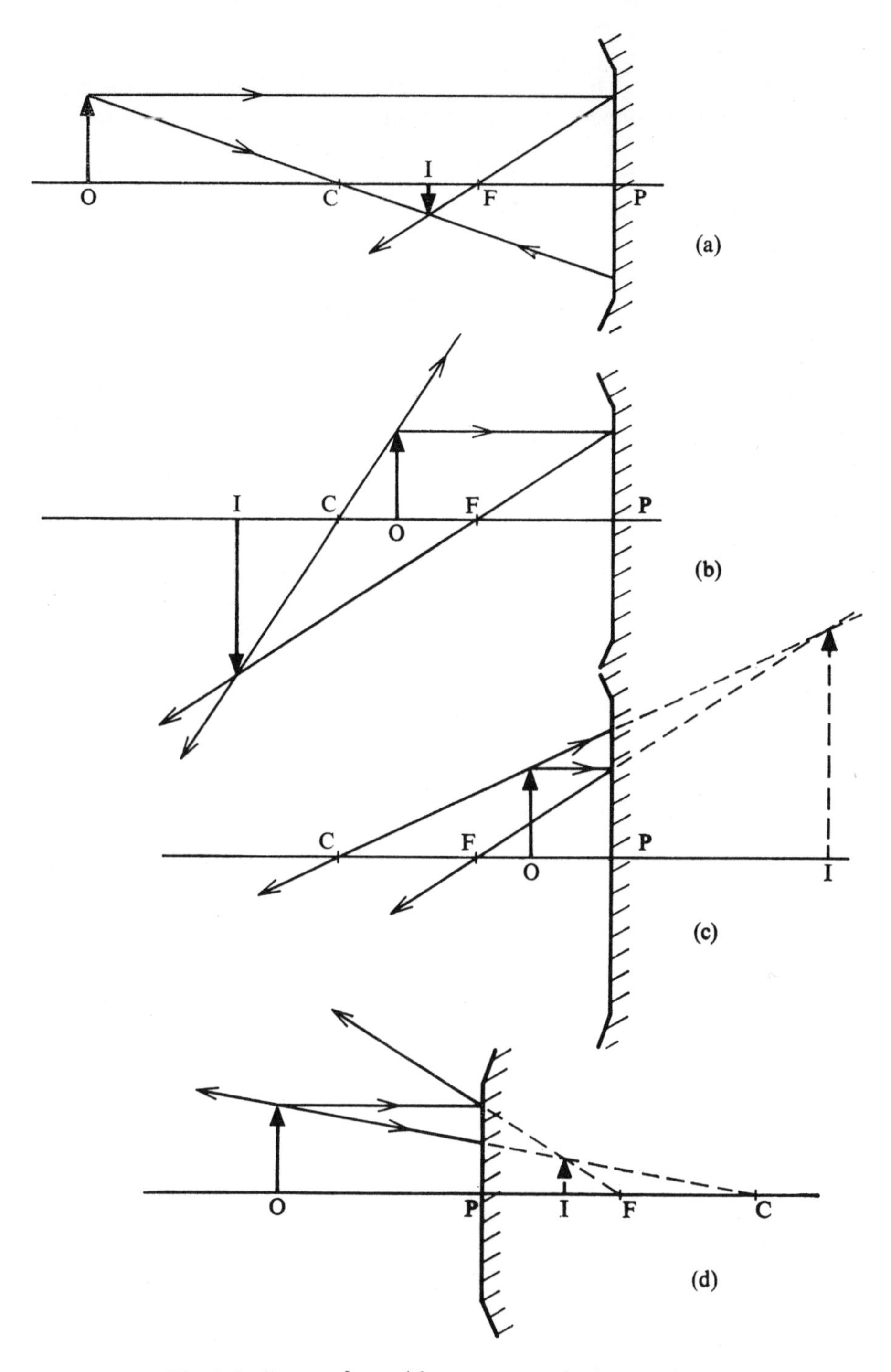


Fig. 2-5 Images formed by concave and convex mirrors

to be at the positions indicated. They cannot be shown on a screen. These images are therefore said to be ***virtual.*** Virtual images are always erect, and real images always inverted.

The images formed by a concave mirror may be diminished, the same size, or magnified, depending on the position of the object, but a convex mirror always forms a diminished image. Car driving mirrors are convex in order to condense a large field of view into a small image. Shaving mirrors are concave and so form a magnified erect image as in Fig. 2-5(c).

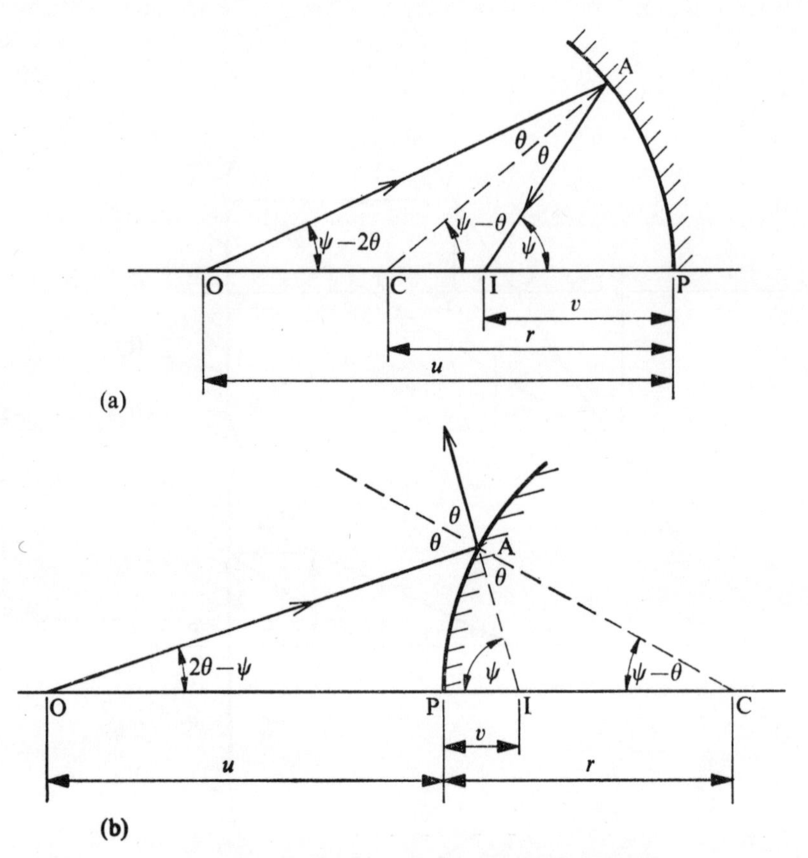


Fig. 2-6 Reflection from a spherical surface

## 2-3 Mirror formulae

Figure 2-6 shows the path of one ray between an on-axis point object and its corresponding image. The reflected ray makes an angle $\psi$ with the principal axis, and $\theta$ is the angle of incidence and reflection. The other angles are as shown in the figure. Let the object be at a distance $u$ from the pole, the image a distance $v$, and the centre of curvature a distance $r$. Since the apertures are small, the rays are very near the axis and are therefore termed *paraxial* rays, and consequently the angles are all small. Therefore, approximately, for the *concave mirror:*

$$\psi = \frac{AP}{v} \tag{2-2}$$

$$\psi - \theta = \frac{AP}{r} \tag{2-3}$$

$$\psi - 2\theta = \frac{AP}{u} \tag{2-4}$$

Adding Eqns (2-2) and (2-4)

$$2\psi - 2\theta = \frac{AP}{u} + \frac{AP}{v} \tag{2-5}$$

and multiplying Eqn (2-3) by 2

$$2\psi - 2\theta = 2\frac{AP}{r} \tag{2-6}$$

Equating Eqn (2-5) with (2-6) gives

$$\frac{AP}{u} + \frac{AP}{v} = 2\frac{AP}{r}$$

i.e.,

$$\frac{1}{u} + \frac{1}{v} = \frac{2}{r} = \frac{1}{f} \tag{2-7}$$

Similarly for the *convex mirror:*

$$\psi = \frac{AP}{v} \tag{2-8}$$

$$\psi - \theta = \frac{AP}{r} \tag{2-9}$$

$$2\theta - \psi = \frac{AP}{u} \tag{2-10}$$

Subtracting Eqn (2-8) from (2-10)

$$2\theta - 2\psi = \frac{AP}{u} - \frac{AP}{v} \tag{2-11}$$

and multiplying Eqn (2-9) by $-2$

$$2\theta - 2\psi = -2\frac{AP}{r} \tag{2-12}$$

Equating Eqn (2-11) with (2-12) gives

$$\frac{AP}{u} - \frac{AP}{v} = -2\frac{AP}{r}$$

i.e.,

$$\frac{1}{u} - \frac{1}{v} = -\frac{2}{r} = -\frac{1}{f} \tag{2-13}$$

Thus two equations are needed to cover these reflections. This is inconvenient, especially when reflections between concave and convex mirrors are both involved in the same problem. However, by a suitable choice of sign convention, this difficulty may be removed and one formula used to cover all cases of reflection from a spherical surface.

## 2-4 Sign conventions

A number of sign conventions have been used in the past, but most of these are unsatisfactory for various reasons. For some systems we need to know whether the final image is real or virtual. This is easy in the case of a single mirror, but in a more complicated optical system it may be difficult. Another system requires a different convention for mirrors than for lenses; although the same formula may be used for both mirrors and lenses, in a mixed system of mirrors and lenses any advantage is lost by the confusion that arises in the allocation of the signs.

In this book we shall use the *New Cartesian* sign convention for both mirrors and lenses; this has the advantage that no prior knowledge is required concerning the final image. In this convention,

> *the pole of the lens or mirror is taken as the centre of a co-ordinate system. All distances measured to the left or downwards are taken as negative distances, and all distances to the right or upwards are taken as positive.*

(See Fig. 2-7.) The direction of the incoming light ray must be from left to right. Usually this just involves starting with the object on the left.

Thus in Fig. 2-6(a), the light is travelling from left to right and, taking the pole P as the centre of a co-ordinate system, the distances $u$, $v$, and $r$ are all on the left and therefore negative. Substituting $-u$, $-v$, and $-r$ for $u$, $v$, and $r$ in Eqn (2-7) gives

$$\frac{1}{-u} + \frac{1}{-v} = \frac{2}{-r} = \frac{1}{-f}$$

i.e.,

$$\frac{1}{u} + \frac{1}{v} = \frac{2}{r} = \frac{1}{f}$$

Similarly, in Fig. 2-6(b) the convex mirror is facing the oncoming light, which is travelling from left to right, and thus $u$, on the left, is negative; but the centre of curvature (and thus $r$) and the image are on the right, so that $r$ and $v$ are both positive. Substituting $-u$, $+v$, and $+r$ in Eqn (2-13) for a convex mirror gives

$$\frac{1}{-u} - \frac{1}{+v} = \frac{-2}{+r} = \frac{-1}{+f}$$

i.e.,

$$\frac{1}{u} + \frac{1}{v} = \frac{2}{r} = \frac{1}{f} \qquad (2\text{-}14)$$

Thus, one formula is applicable to all spherical mirrors.

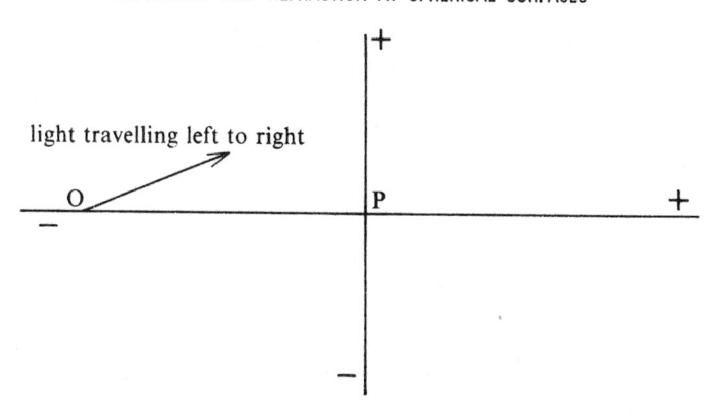


Fig. 2-7 New Cartesian sign convention

The lateral magnification may also be simply calculated. In Fig. 2-8, since the angle of incidence is equal to the angle of reflection, the triangles OO′P, II′P are similar. Thus, the *lateral magnification M*, defined as the ratio of the size of the image to that of the object, is given by

$$M = \frac{y'}{y} = \frac{v}{u}$$

If we apply the sign convention and write the magnification as

$$M = \frac{y'}{y} = -\frac{v}{u} \tag{2-15}$$

where $y$ and $y'$ are the actual physical sizes of the object and image, then, on applying the correct signs for $v$ and $u$, a negative magnification indicates an inverted image and a positive sign an erect image. The sign is part of our sign convention; the magnification is only the ratio of the sizes.

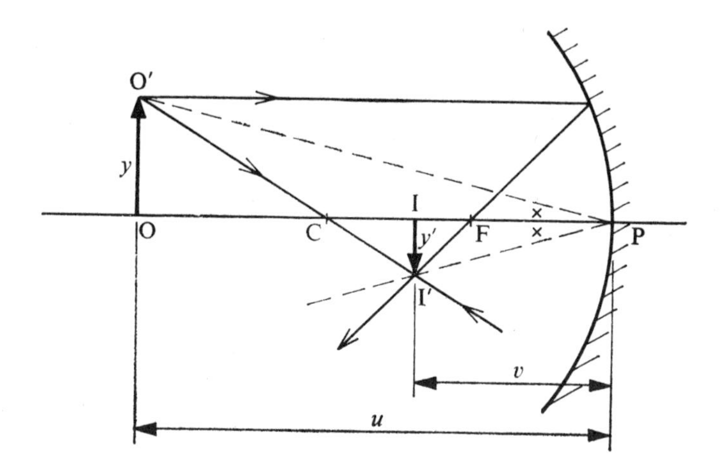


Fig. 2-8 Magnification by a spherical mirror

Thus, to summarize, for any spherical mirror, applying the New Cartesian sign convention,

$$\frac{1}{u} + \frac{1}{v} = \frac{1}{f} = \frac{2}{r}$$

$$M = \frac{\text{image size}}{\text{object size}} = -\frac{v}{u}$$

**Example 2-1** An object 4 cm long is placed 45 cm from a convex mirror whose radius of curvature is 30 cm. Find the position and size of the image.

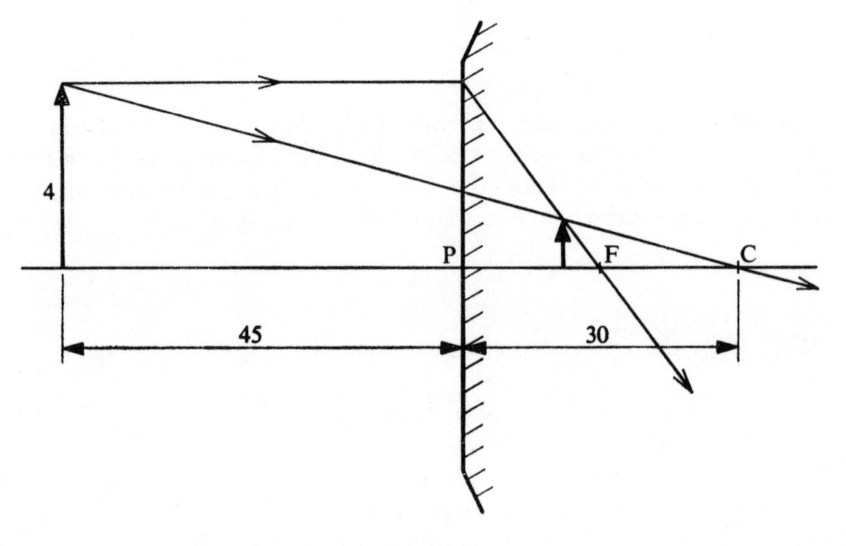


Fig. 2-9

Applying the sign convention

$$r = +30 \text{ cm}$$
$$u = -45 \text{ cm} \quad \text{(Fig. 2-9)}$$

Substituting in Eqn (2-14)

$$\frac{1}{u} + \frac{1}{v} = \frac{2}{r}$$

i.e.,

$$\frac{1}{-45} + \frac{1}{v} = \frac{2}{+30}$$

$$\therefore \frac{1}{v} = \frac{1}{45} + \frac{2}{30} = +\frac{4}{45}$$

$$\therefore v = +\frac{45}{4} = +11{\cdot}25 \text{ cm}$$

Thus the image is 11·25 cm behind the mirror—that is, the image is virtual. The sign arises from our artificial convention allowing us to calculate the correct answer. Having found the image position, it has no further meaning. Thus it is important to describe in words the image position with reference to the reflecting surface.

Also

$$M = \frac{\text{image size}}{\text{object size}} = -\frac{v}{u}$$

$$\begin{aligned} \text{image size} &= -\frac{v}{u} \times \text{object size} \\ &= -\left(\frac{+45/4}{-45}\right)4 \\ &= +\frac{1}{4} \times 4 \\ &= +1 \text{ cm} \end{aligned}$$

Therefore the image is 1 cm long and the positive sign indicates that it is erect. The actual magnification, if required, would be written as $\times\frac{1}{4}$, indicating a diminished image.

Alternatively, these results could have been obtained by an accurate scale drawing. In any case, a rough drawing should always be made as an approximate check on the calculation.

**Example 2-2** A concave mirror is used to focus an image of an object onto a screen 10 m from the object. If the lateral magnification is $\times 6$, calculate the radius of curvature of the mirror.

Since the image is formed on a screen we know it must be real and therefore inverted (see Fig. 2-5). Thus, the magnification may be written as

$$M = -\frac{v}{u} = -6$$

The *minus* 6 indicates an inverted image.

$$\therefore\ v = 6u$$

Also

$$v = 10 + u \quad \text{(Fig. 2-10)}$$

$$\begin{aligned} \therefore\ 6u &= 10 + u \\ u &= \frac{10}{5} = 2 \text{ m} \\ \therefore\ v &= 6u = 12 \text{ m} \end{aligned}$$

Applying the sign convention

$$\begin{aligned} u &= -2 \\ v &= -12 \end{aligned}$$

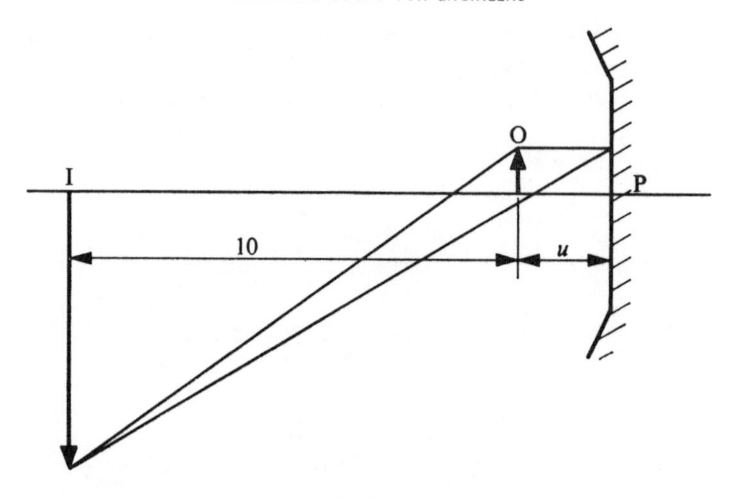


Fig. 2-10

Substituting in

$$\frac{2}{r} = \frac{1}{u} + \frac{1}{v}$$

$$\frac{2}{r} = \frac{1}{-2} + \frac{1}{-12}$$

$$= -\left(\frac{1}{2} + \frac{1}{12}\right)$$

$$= -\frac{7}{12}$$

$$\therefore\ r = -\frac{12}{7} \times 2$$

$$= -3{\cdot}43 \text{ m}$$

The required radius of curvature is thus 3·43 m and the negative sign indicates that the centre of curvature is on the left—that is, on the same side as the object.

## 2-5 Thin lenses

Thin lenses may be dealt with in a similar way to spherical mirrors. The same restrictions as to small aperture apply, otherwise a sharply focused image is not formed. A lens is termed *thin* when its thickness is negligible compared with all other distances. A principal focus may be defined in the same way as for a mirror. In Fig. 2-11, rays parallel to the axis are shown to converge towards, or diverge from, the principal foci, F. Also, parallel rays inclined to the axis come to a focus in an off-axis point lying on the focal plane through F, as for a spherical mirror.

Every lens has two principal focal points. The focal length $f$ is defined as the distance between the principal focus and the pole P, assuming that the lens has negligible thickness. An addition to our sign convention is now required:

*all converging lenses are considered to have a positive focal length and all diverging lenses a negative focal length.*

Applying the sign convention as for a mirror, a formula connecting image and object distances with focal length may be derived. But, as before, let us first determine the object and corresponding image positions by graphical means.

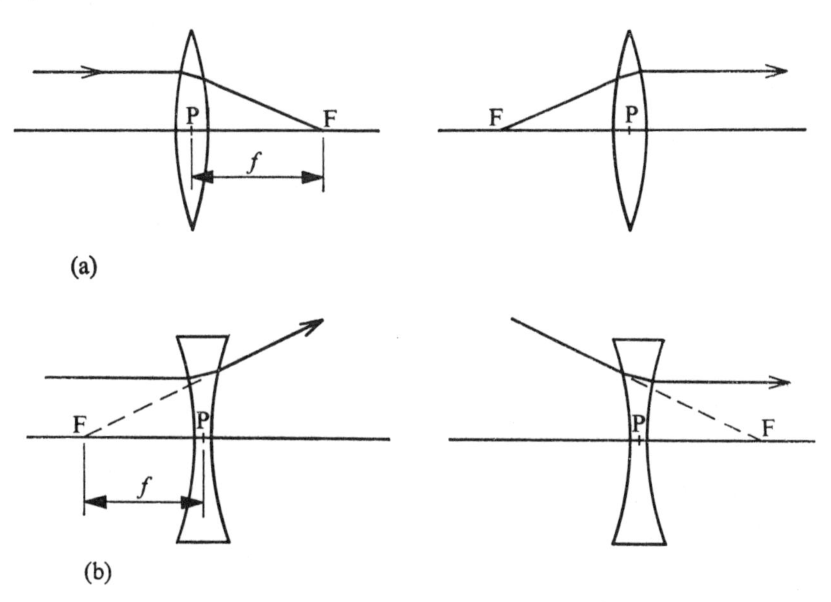


Fig. 2-11 Principal foci of (a) converging lens and (b) diverging lens

## 2-6 Graphical construction for thin lenses

In the same manner as for a mirror, rays may be considered as diverging from every point of an object and being brought to a focus at a corresponding point of the image. In particular, we know the paths of two rays. Thus a ray from the object parallel to the axis, by definition, converges to, or diverges from, the appropriate principal focus; also, a ray from the object passing through the pole of the lens remains undeviated, because the lens has negligible thickness and the two refracting surfaces may be considered parallel at the centre of the lens.

Figure 2-12 illustrates the use of these two rays in determining the size and position of the image. It must be emphasized that these are only two of the infinite number of rays which go to make up the image. Figure 2-12(a) shows that, when the object is at a distance greater than the focal length from a converging lens, a real inverted image is formed which may be magnified or diminished depending on whether the object distance is less or greater than twice the focal length. The object and image may be interchanged and are therefore said to be conjugate. For an object distance less than the focal length, or for a diverging lens, the refracted rays appear to diverge from an image, which is thus virtual and is also always erect.

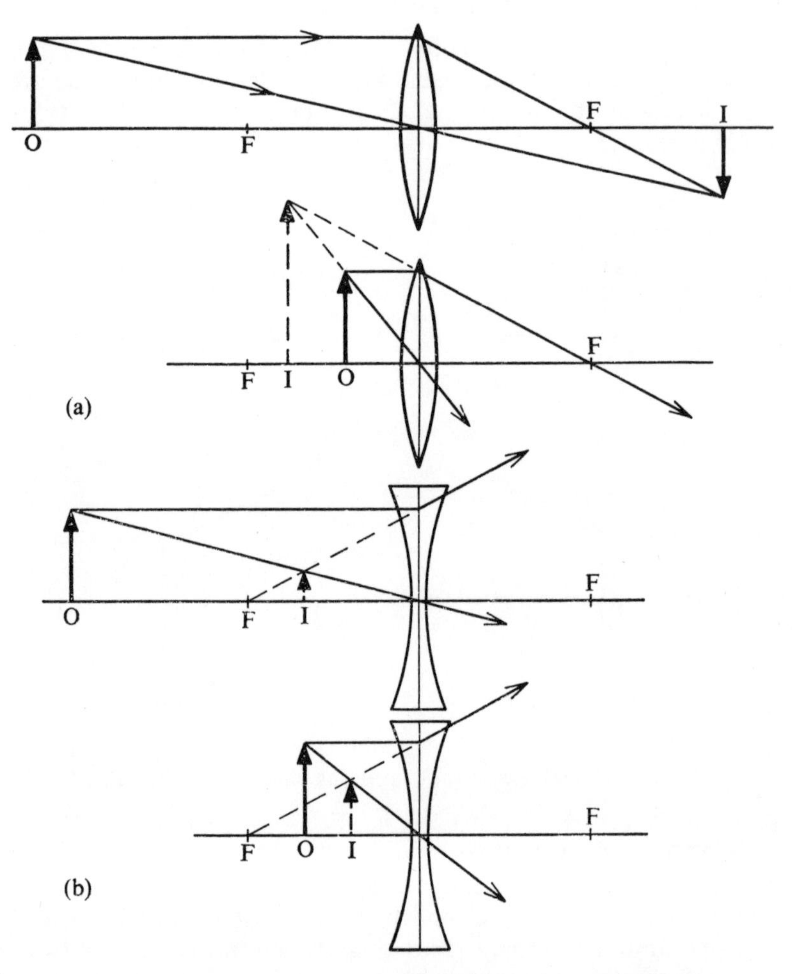


Fig. 2-12 Images formed by (a) converging and (b) diverging lenses

For a converging lens, the virtual image is magnified and, in fact, the lens is being used as a simple magnifying glass; but the image formed by a diverging lens is always diminished.

## 2-7 Thin-lens formulae

Consider a real image formed by a converging lens as in Fig. 2-13. Two rays passing through the principal foci define the image position. Let the object OO′ and the corresponding image II′ be of size $y$ and $y'$ and distance

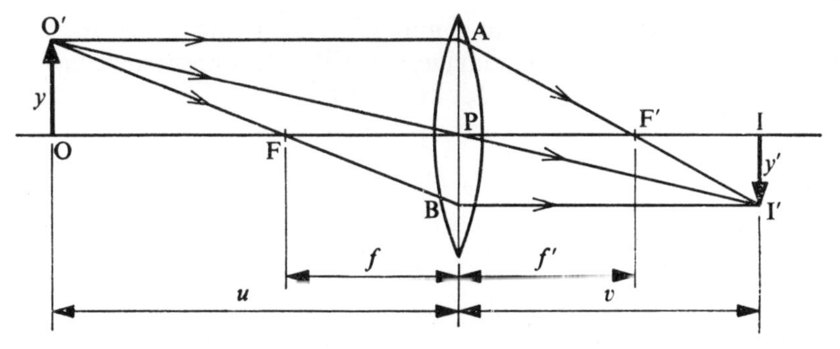


Fig. 2-13 Refraction by a thin lens

$u$ and $v$ from the pole respectively. Let the lens have a focal length $f = f'$. Then, in the image space, similar triangles ABI′, APF′ are formed.

$$\therefore \frac{AB}{BI'} = \frac{AP}{PF'}$$

Applying the sign convention, this becomes

$$\frac{y - y'}{v} = \frac{y}{f'} \tag{2-16}$$

since all distances to the left and downwards are classed as negative.

In the object space, triangles BAO′, BPF are similar

$$\therefore \frac{BA}{AO'} = \frac{BP}{PF}$$

Applying the sign convention again, this becomes

$$\frac{y - y'}{-u} = \frac{-y'}{f} \tag{2-17}$$

since the convention is that the focal length of a converging lens is positive.

Therefore, adding Eqns (2-16) and (2-17) and remembering that $f = f'$ in this case,

$$\frac{y - y'}{v} + \frac{y - y'}{-u} = \frac{y - y'}{f}$$

i.e.,

$$\frac{1}{v} - \frac{1}{u} = \frac{1}{f} \tag{2-18}$$

This is a general equation applicable to all thin lenses, providing the sign convention is applied.

An expression for the lateral magnification may also be simply deduced. From Fig. 2-13, since APO′ and BPI′ are similar triangles, the lateral magnification $M$ is given by

$$M = \frac{\text{size of image}}{\text{size of object}} = \frac{y'}{y} = \frac{v}{u} \tag{2-19}$$

Applying the correct signs to $v$ and $u$, where $y$ and $y'$ are the actual physical sizes of the object and image, $v/u$ will have a positive value for an erect image and a negative value for an inverted image. This equation, with

$$\frac{1}{v} - \frac{1}{u} = \frac{1}{f}$$

enables the size, position, and nature of the image to be determined for any thin lens.

**Example 2-3** Find the position and magnification of the image of an object 40 cm from a concave lens of focal length 20 cm.

This can be determined by a scale drawing as in Fig. 2-14, or calculated.

Applying the sign convention, $u = -40$ cm since the light must travel from left to right, and hence the object is on the left; $f = -20$ cm since

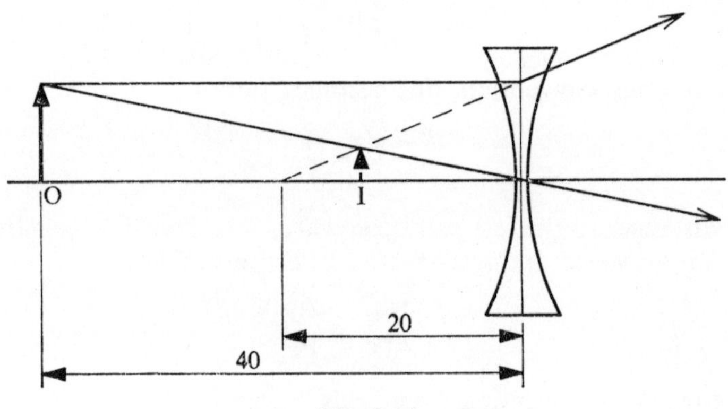


Fig. 2-14

the lens is a diverging one, and hence is considered to have a negative focal length. Substituting in Eqn (2-18)

$$\frac{1}{v} - \frac{1}{u} = \frac{1}{f}$$

$$\frac{1}{v} - \frac{1}{-40} = \frac{1}{-20}$$

i.e.,

$$\frac{1}{v} = -\left(\frac{1}{40} + \frac{1}{20}\right)$$

$$= -\frac{3}{40}$$

$$\therefore\ v = -\frac{40}{3}\ \text{cm}$$

Thus, the image is formed 13·3 cm from the lens on the same side of the lens as the object and is therefore a virtual image.

The magnification is given by

$$M = \frac{v}{u} = \frac{-40/3}{-40} = +\frac{1}{3}$$

that is, the magnification is $\times\frac{1}{3}$ and the positive sign indicates that the image is erect.

Even if a scale drawing is not made, a rough sketch should be done to give an approximate check that the calculation is correct.

**Example 2-4** A two-times magnified image of an object is produced by a convex lens of focal length 4 cm. Find the positions of object and image.

This is possible at two positions of the object as shown in Fig. 2-15.

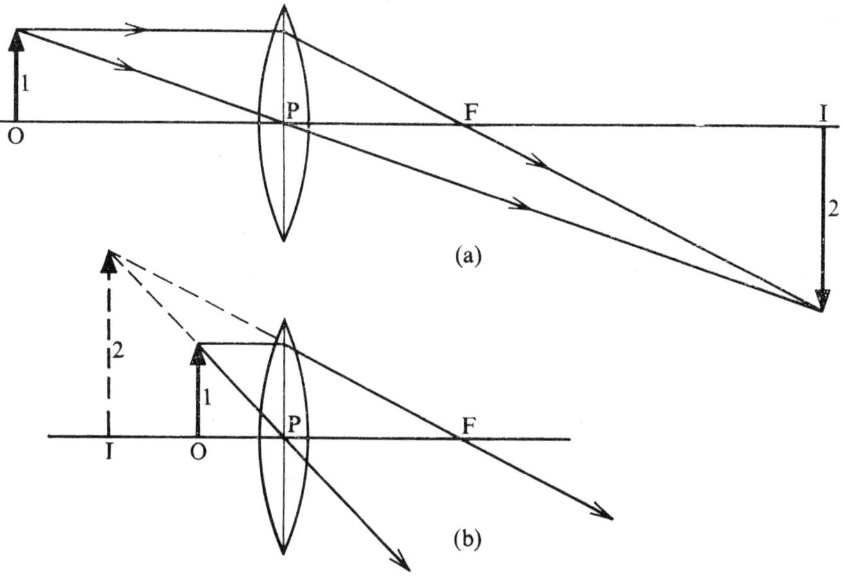


Fig. 2-15

(a) A real image is inverted, hence

$$M = \frac{v}{u} = -2$$

the negative sign indicating the inversion.

$$\therefore\ v = -2u$$

Substituting in

$$\frac{1}{v} - \frac{1}{u} = \frac{1}{f}$$

$$\frac{1}{-2u} - \frac{1}{u} = \frac{1}{+4}$$

i.e.,

$$-\frac{3}{2u} = \frac{1}{4}$$

$$u = -6$$

that is, the object is 6 cm to the left of the lens, as it should be for light travelling from left to right. Also,

$$v = -2u = +12$$

Thus a two-times magnified real inverted image is formed 12 cm from the lens by an object 6 cm from the lens.

(b) Similarly, a virtual image is erect, hence

$$M = \frac{v}{u} = +2$$

$$\therefore\ v = +2u$$

Substituting in

$$\frac{1}{v} - \frac{1}{u} = \frac{1}{f}$$

$$\frac{1}{+2u} - \frac{1}{u} = \frac{1}{+4}$$

hence

$$u = -2$$

$$v = +2u = -4$$

that is, an erect virtual image of an object 2 cm from the lens is formed 4 cm from the lens.

## 2-8 Thin lenses in contact

If two or more thin lenses are in contact, they may be replaced by one lens of a suitable focal length. In Fig. 2-16, two lenses of focal length $f_1$ and $f_2$ are in contact. If the lens of focal length $f_1$ were the only lens, an image of O would be formed at I′ to satisfy the equation

$$\frac{1}{v'} - \frac{1}{u} = \frac{1}{f_1} \tag{2-20}$$

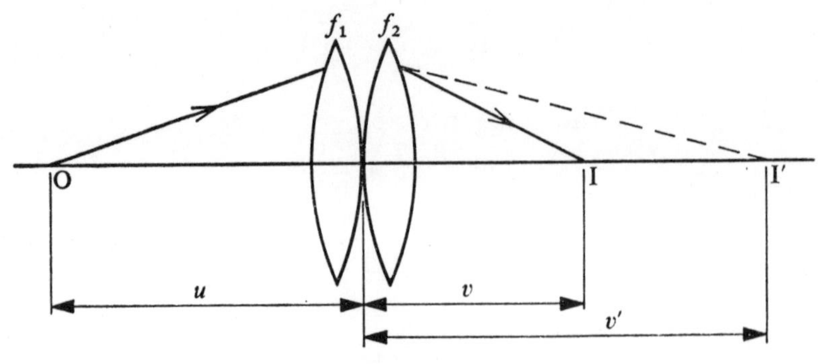


Fig. 2-16 Thin lenses in contact

With the second lens in position, light that would have travelled towards I′ now travels towards I. Thus it may be considered that the second lens forms an image, I, of an object at I′. The positions would now satisfy

$$\frac{1}{v} - \frac{1}{v'} = \frac{1}{f_2} \tag{2-21}$$

since $+v'$ is the object distance. The light is still travelling from left to right.

In practice, of course, no image is formed at I′, and so eliminating $v'$ by adding Eqns (2-20) and (2-21) gives

$$\frac{1}{v} - \frac{1}{u} = \frac{1}{f_1} + \frac{1}{f_2}$$

But the two lenses may be replaced by one lens of focal length $f$ such that

$$\frac{1}{v} - \frac{1}{u} = \frac{1}{f}$$

and therefore

$$\frac{1}{f_1} + \frac{1}{f_2} = \frac{1}{f} \tag{2-22}$$

The sign convention must still be applied when using this equation—that is, a focal length is considered positive for a converging lens and negative for a diverging one.

## 2-9 Refraction at a spherical surface

So far, we have not considered either the actual shape or refractive index of the lens. In Fig. 2-17, a ray from a point O is incident on the spherical surface at a point A. It is then refracted and appears to come from a point I. If these rays make angles $\theta$ and $\theta'$ with the normal to the surface and the object ray makes an initial angle $\psi$ with the axis, the other angles involved are as shown in the figure.

Applying the sign convention and still assuming the aperture is small, so that all the angles considered are small, then

$$\psi = \frac{\text{AP}}{-u} \tag{2-23}$$

$$\psi - \theta = \frac{\text{AP}}{-r} \quad \left(\text{or} \quad \theta - \psi = \frac{\text{AP}}{+r}\right) \tag{2-24}$$

$$\theta' - \theta + \psi = \frac{\text{AP}}{-v} \tag{2-25}$$

Subtracting Eqn (2-24) from (2-23) and (2-25) in turn gives

$$\theta = -\frac{\text{AP}}{u} + \frac{\text{AP}}{r} \tag{2-26}$$

$$\theta' = -\frac{\text{AP}}{v} + \frac{\text{AP}}{r} \tag{2-27}$$

If $n$ is the refractive index of the medium to the left of the surface and $n'$ that of the medium to the right, then

$$\frac{\sin\theta}{\sin\theta'} = \frac{n'}{n}$$

i.e., $$n\theta = n'\theta' \quad \text{for small angles}$$

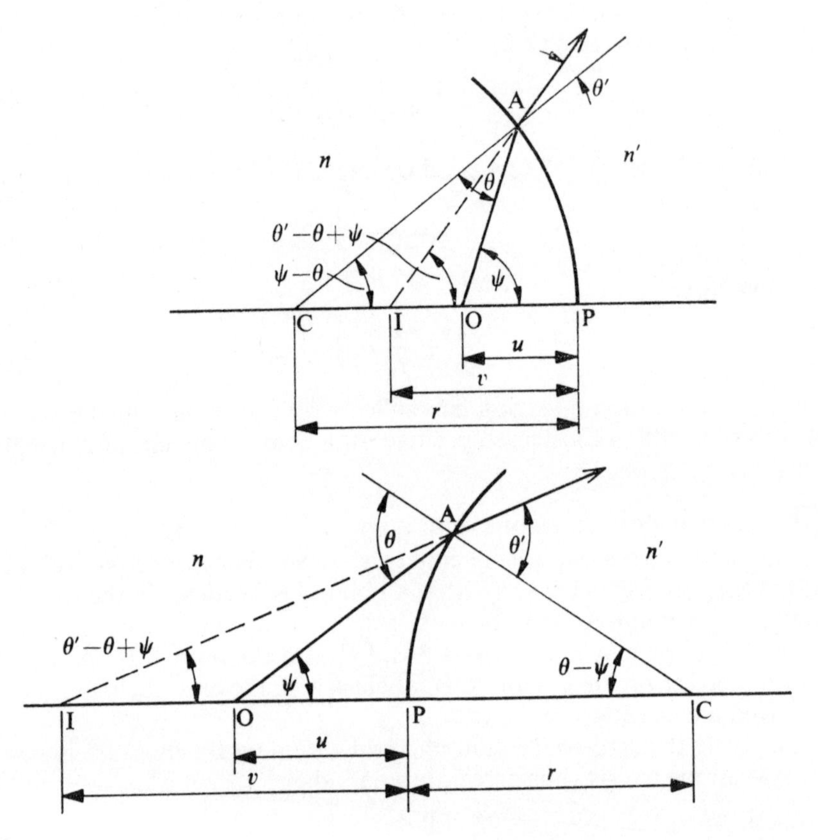


Fig. 2-17 Refraction at a spherical surface

Substituting for $\theta$ and $\theta'$ gives

$$n\left(\frac{\text{AP}}{r} - \frac{\text{AP}}{u}\right) = n'\left(\frac{\text{AP}}{r} - \frac{\text{AP}}{v}\right)$$

$$\therefore \frac{n'}{v} - \frac{n}{u} = \frac{n'-n}{r} \tag{2-28}$$

**Example 2-5** A glass sphere of radius 20 cm and refractive index 1·6 has a small bubble 10 cm in from the surface. Where will this bubble appear to be when viewed from the point nearest to the bubble, as in Fig. 2-18?

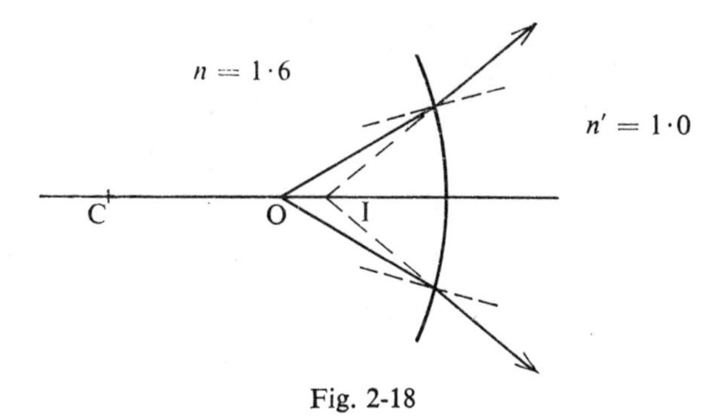

Fig. 2-18

Applying the sign convention, the object distance is −10 cm and the radius of curvature is −20 cm. Thus, substituting in Eqn (2-28)

$$\frac{n'}{v} - \frac{n}{u} = \frac{n' - n}{r}$$

$$\frac{1}{v} - \frac{1{\cdot}6}{-10} = \frac{1 - 1{\cdot}6}{-20}$$

No signs are assigned to the refractive indices, but $n$ refers to the refractive index of the medium to the left of the surface and $n'$ to the medium to the right, remembering that the light must travel from left to right.

$$\therefore \frac{1}{v} = +\frac{0{\cdot}6}{20} - \frac{1{\cdot}6}{10}$$

hence $$v = -7{\cdot}69 \text{ cm}$$

That is, the bubble appears to be 7·69 cm in from the surface instead of 10 cm.

**Example 2-6** A glass sphere of radius 3 cm and refractive index 1·5 is used as an equi-convex lens. One side of the lens is in air and the other side is in contact with water, refractive index 1·33. Calculate the position and magnification of the image of an object located 8 cm from the lens in air (Fig. 2-19).

Although the lens has a thickness of 6 cm, we can still apply thin-lens formulae because the aperture, and hence the angles involved, are small.

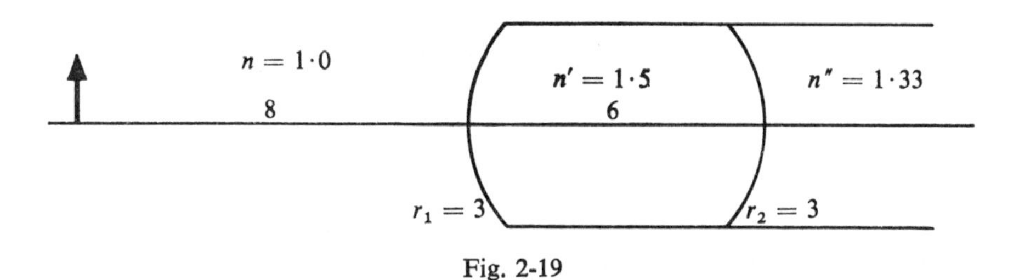

Fig. 2-19

Then, applying Eqn (2-28) to each surface in turn:

*First surface*

$$\frac{1\cdot5}{v} - \frac{1}{-8} = \frac{1\cdot5 - 1}{+3}$$

$$\frac{1\cdot5}{v} = \frac{0\cdot5}{3} - \frac{1}{8} = \frac{1}{24}$$

hence $$v = +36 \text{ cm}$$

Also, $$\text{magnification} = +\frac{v}{u} = \frac{+36}{-8} = -4\cdot5$$

*Second surface*

The image produced by the first surface now becomes the object for the second surface and, since the light is still travelling from left to right, this object distance is $+(36 - 6)$, that is, $+30$ cm, allowing for the thickness of the lens. Then,

$$\frac{1\cdot33}{v} - \frac{1\cdot5}{+30} = \frac{1\cdot33 - 1\cdot5}{-3}$$

i.e., $$\frac{1\cdot33}{v} = \frac{0\cdot17}{3} + \frac{1\cdot5}{30} = \frac{3\cdot2}{30}$$

hence $$v = +12\cdot5 \text{ cm}$$

Also, $$\text{magnification} = +\frac{v}{u} = \frac{+12\cdot5}{+30} = +0\cdot42$$

The total magnification is the product of the separate magnifications, that is,

$$\begin{aligned}\text{magnification} &= (-4\cdot5)(+0\cdot42) \\ &= -1\cdot89\end{aligned}$$

Thus, the image is produced in the water 12·5 cm from the second glass surface, is magnified $\times 1\cdot89$, and inverted.

## 2-10 Refraction at two spherical surfaces

The formula derived for refraction at one surface may be extended to include refraction at two surfaces, as in a lens. It must still be assumed that the lens is of negligible thickness compared with other dimensions involved and that only a small aperture is used so that the refraction angles are small.

In Fig. 2-20, rays, after refraction, appear to converge to form an image at I. For convenience, the figure is drawn to give all distances positive when applying the sign convention. For refraction by the first surface alone, the image would be at I′, given by

$$\frac{n'}{v'} - \frac{n}{u} = \frac{n' - n}{r_1} \qquad (2\text{-}29)$$

where the refractive indices are as indicated in the figure.

For refraction at the second surface, a final image is formed at I of an imaginary object which may be considered to be at I′. Thus, the object distance is now $v'$, the light still travelling from left to right. Applying the formula again to the second surface,

$$\frac{n''}{v} - \frac{n'}{v'} = \frac{n'' - n'}{r_2}$$

$$= -\left(\frac{n' - n''}{r_2}\right) \qquad (2\text{-}30)$$

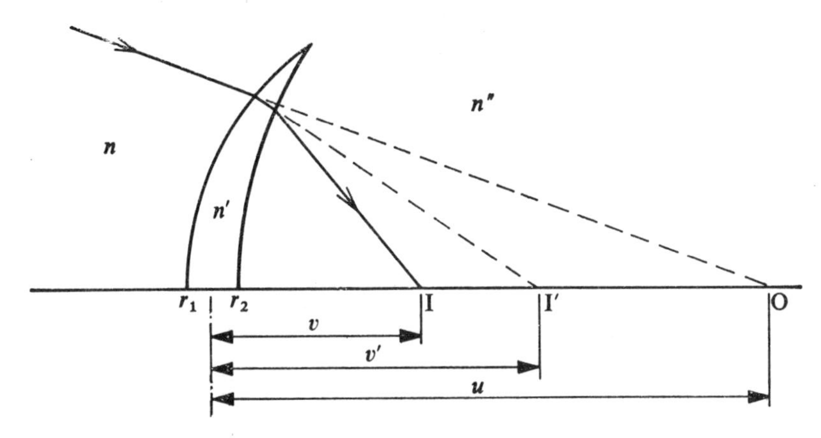


Fig. 2-20 Refraction by two spherical surfaces

Eliminating the constructional distance $v'$ by adding Eqns (2-29) and (2-30) gives

$$\frac{n''}{v} - \frac{n}{u} = \frac{n' - n}{r_1} - \frac{n' - n''}{r_2} \qquad (2\text{-}31)$$

For the usual case where the lens is in one medium, $n = n''$ and therefore Eqn (2-31) simplifies to

$$\frac{n}{v} - \frac{n}{u} = (n' - n)\left(\frac{1}{r_1} - \frac{1}{r_2}\right) = \frac{n}{f} \qquad (2\text{-}32)$$

When the surrounding medium is air, $n = 1$, and

$$\frac{1}{v} - \frac{1}{u} = (n' - 1)\left(\frac{1}{r_1} - \frac{1}{r_2}\right) = \frac{1}{f} \qquad (2\text{-}33)$$

**Example 2-7** Calculate the focal length of a double-convex lens whose refractive index is 1·5 and whose surfaces have radii of curvature of 20 cm and 15 cm (Fig. 2-21).

Substituting in the formula, Eqn (2-33),

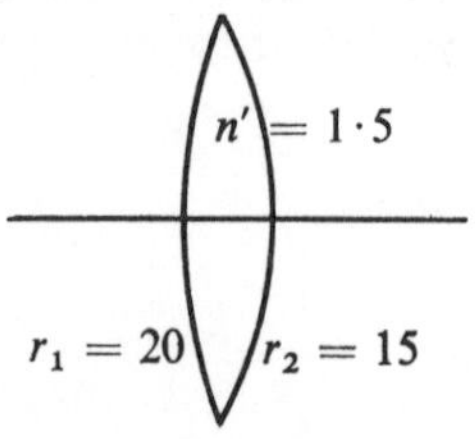

$n' = 1{\cdot}5$

$r_1 = 20 \quad r_2 = 15$

Fig. 2-21

$$\frac{1}{f} = (n' - 1)\left(\frac{1}{r_1} - \frac{1}{r_2}\right)$$

$$= (1{\cdot}5 - 1)\left(\frac{1}{+20} - \frac{1}{-15}\right)$$

$$= \frac{1}{2}\left(\frac{1}{20} + \frac{1}{15}\right) = +\frac{7}{120}$$

Hence $f = +17\tfrac{1}{7}$ cm

That the focal length is positive agrees with the sign convention for a converging lens.

## 2-11 Power notation

It is often more convenient to deal with the power of a lens rather than with its focal length. The power of a lens, measured in diopters, is given by the reciprocal of the focal length as measured in air, expressed in metres, that is,

$$P = \frac{1}{f} \tag{2-34}$$

or diopters = 1/focal length in metres

= 100/focal length in centimetres

Thus, a *converging* lens of focal length 20 cm may be described as a +5 diopter lens, and a *diverging* lens of focal length 25 cm as a −4 diopter lens, or simply as a −4 D lens.

This method of notation has the advantage that powers of individual lenses may be simply added. Thus the equation giving the total focal length of a combination of lenses of different focal lengths [Eqn (2-22)]

$$\frac{1}{f} = \frac{1}{f_1} + \frac{1}{f_2}$$

may be simplified to

$$P = P_1 + P_2 \tag{2-35}$$

This may be extended to any number of thin lenses in contact, since, in general, the power of the combination is equal to the sum of the powers of the individual lenses.

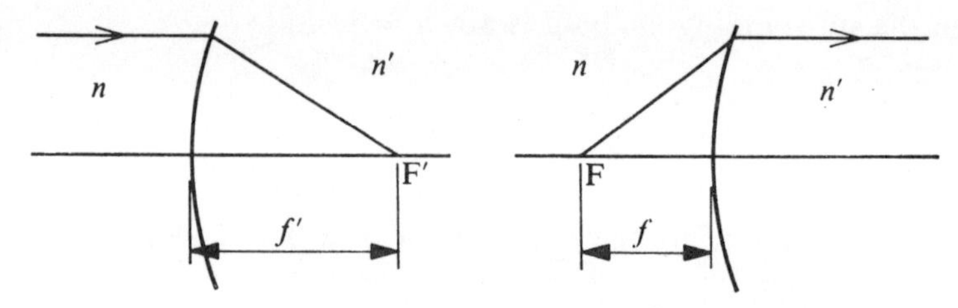


Fig. 2-22 Principal foci of a refracting surface

Further, by Eqn (2-28)

$$\frac{n'}{v} - \frac{n}{u} = \frac{n' - n}{r}$$

a ray parallel to the axis (Fig. 2-22) in a medium of refractive index $n$, by definition, is brought to a principal focus $F'$ in the second medium of refractive index $n'$, such that

$$\frac{n'}{f'} - \frac{n}{\infty} = \frac{n' - n}{r}$$

i.e.,

$$\frac{n'}{f'} = \frac{n' - n}{r}$$

Similarly, a ray diverging from the principal focus F in the first medium is refracted parallel to the axis, thus

$$\frac{n'}{\infty} - \frac{n}{-f} = \frac{n' - n}{r}$$

i.e.,

$$\frac{n}{f} = \frac{n' - n}{r}$$

Therefore, for a single surface of radius $r$, the power is also given by

$$P = \frac{n}{f} = \frac{n'}{f'} = \frac{n' - n}{r} \tag{2-36}$$

## 2-12 Newton's formula

An alternative formula for determining object and image positions for lenses was derived by Newton. So far, all measurements have been made from the pole of the lens but, using Newton's method, all distances are measured from the principal foci. In Fig. 2-23, an image is formed at I of

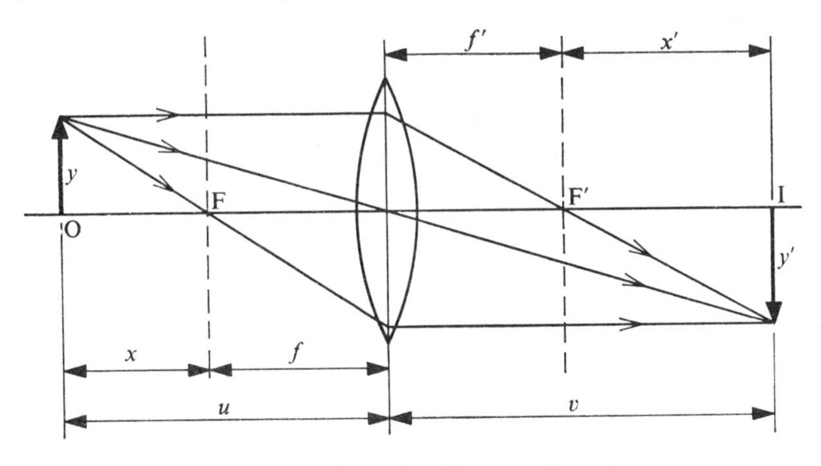


Fig. 2-23 Illustrating Newton's formula

an object at O. If the object is a distance $x$ from F and the image a distance $x'$ from F′, and applying the usual sign convention but with object distances measured on a co-ordinate system centred at F and image distances on a co-ordinate system centred at F′, then object distance

$$u = -(-x + f) = x - f$$

and image distance

$$v = +(f' + x') = x' + f$$

since $f = f'$ for a lens in air, and is regarded as positive for a converging lens.

Substituting in Eqn (2-18)

$$\frac{1}{v} - \frac{1}{u} = \frac{1}{f}$$

$$\frac{1}{x' + f} - \frac{1}{x - f} = \frac{1}{f}$$

i.e.,

$$f(x - 2f - x') = (x' + f)(x - f)$$

hence

$$xx' = -f^2 \tag{2-37}$$

This may also be deduced from a consideration of similar triangles formed, as for example in Fig. 2-23.

To use Newton's formula for a converging lens it is necessary to consider the principal foci, F and F′, as *crossed*. That is, they are reversed in position to that shown in Fig. 2-23, with the distances $x$ and $x'$ now measured from these new positions.

Also from Fig. 2-23,

$$\text{magnification} = \frac{y'}{y} = \frac{f}{x} = -\frac{x'}{f'} \tag{2-38}$$

As usual, a negative value indicates an inverted image.

## 2-13 Summary of formulae

The main formulae derived in this chapter are collected here for convenience.

*Spherical Mirrors*

Eqn (2-14) $\quad \dfrac{1}{v} + \dfrac{1}{u} = \dfrac{1}{f} = \dfrac{2}{r}$

(2-15) $\quad M = \dfrac{y'}{y} = -\dfrac{v}{u}$

*Thin Lenses*

Eqn (2-18) $\quad \dfrac{1}{v} - \dfrac{1}{u} = \dfrac{1}{f}$

(2-19) $\quad M = \dfrac{y'}{y} = +\dfrac{v}{u}$

(2-22) $$\frac{1}{f_1} + \frac{1}{f_2} = \frac{1}{f}$$

(2-28) $$\frac{n'}{v} - \frac{n}{u} = \frac{n' - n}{r}$$

(2-32) $$\frac{n}{v} - \frac{n}{u} = (n' - n)\left(\frac{1}{r_1} - \frac{1}{r_2}\right) = \frac{n}{f}$$

where $n = 1$ for a lens in air.

(2-36) $$\text{Dioptric power} = P = \frac{n}{f} = \frac{n'}{f'} = \frac{n' - n}{r}$$

where distances are measured in metres.

*Sign Convention*

The incident ray travels from left to right.

All distances are measured from the pole of the lens or mirror
—distances to the left are negative
—distances to the right are positive.

A convex lens has a positive focal length.

A concave lens has a negative focal length.

When $M$ is negative, the image is inverted, and conversely.

## PROBLEMS

**2-1** Find graphically and by calculation the image distance, magnification, and nature of the image of an object located 15 cm in front of a concave mirror of radius of curvature 10 cm. What would the results be if the mirror were convex? [$7\frac{1}{2}$ cm, $\times\frac{1}{2}$, real, inverted; $3\frac{3}{4}$ cm, $\times\frac{1}{4}$, virtual, erect]

**2-2** An object 5 mm high is located 5 cm in front of a concave mirror of focal length 10 cm. What is the position, size, and nature of the image? What would the results be if the mirror were convex?
[10 cm, 1 cm, virtual, erect; $3\frac{1}{3}$ cm, $\frac{1}{3}$ cm, virtual, erect]

**2-3** A plane mirror and a convex mirror are arranged 30 cm apart and facing each other. A small object is located midway between them and on the principal axis of the convex mirror. Explain why on looking into the plane mirror two images are seen. If the image formed by two reflections is located 40 cm behind the plane mirror, calculate the radius of curvature of the convex mirror.
[60 cm]

**2-4** A convex and a concave mirror of radii 20 cm and 40 cm respectively are arranged concentrically. A small object is located midway between them and on their common axis. Calculate the image position and magnification resulting from two reflections, the first occurring at the concave mirror.
[8 cm behind convex mirror; $\times\frac{2}{5}$]

**2-5** An object is located (a) 10 cm, (b) 20 cm from a convex lens of focal length 15 cm. Find graphically and by calculation the position, magnification, and nature of the image formed in the two cases.
[(a) 30 cm, $\times 3$, virtual, erect; (b) 60 cm, $\times 3$, real, inverted]

**2-6** A concave lens forms a virtual image 10 cm from the object. If the magnification is $\times\frac{2}{3}$, calculate the focal length of the lens. [60 cm]

**2-7** Show that the minimum distance between an object and its real image as formed by a lens is equal to four times the focal length of the lens. (Hint: distance $= l = v - u = [uf/(u + f)] - u$; then differentiate for a minimum.)

**2-8** Show that the image of a small object, distance $r/n$ from the centre of a glass sphere of radius $r$ and refractive index $n$, is formed at a distance $rn$ from the centre of the sphere when viewed from the side of the sphere remote from the object.

**2-9** A lens is used to produce a four times magnified image on a screen placed 25 cm from the object. Calculate the focal length of the lens. [4 cm]

**2-10** Light from a small luminous object on the axis of a convex lens traverses the lens and is then reflected back through the lens by a plane mirror to form an image coincident with the object. Show that the image may either be erect or inverted.

**2-11** A small bubble is enclosed in a solid sphere of glass of radius 12 cm and refractive index 1·6. If the object is 4 cm from the centre of the sphere, find how much further from the centre its image will be when the object is viewed from the side to which it is (a) nearest, (b) most distant.

[(a) $1\frac{1}{3}$ cm, (b) 4 cm]

**2-12** If the refractive index from air to glass is 3/2, and from air to water 4/3, find the ratio of the focal lengths of a glass lens in water and in air. [4:1]

**2-13** A convex and a concave lens, each of 10 cm focal length, are arranged coaxially 3 cm apart. Find the distance between an object and its final image when the object is on the axis and 15 cm beyond (a) the convex, (b) the concave lens. [(a) 2·1 cm, 39·7 cm; (b) 72 cm, 2·9 cm]

**2-14** Calculate the possible focal lengths of a lens of refractive index 1·5 made by combining two surfaces whose radii of curvature are 10 and 15 cm.

[$\pm$12 cm; $\pm$60 cm]

**2-15** A parallel beam of light is incident onto a glass sphere of radius 2 cm and refractive index 1·61 submerged in water, refractive index 1·33. Calculate the distance beyond the sphere at which the rays come to a focus. [3·75 cm]

# 3 Thick lenses

## 3-1 Introduction

In the previous chapter, equations were derived connecting object and image positions with the focal length of a lens, and with its radii of curvature and refractive index. For many purposes these equations are sufficiently accurate, but it must be remembered that they have two important limitations: the aperture is restricted, that is, the equations are only true for paraxial rays, involving small angles of incidence and refraction; and the thickness is considered negligible in comparison with the focal length. The first of these restrictions is an important one in so far as marginal rays are brought to a slightly different focus than paraxial rays, thus causing spherical aberration. In many cases, however, this will cause only a small error in the calculated positions of the object and image. If such small errors do become important in a particular application, and especially in actual lens computation work, then they can only be dealt with by calculating the paths of individual rays through the system (Section 3-7). The restriction on the thickness is of more practical concern. Many lenses do in fact have a thickness which is quite an appreciable fraction of the focal length. The category of thick lenses includes not only two separated spherical surfaces, but also two or more thin lenses which are not in contact but spaced out to form a coaxial lens system. If we apply the thin-lens formulae directly to such thick-lens systems considerable errors can result. As was shown in Example 2-6, the problem can be dealt with surface by surface, but an alternative approach is to make use of the *cardinal points* of the lens.

## 3-2 Cardinal points and planes

A lens has six cardinal points. These are two principal points, two focal points (principal foci), and two nodal points. Planes normal to the optic axis through these points are the corresponding principal, focal, and nodal planes.

In using a thin-lens formula for a thick lens, the problem is to decide from where the object, image, and focal distances should be measured in order to obtain the correct object and image positions. In a thin lens, of course, these would be measured from the centre, which is considered to be a point that also coincides with both the lens surfaces.

Figure 3-1 shows that the single refracting plane of a thin lens in air

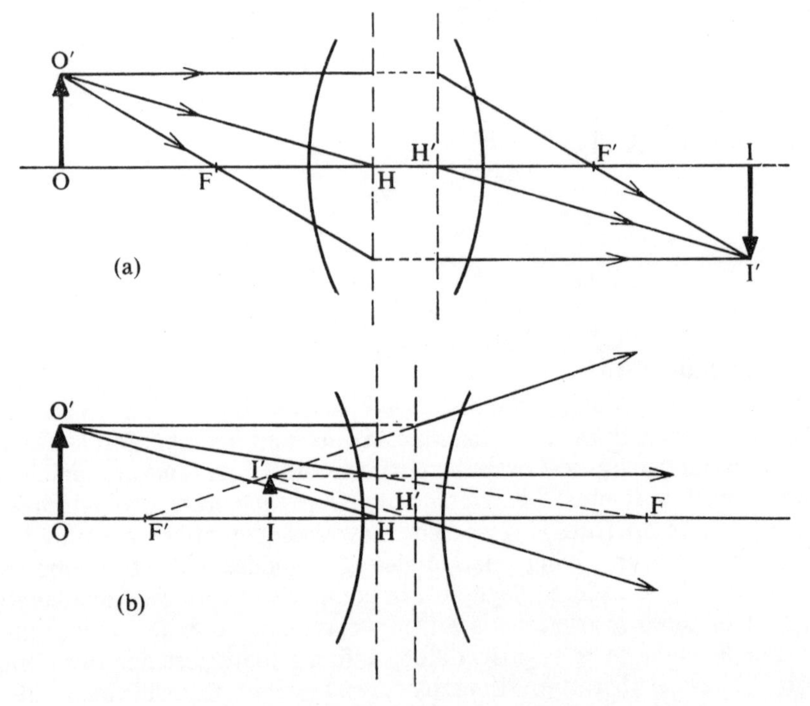


Fig. 3-1 Principal planes of a thick lens

may be replaced by two imaginary refracting planes, the *principal planes*, passing through the *principal points*, H and H′. Thus the two points, H and H′, have replaced the one point at the centre of the thin lens; so O′H and H′I′ are now parallel lines instead of one line passing through the optical centre, as for example in Fig. 2-13. The position of the principal planes is such that if the object, image, and focal distances are measured from H and H′, then $u$, $v$, and $f$ are correctly related by the thin-lens formulae of the previous chapter. For a symmetrical lens, the principal planes are symmetrically located within the lens, but this is not so for an asymmetrical lens (Fig. 3-2). Thus, the principal planes for a lens of given power may be shifted by 'bending' the lens.

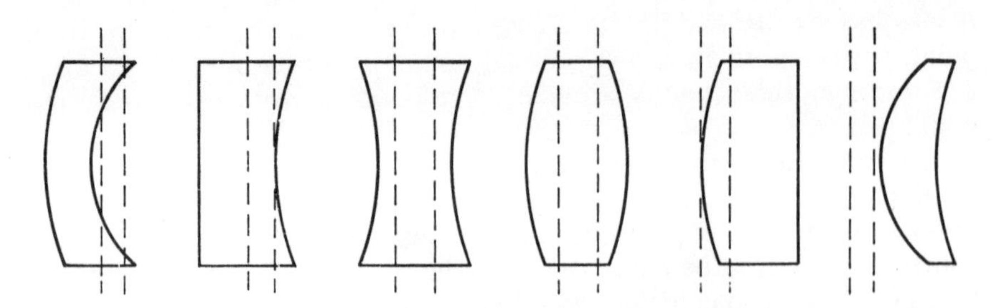

Fig. 3-2 Positions of principal planes

In Fig. 3-1, H is the *first principal plane* and H′ the *second*, although in certain instances the planes may be crossed and occur in the reverse order to that shown. It will also be seen that these are conjugate planes with a one-to-one relationship, thus giving unit magnification. They are, therefore, also termed *unit planes*.

The remaining cardinal points are the principal foci and the nodal points. The *focal points*, F and F′, are by definition the points to which rays parallel to the axis converge after traversing a positive lens system, or appear to diverge from in a negative system. Thus a ray travelling from left to right and parallel to the axis is refracted towards the second principal focus F′ for a positive lens. For a negative lens, the ray is refracted away from the second principal focus, that is, F′ is on the left of the lens.

The position of the *nodal points* is illustrated in Fig. 3-3. A certain incident ray gives rise to a parallel emergent ray. Then the points where

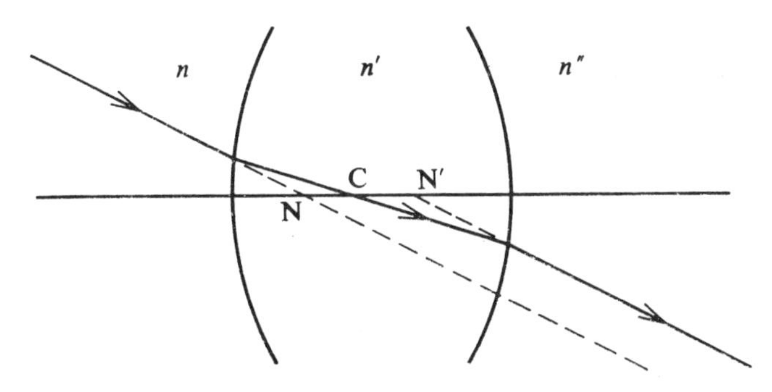


Fig. 3-3 Nodal points of a thick lens

these rays intersect the principal axis are the first and second nodal points, N and N′. These are also conjugate points giving unit angular magnification.

The planes normal to the principal axis through the focal and nodal points are the *focal and nodal planes* respectively.

If the lens is in a homogeneous medium such as air, so that $n = n''$ in Fig. 3-3, then the nodal planes and principal planes coincide. If the lens in Fig. 3-1 were not in a homogeneous medium, the rays O′H and H′I′ would not be parallel and the construction would have to be made by means of the rays passing through the principal focal points only.

Figure 3-3 also defines the position of the *optical centre*, C, of the lens. Incident rays drawn through this point emerge undeviated. For a thin lens N, N′, H, H′, and C coincide and are considered to be at the lens surface.

## 3-3 Graphical construction for thick lenses

The corresponding object and image positions may be determined in exactly the same manner as that used for thin lenses, providing that the defining rays are drawn parallel to the axis between their intersections with the principal planes (Fig. 3-1). To do this, we must first find the positions of the principal planes.

Consider the lens in Fig. 3-4. Let $F_1$ and $F_1'$ be the principal foci for the first surface, and $F_2$ and $F_2'$ those for the second surface. If these positions are not known they may be calculated using Eqn (2-36). Let $C_1$ and $C_2$ be the respective centres of curvature of the two surfaces.

A ray from any point P, on the first focal plane of the first surface, to the centre of curvature $C_1$ of this surface is, of course, a normal to the surface and therefore undeviated. Also, a ray parallel to $PC_1$ and passing through $F_2$, the principal focus of the second surface, is refracted by the second surface at R and emerges as ray RS parallel to the principal axis. Let $F_2R$ intersect the first surface at Q; then since P is a point on the first focal plane

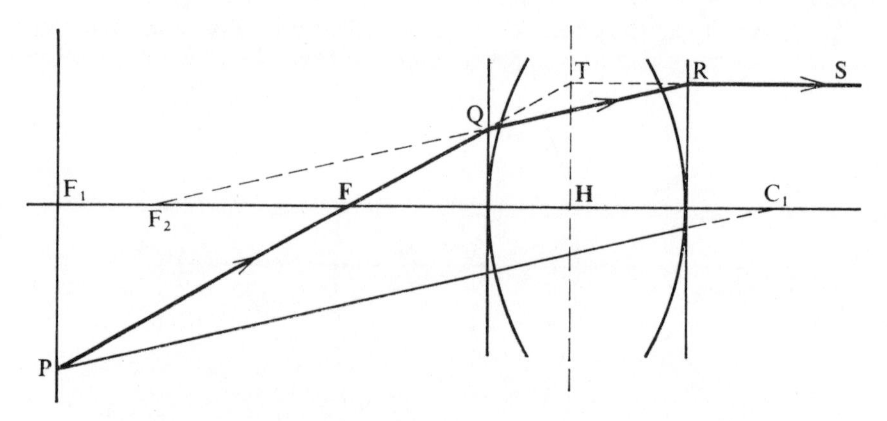


Fig. 3-4 Location of the first principal and focal points of a lens

of the first surface a ray PQ is refracted along QR, parallel to $PC_1$. If PQ intersects the principal axis at F, then, as is seen from the figure, a ray from F is refracted at Q and R and emerges along RS parallel to the principal axis. Thus the point F, by definition, becomes the first principal focus for the complete lens. Also from the figure, all the refraction could be considered as taking place at a point T, thus defining the first principal plane and with it the first principal point, H.

So to establish the position of the first principal and focal points of the lens, rays are drawn as in the figure. First $F_2QR$ is drawn parallel to $PC_1$ where $F_2$ and $C_1$ are fixed positions. Then P and Q are joined and the line produced, and RS is drawn parallel to the axis and produced back to locate point T. In this way the positions of F and H are defined. The lens may then be imagined reversed to locate F′ and H′, the second focal and principal points, in the same manner. Having located these points, thin-lens formulae may be used, with *all distances measured from* H *and* H′.

Although equations may be derived for a single thick lens giving the positions of the cardinal points with respect to the lens surfaces, they are somewhat cumbersome and of little interest. In practice single thick lenses are rarely used. Where they are—for example, in microscope objectives—the formulae are not sufficiently accurate since they consider only paraxial rays and hence make the assumption that $\sin\theta/\sin\theta' = \theta/\theta'$. Thus in the few practical applications where it is necessary to use single thick lenses, it is usually necessary to trace the actual paths of individual rays through the lens. However, if it is required to locate image positions for paraxial rays, it is best to use the methods outlined in the previous chapter, applying Eqn (2-28) to each surface in turn, allowing for thickness. For the practically more important coaxial systems of separated thin lenses, however, the paraxial formulae are often sufficiently accurate to be of practical use, although, since these systems are really 'thick', for the highest accuracy it is still necessary to ray trace so as to obtain the paths of non-paraxial rays.

## 3-4 Two separated thin lenses

Consider first a ray through a single thin lens as shown in Fig. 3-5. The lens is depicted as a line normal to the axis, the principal planes, HH′,

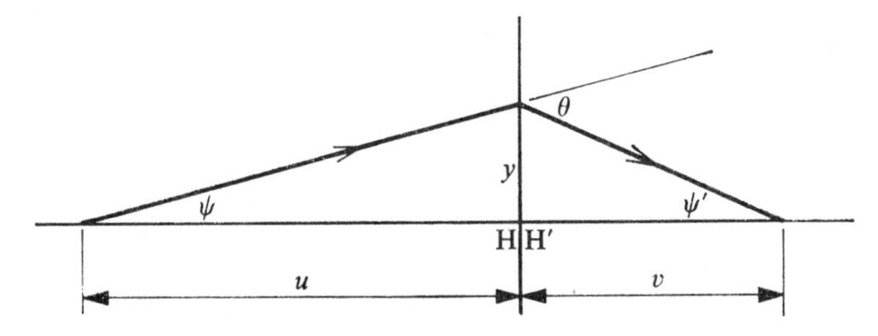


Fig. 3-5 Deviation of a ray by a thin lens

being coincident. Then from the figure, a ray incident to the lens at a height $y$ is deviated through a small angle $\theta$ such that

$$\theta = \psi + \psi' \tag{3-1}$$

$$= \frac{y}{-u} + \frac{y}{v}$$

$$\therefore \frac{\theta}{y} = \frac{1}{v} - \frac{1}{u} = \frac{1}{f} \tag{3-2}$$

i.e., the deviation, $\theta$, of the ray is given by $y/f$.

Consider now two thin lenses in air with principal planes $H_1H_1'$ and $H_2H_2'$, focal lengths $f_1$ and $f_2$, and separated by a distance $d$ (Fig. 3-6). A ray incident at a height $y_1$ onto the first lens is deflected through an angle $\theta_1$ where

$$\theta_1 = \frac{y_1}{f_1} \tag{3-3}$$

This ray then meets the second lens at some height $y_2$ and is deflected through $\theta_2$, where

$$\theta_2 = \frac{y_2}{f_2} \tag{3-4}$$

Eventually it crosses the principal axis at a point F′, which by definition is the second principal focus for the total system. The total deflection of the ray is thus

$$\theta_1 + \theta_2 = \frac{y_1}{f_1} + \frac{y_2}{f_2} \tag{3-5}$$

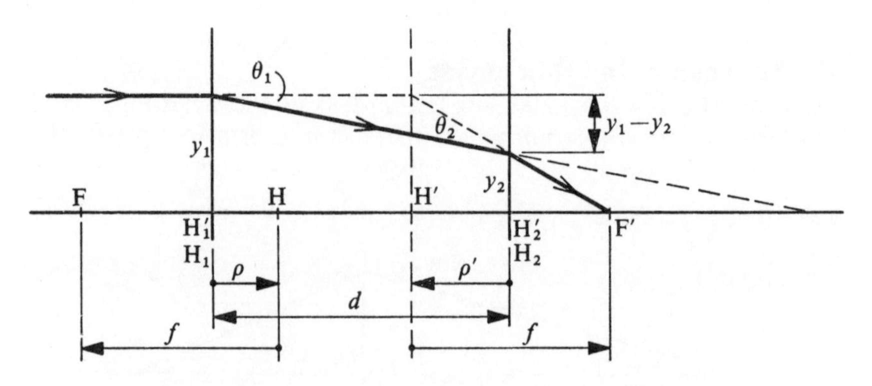


Fig. 3-6 Deviation of a ray by two separated thin lenses

Also, from the figure

$$y_1 - y_2 = \theta_1 d$$

i.e.,

$$y_2 = y_1 - \theta_1 d$$

$$= y_1 - \frac{y_1}{f_1} d$$

$$= y_1\left(1 - \frac{d}{f_1}\right) \tag{3-6}$$

All these deviations may also be considered as taking place in the plane through H′, which is thus the second principal plane for the whole system. Let this plane be a distance $\rho'$ from the second lens. In the figure, H′ is to the left of the second lens and $\rho'$ may thus be regarded as a negative distance. Then, from the figure

$$\rho' = -\frac{y_1 - y_2}{\theta_1 + \theta_2} \tag{3-7}$$

Substituting from Eqns (3-5) and (3-6)

$$\rho' = -\frac{\dfrac{y_1}{f_1}d}{\dfrac{y_1}{f_1} + \dfrac{y_2}{f_2}}$$

$$= -\frac{\dfrac{y_1}{f_1}d}{\dfrac{y_1}{f_1} + \dfrac{y_1}{f_2}\left(1 - \dfrac{d}{f_1}\right)}$$

$$= -\frac{f_2 d}{f_1 + f_2 - d} \qquad (3\text{-}8)$$

The separation $d$ is always regarded as positive. Similarly, if the first principal plane of the combined system passes through H and is situated a distance $\rho$ to the right of the first lens, that is, a positive distance, then

$$\rho = +\frac{f_1 d}{f_1 + f_2 - d} \qquad (3\text{-}9)$$

Similarly from the figure, the focal length, $f$, of the combined system is the distance between H′ and F′ (or H and F since the system is in air). Then

$$f = \frac{y_1}{\theta_1 + \theta_2}$$

Substituting as before

$$f = \frac{y_1}{\dfrac{y_1}{f_1} + \dfrac{y_1}{f_2}\left(1 - \dfrac{d}{f_1}\right)}$$

$$= \frac{f_1 f_2}{f_1 + f_2 - d} \qquad (3\text{-}10)$$

or rearranging

$$\frac{1}{f} = \frac{1}{f_1} + \frac{1}{f_2} - \frac{d}{f_1 f_2} \qquad (3\text{-}11)$$

This may also be written in power notation, giving the combined power $P$ for the system as

$$P = P_1 + P_2 - P_1 P_2 d \qquad (3\text{-}12)$$

In particular, for two thin lenses in contact so that $d = 0$, Eqn (3-11) simplifies to

$$\frac{1}{f} = \frac{1}{f_1} + \frac{1}{f_2}$$

as has previously been shown [Eqn (2-22)].

Thus Eqns (3-8), (3-9), and (3-10) may be used to find the positions of the principal planes and the focal length of the combined system. Knowing these, object and corresponding image positions may be determined, using the thin-lens formula, $1/v - 1/u = 1/f$, where the object distance $u$ is measured from the first principal point, H, and the image distance $v$ is measured from the second principal point, H′. Similarly, the first and second principal focal distances (both $f$) are also measured from H and H′.

The *equivalent lens* of the combination is a thin lens of the same focal length as the combination and placed at the position of the first principal plane of the system. Such a lens gives the same lateral magnification as the system but the image distance is measured from the thin lens and not from the second principal plane, giving an incorrect image position. Thus the equivalent lens is only equivalent as far as the lateral magnification is concerned.

**Example 3-1** Consider, as an example of the use of the equations developed, a system consisting of a concave lens, focal length 15 cm, and a convex lens, focal length 10 cm, mounted coaxially 5 cm apart in air (Fig. 3-7).

Then, applying the correct signs to the focal lengths and, as always, counting the separation $d$ as a positive number,

$$\rho = +\frac{f_1 d}{f_1 + f_2 - d} = +\frac{(-15)5}{(-15) + (+10) - 5}$$

$$= +\frac{-75}{-10}$$

$$= +7{\cdot}5 \text{ cm}$$

$$\rho' = -\frac{f_2 d}{f_1 + f_2 - d} = -\frac{(+10)5}{(-15) + (+10) - 5}$$

$$= -\frac{+50}{-10}$$

$$= +5 \text{ cm}$$

$$f = \frac{f_1 f_2}{f_1 + f_2 - d} = \frac{(-15)(+10)}{(-15) + (+10) - 5}$$

$$= \frac{-150}{-10}$$

$$= +15 \text{ cm}$$

Thus, since $\rho = +7{\cdot}5$ cm, the first principal point H is located 7·5 cm to the right of the first lens. Similarly, as $\rho' = +5$ cm, the second principal point H′ is 5 cm to the right of the second lens. Also, the principal focal points of the system are located at F and F′, 15 cm to the left and right of

H and H′ respectively. Also, since $f = +15$ cm, the system is a converging one.

If in a particular problem the focal lengths of the individual lenses were not known, they could be calculated using, for example, Eqn (2-33). If in the above example an object were located a distance 20 cm from the first

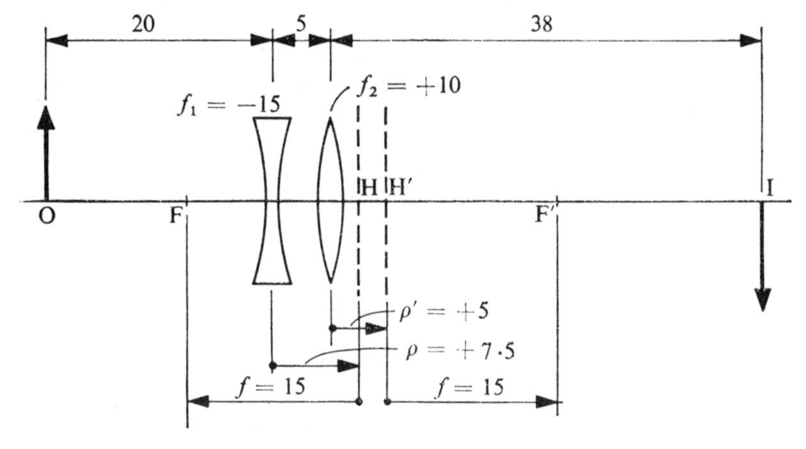


Fig. 3-7

lens, as in the figure, then the object distance $u$, measured from the first principal plane H, would be $u = -(20 + 7{\cdot}5) = -27{\cdot}5$ cm. Since also $f = +15$ cm

$$\frac{1}{v} - \frac{1}{u} = \frac{1}{f}$$

becomes

$$\frac{1}{v} - \frac{1}{-27{\cdot}5} = \frac{1}{+15}$$

from which $$v = +33 \text{ cm}$$

Thus the image is real and located 33 cm to the right of the second principal point H′, as in the figure. That is, it is $(33 + 5) = 38$ cm from the second lens. The lateral magnification is

$$M = \frac{v}{u} = \frac{+33}{-27{\cdot}5}$$

$$= -1{\cdot}2$$

That is, the image is inverted and magnified $\times 1{\cdot}2$.

**Example 3-2** The principal points H and H′ may be in reverse order, that is, the planes may be said to be crossed as in the following example.

Consider two converging lenses of focal length 15 cm and 20 cm, mounted coaxially 10 cm apart (Fig. 3-8). Then

$$d = +10 \text{ cm}$$
$$f_1 = +15 \text{ cm}$$
$$f_2 = +20 \text{ cm}$$

$$\rho = +\frac{f_1 d}{f_1 + f_2 - d} = +\frac{15.10}{15 + 20 - 10} = +\frac{150}{25}$$
$$= +6 \text{ cm}$$

$$\rho' = -\frac{f_2 d}{f_1 + f_2 - d} = -\frac{20.10}{15 + 20 - 10} = -\frac{200}{25}$$
$$= -8 \text{ cm}$$

$$f = +\frac{f_1 f_2}{f_1 + f_2 - d} = +\frac{15.20}{15 + 20 - 10} = +\frac{300}{25}$$
$$= +12 \text{ cm}$$

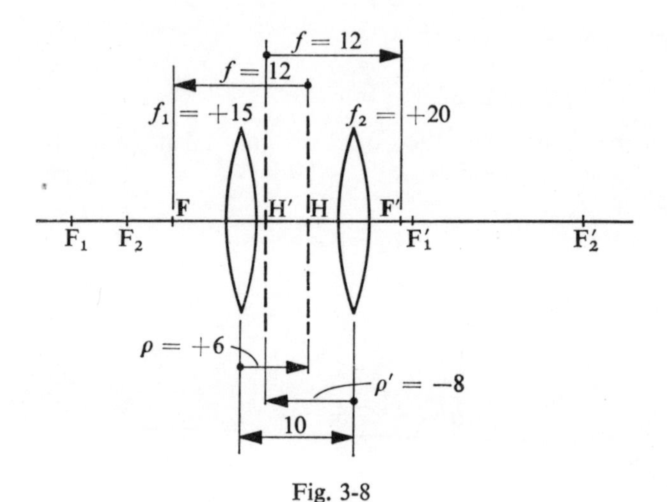


Fig. 3-8

Hence a converging system of focal length 12 cm is formed with the positions of the principal planes as shown in the figure. Since the planes at H and H′ are crossed, care must be exercised in applying the thin-lens formula, $1/v - 1/u = 1/f$, so that the object distance $u$ is measured from H and the image distance $v$ from H′.

## 3-5 Two separated thin lenses—alternative method

Newton's method, discussed in Section 2-12, may be used to provide a completely different approach to the problem of two thin lenses. By measuring from the focal points, formulae may be derived that involve distances measured from the principal foci of the system, rather than from the principal points. Knowing the positions of the principal foci and the focal length of the system, the principal points H and H′ are readily determined, as were the positions of F and F′ once H and H′ were known in the previous method. It is a matter of personal choice which method is used in a particular problem; but the method of the previous section is often simpler to apply in practice.

In Fig. 3-9 the principal foci of the two individual thin lenses of focal length $f_1$ and $f_2$ are located at $F_1$, $F_1'$ and $F_2$, $F_2'$ respectively. It is important to see that the principal foci are correctly labelled. Thus the first principal focus is to the left of a positive lens but to the right of a negative lens. F and F′ are the principal foci of the combined system. $\Delta$, the separation of the two systems, is the distance $F_1'F_2$ and is considered positive if $F_2$ is to the right of $F_1'$, as drawn.

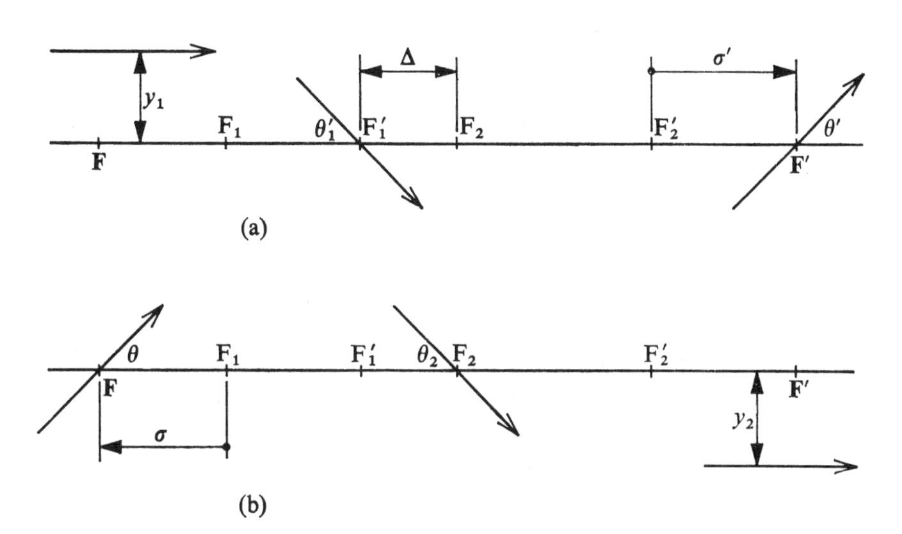


Fig. 3-9 Rays through a system of two separated thin lenses

A ray parallel to the axis and incident from the left at a height $y_1$ onto the first lens surface is refracted to pass through $F_1'$, but it also passes through F′, the second principal focus for the combined system. Thus for the second lens, $F_1'$ and F′ are conjugate points. Hence, applying Newton's formula, $xx' = -f^2$ [Eqn (2-37)], to this lens and remembering that all distances are measured from the appropriate focal points and with the appropriate signs,

$$-\Delta . \sigma' = -f_2^2$$

where $f_2$ is the focal length of the second lens and $\sigma'$ is as indicated in the figure. No sign is given to $\sigma'$ as it is not yet located.

$$\therefore \ \sigma' = +\frac{f_2^2}{\Delta} \tag{3-13}$$

Similarly from Fig. 3-9(b)

$$\sigma = -\frac{f_1^2}{\Delta} \tag{3-14}$$

since the position of F is to be determined and hence no sign is yet given to $\sigma$. A negative value of $\sigma$ will indicate that F is to the left of $F_1$ and vice versa. The same applies to $\sigma'$ measured from $F_2'$ to give the position of F′.

For small angles, by Eqn (3-2),

$$\begin{aligned} y_2 &= f_2\theta_2 \quad \text{for second lens} \\ &= f.\theta \quad \text{for combined system} \end{aligned}$$

i.e.,

$$\frac{\theta}{\theta_2} = \frac{f_2}{f} \tag{3-15}$$

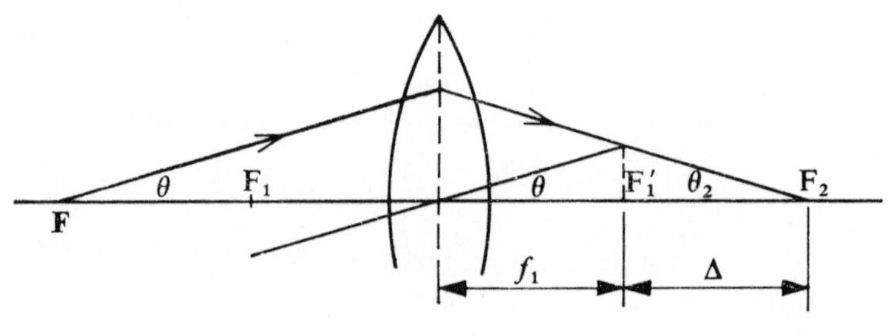


Fig. 3-10

Considering rays through the first lens in more detail, as in Fig. 3-10, it will also be seen that

$$\frac{\theta}{\theta_2} = \frac{\Delta}{-f_1} \tag{3-16}$$

Therefore, from Eqns (3-15) and (3-16)

$$f = -\frac{f_1 f_2}{\Delta} \tag{3-17}$$

The convention is applied that $\Delta$ is a positive distance if $F_2$ is to the right of $F_1'$. It will also be seen by comparing Eqns (3-17) and (3-10) that $\Delta = -(f_1 + f_2 - d)$, where the separation $d$ is always regarded as positive. Hence, using Eqns (3-13) and (3-14), the positions of the principal foci may be found for the combined system and its focal length from Eqn (3-17). Knowing these, object and corresponding image positions may be determined using Newton's formula $xx' = -f^2$, all measurements of $x$ and $x'$ being made from F and F′.

Alternatively, since the positions of the principal foci are known as well as the focal length of the system, the positions of the principal points H and H′ are known. Knowing these, the thin-lens formula, $1/v - 1/u = 1/f$, may be used as before.

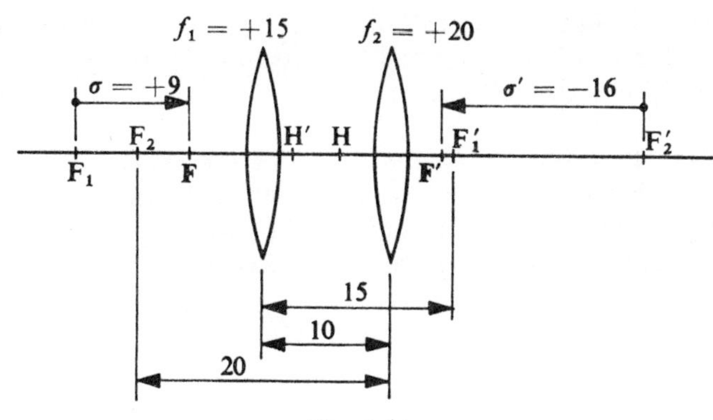


Fig. 3-11

**Example 3-3** As an example of the use of the formulae, consider the same example as dealt with in Example 3-2 and Fig. 3-8.

Since $F_2$ is to the left of $F'_1$, the separation $\Delta$ is a negative distance (see Fig. 3-11),

i.e.,
$$\begin{aligned} \Delta = F'_1F_2 &= -(f_1 + f_2 - d) \\ &= -(15 + 20 - 10) \\ &= -25 \text{ cm} \end{aligned}$$

$$\sigma = -\frac{f_1^2}{\Delta} = -\frac{15^2}{-25} = +9 \text{ cm}$$

$$\sigma' = +\frac{f_2^2}{\Delta} = +\frac{20^2}{-25} = -16 \text{ cm}$$

$$\begin{aligned} f = -\frac{f_1 f_2}{\Delta} &= -\frac{15.20}{-25} \\ &= +12 \text{ cm} \end{aligned}$$

Hence a converging system is formed of focal length 12 cm with $\sigma$ measured from $F_1$ and $\sigma'$ measured from $F'_2$, giving the positions of the

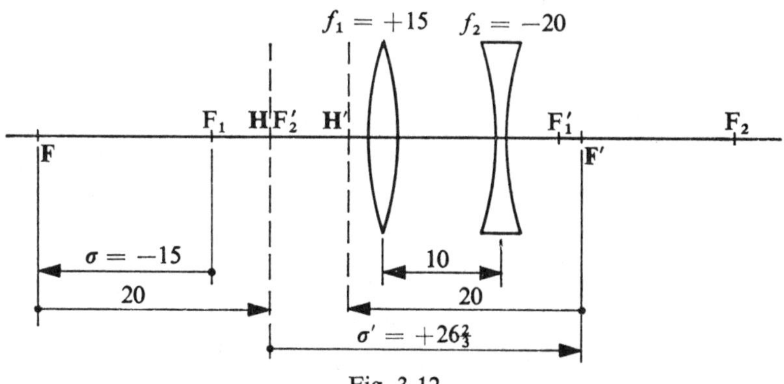


Fig. 3-12

principal foci, F and F′, as in Fig. 3-11, which should also be compared with Fig. 3-8. Since F and F′ are 12 cm from the principal points H and H′, their position is also established.

**Example 3-4** It is most important to label correctly the principal foci in this method. Thus if in the previous example the second lens were negative instead of positive, the positions of $F_2$ and $F'_2$ would be reversed. The first principal focus is now on the right of the lens and the second on the left. This is shown in Fig. 3-12.

$F_2$ is to the right of $F'_1$ and therefore $\Delta$ is positive.

$$\begin{aligned}\Delta = F'_1F_2 &= -(f_1 + f_2 - d)\\ &= -(+15 - 20 - 10)\\ &= +15 \text{ cm}\end{aligned}$$

$$\sigma = -\frac{f_1^2}{\Delta} = -\frac{+15^2}{+15} = -15 \text{ cm}$$

$$\sigma' = +\frac{f_2^2}{\Delta} = +\frac{(-20)^2}{+15} = +26\tfrac{2}{3} \text{ cm}$$

$$f = -\frac{f_1 f_2}{\Delta} = -\frac{15(-20)}{+15} = +20 \text{ cm}$$

Hence this is a converging system of focal length 20 cm. The principal foci and principal points are shown in the figure, remembering that H and H′ are 20 cm from F and F′.

## 3-6 Practical applications

Practical uses of these formulae are shown in Chapter 5 when the design and operation of a number of optical instruments are discussed but, as an immediate example, we may consider the operation of a *telephoto objective*. This is used to photograph a small distant object and obtain a large image. Since the size of the image of such a distant object is directly proportional to the focal length of the lens or system of lenses, a system of long focal length is required. This could involve an inconveniently long camera or a camera with a very large bellows movement. However, a telephoto lens overcomes this inconvenience and provides a compact system. As shown in Fig. 3-13, the objective is made up of a positive and a negative component with a large separation, and so the second principal point H′ is located well in front of the objective. This means that the focal length of the combination, as measured from H′, is quite large, but the distance from the back lens to the second principal focus at F′, termed the *back focal length*, is relatively short and so we have a compact system with a long focal length.

In practice, each of the lenses of the combination is a doublet to give the necessary aberration corrections, as discussed in the next chapter. By

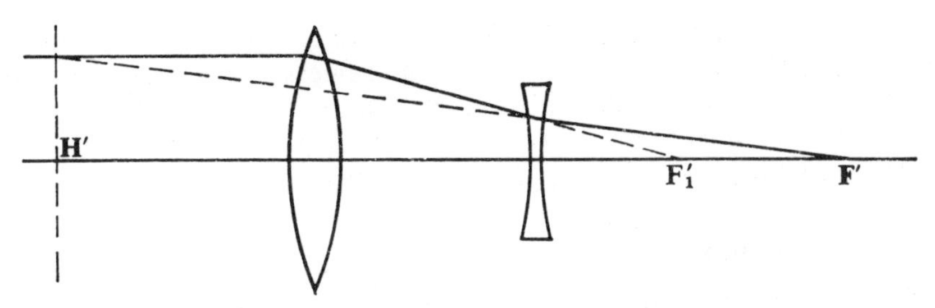


Fig. 3-13 Telephoto lens system

varying the separation of the component lenses, it is possible to change the focal length of the combination and hence the magnification. However, this is difficult and expensive to accomplish, because complex motions are needed to keep the image in focus on the film. Also, to maintain satisfactory aberration corrections with the varying focal length more lens elements are required, thus adding to the complexity of the motions and to the cost of the system. A reasonable compromise is to use several objectives each with a different fixed focal length.

Section 3-2 defined the nodal points of a lens as the points at which parallel incident and emergent rays cut the principal axis. Although these points are of only limited use, they have an important practical application in *panoramic cameras*. Given a lens of focal length $f$ and with nodal points at N and N′ then an incident ray through N emerges parallel through N′, by definition. Figure 3-14 shows that if a film is curved into an arc, of

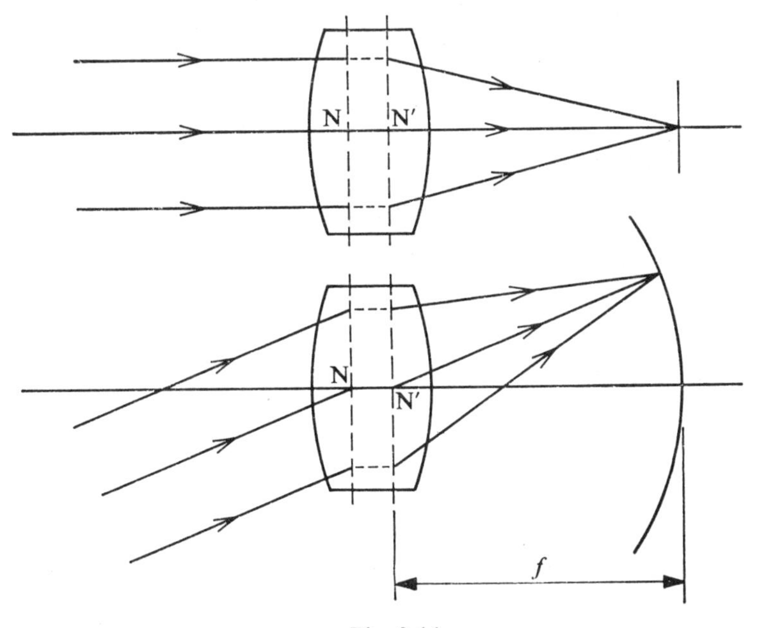


Fig. 3-14

radius $f$ centred on N′, incident parallel rays are brought to a focus on the film. To use this principle in a camera, the camera lens is made to pivot about N′. Usually a slit shutter moves in front of the curved film so that incident rays parallel to the lens axis are focused at each point on the film. By dealing with mainly axial rays, the lens aberrations are minimized.

## 3-7 Ray tracing

So far this chapter has dealt only with paraxial rays—that is, rays that are close to the axis and hence involve only small angles of incidence and refraction at lens surfaces. This allowed us to assume that $\sin \theta/\sin \theta' = \theta/\theta'$ and so use the cardinal points, to derive a valuable set of formulae. These formulae are accurate enough for all practical purposes, provided only the paraxial rays are involved. If this is not so, small errors occur in the calculated distances. These errors increase as the radius of curvature becomes shorter and as the point of refraction is moved farther from the axis. That is, the error increases as $\sin \theta/\sin \theta'$ diverges from $\theta/\theta'$. Where accuracy is necessary in such conditions—for example, in the design of microscope objectives—we must trace the paths of individual rays through the system. Figure 3-15 shows the arrangement of lenses in a typical microscope objective designed to give a magnification of $\times 40$, showing the large refraction angles involved. To deal with this type of problem, a new series of formulae are required.

Consider now a ray, not paraxial, incident at a height $h$ onto a refracting surface of radius $r$, as in Fig. 3-16. Let the incident ray be deviated to cut the axis at a point I, instead of at O. The distances and angles are as indicated in the figure. As drawn, all the distances are positive.

Then, applying the sine rule to the triangle AOC

$$\sin \theta = \frac{(u - r)}{r} \sin \psi \tag{3-18}$$

Also, by Snell's law

$$\sin \theta' = \frac{n}{n'} \sin \theta \tag{3-19}$$

In the triangle AIC

$$\psi' = \psi + \theta - \theta' \tag{3-20}$$

and applying the sine rule

$$v - r = \sin \theta' \frac{r}{\sin \psi'} \tag{3-21}$$

Hence

$$v = (v - r) + r \tag{3-22}$$

Thus, by application of Eqns (3-18) to (3-22), the position of the image due to the first refraction is computed. This then becomes the object for refraction by the second surface and $\psi'$ becomes the new value of $\psi$. The

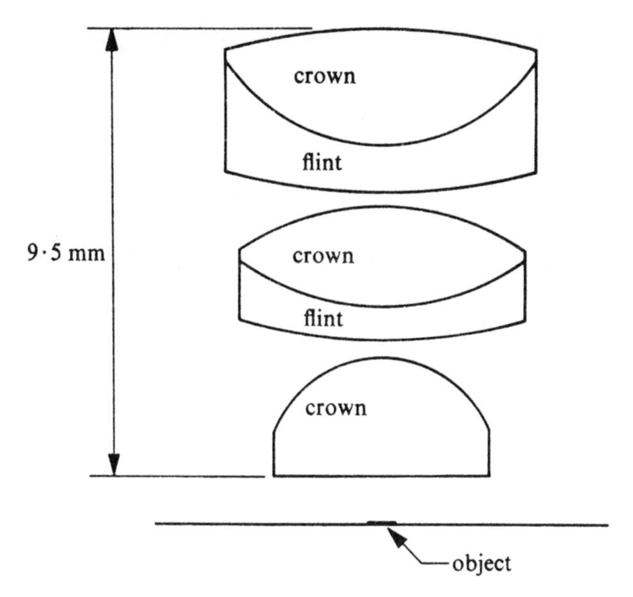


Fig. 3-15 Arrangement of lenses of ×40 microscope objective

computation is then repeated for the second surface. If the incident ray is parallel to the axis, $\psi = 0$, and the computation starts at Eqn (3-19) after having first calculated $\theta$ corresponding to the incident height $h$.

For paraxial rays, the same equations may be used, but now, since the angles involved are small, they are expressed in radians and are equal to the sines. Equations may similarly be derived for refraction at plane surfaces and for reflection by spherical mirrors.

We must remember to apply the sign convention. All distances measured to the right of the pole P are regarded as positive and all distances to the

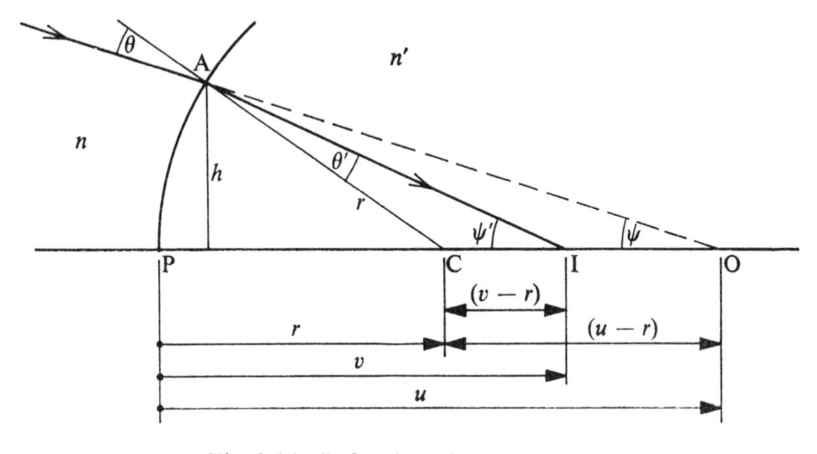


Fig. 3-16 Refraction of a marginal ray

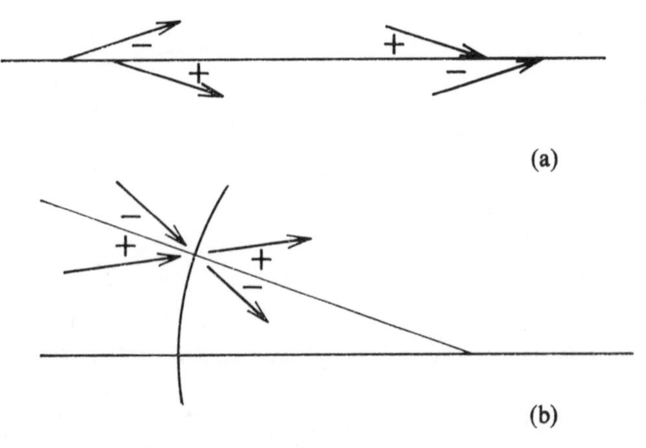


Fig. 3-17 Angle sign convention

left as negative, with the light travelling from left to right. An extra convention is also required for angles:

(*a*) *for rays meeting the axis, measure from axis to ray,*
(*b*) *for rays meeting a surface away from the axis, measure from ray to radius.*

*In both instances, a clockwise rotation gives a positive angle and an anti-clockwise rotation a negative angle.*

This is illustrated in Fig. 3-17. Thus in Fig. 3-16 all the angles named were positive.

**Example 3-5** Consider the path of a ray through the first surface of the lens depicted in Fig. 3-18. Let this lens have a diameter of 2·5 cm with the radius of curvature of the convex first surface of 4 cm and let the lens be made from a glass of refractive index 1·6120 for sodium yellow light. Consider the path of a marginal ray from an axial point 20 cm from this first surface, assuming the lens to be in air.

Allowing for a bezel to hold the lens, consider this marginal ray to be

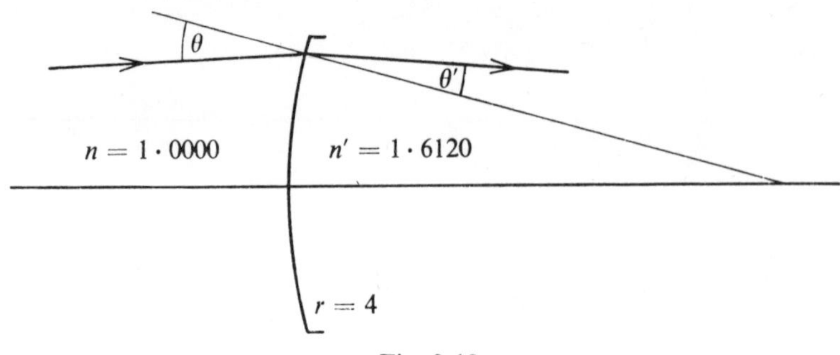


Fig. 3-18

incident on the convex surface 1 cm from the axis. Then the sag, $s$, of the curved surface is given by

$$(2r - s)s = h^2$$

(Fig. 3-19—rectangular property of circle theorem).

$$\therefore \quad -2rs + s^2 = -h^2$$

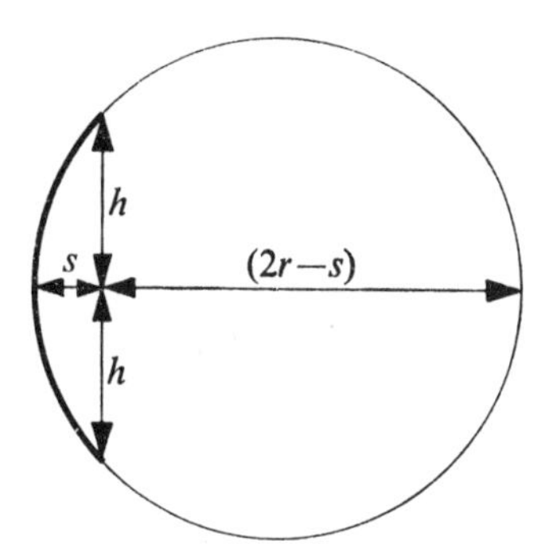


Fig. 3-19 Sag of a curve

Completing the square

$$r^2 - 2rs + s^2 = r^2 - h^2$$

$$\therefore (r - s)^2 = r^2 - h^2$$

and

$$s = r - (r^2 - h^2)^{1/2} \qquad (3\text{-}23)$$

Hence

$$\begin{aligned} s &= 4 - (4^2 - 1^2)^{1/2} \\ &= 4 - (15)^{1/2} \\ &= 4 - 3{\cdot}8730 \\ &= 0{\cdot}127 \text{ cm} \end{aligned}$$

Therefore the distance of the object from the plane of intersection of the marginal ray with the lens is $(20 + 0{\cdot}127) = 20{\cdot}127$ cm.

Hence the angle $\psi$ is given by $\tan^{-1}(1/20{\cdot}127)$, that is, $\psi = 2°50'40''$. Obviously for very steep rays the error introduced by neglecting the sag becomes quite appreciable. To compute the further path of this ray we use Eqns (3-18) to (3-22) and apply the correct signs. The equations are best broken down into easily computable steps, as is shown in Table 3-1. The calculations are most easily made by a calculating machine.

Table 3-1 also shows a paraxial trace for comparison. This is not made for a ray of a particular height of incidence, which would be negligible, but for a ray making an arbitrary small angle with the axis, for example, 0·001 radians. The angles then involved, measured in radians, may be taken as equal to the sines of the angles. The computed image distance $v$ is then 15·050 cm for the marginal ray but 15·649 cm for the paraxial ray.

That is an axial difference, termed the *longitudinal spherical aberration*, of 6 mm, about 4% of the image distance. For lenses and rays involving larger angles of incidence and refraction, this aberration can be considerably more and accumulative over several surfaces.

Using the appropriate thin-lens formula [Eqn (2-28)], that is,

$$\frac{n'}{v} - \frac{n}{u} = \frac{n' - n}{r}$$

then
$$\frac{1{\cdot}612}{v} - \frac{1}{-20} = \frac{1{\cdot}612 - 1}{+4}$$

i.e.,
$$\frac{1{\cdot}612}{v} + \frac{1}{20} = \frac{0{\cdot}612}{4}$$

from which
$$v = +15{\cdot}65 \text{ cm}$$

Table 3-1

| *Marginal ray* | | *Paraxial ray* | |
|---|---|---|---|
| $u$ | $-20{\cdot}0000$ | $u$ | $-20{\cdot}00000$ |
| $r$ | $+4{\cdot}0000$ | $r$ | $+4{\cdot}00000$ |
| $u - r$ | $-24{\cdot}0000$ | $u - r$ | $-24{\cdot}00000$ |
| $\frac{u-r}{r}$ | $-6{\cdot}0000$ | $\frac{u-r}{r}$ | $-6{\cdot}00000$ |
| $\psi$ | $-2^\circ\ 50'\ 40''$ | | |
| $\sin\psi$ | $-0{\cdot}0496$ | $\psi$ | $-0{\cdot}00100$ |
| $\frac{u-r}{r}\sin\psi = \sin\theta$ | $+0{\cdot}2977$ | $\frac{u-r}{r}\psi = \theta$ | $+0{\cdot}00600$ |
| $\theta$ | $+17^\circ\ 19'\ 20''$ | | |
| $n$ | $1{\cdot}0000$ | $n$ | $1{\cdot}00000$ |
| $n'$ | $1{\cdot}6120$ | $n'$ | $1{\cdot}61200$ |
| $\frac{n}{n'}$ | $0{\cdot}6203$ | $\frac{n}{n'}$ | $0{\cdot}62034$ |
| $\frac{n}{n'}\sin\theta = \sin\theta'$ | $+0{\cdot}1847$ | $\frac{n}{n'}\theta = \theta'$ | $+0{\cdot}00372$ |
| $\theta'$ | $+10^\circ\ 38'\ 38''$ | | |
| $\psi' = \psi + \theta - \theta'$ | $+3^\circ\ 50'\ 2''$ | | |
| $\sin\psi'$ | $+0{\cdot}0669$ | $\psi' = \psi + \theta - \theta'$ | $+0{\cdot}00128$ |
| $\frac{\sin\theta'}{\sin\psi'}$ | $+2{\cdot}7624$ | $\frac{\theta'}{\psi'}$ | $+2{\cdot}91236$ |
| $r\frac{\sin\theta'}{\sin\psi'} = v - r$ | $+11{\cdot}0496$ | $r\frac{\theta'}{\psi'} = v - r$ | $+11{\cdot}64945$ |
| $v = (v - r) + r$ | $+15{\cdot}0496$ | $v = (v - r) + r$ | $+15{\cdot}64945$ |

This agrees with the previous value for the paraxial trace. Thus the thin-lens formulae, if necessary used in conjunction with the cardinal points, may be regarded as exact formulae, providing their use is confined to paraxial rays. Where angles of incidence and refraction greater than about two degrees are involved, it becomes necessary to trace the path of each particular ray.

It should be noted that due to small differences being cumulative during the ray trace computation, the calculations have to be made to a far greater accuracy than that to which the lens can be made. The traces would also be normally repeated for other refractive indices, corresponding to other wavelengths of light, to determine the longitudinal chromatic aberration.

## PROBLEMS

**3-1** A 10-cm focal length convex lens and a 12-cm focal length concave lens are arranged coaxially 18 cm apart. What is the focal length of the combination and the distance between the principal planes? Find the position of the image and its magnification for an object located 20 cm beyond the convex lens. Check the image position and magnification by applying the thin-lens formula to each lens in turn. [6 cm; 16·2 cm; 2·4 cm beyond concave lens; ×1·2]

**3-2** Two 10-cm focal length convex lenses are arranged coaxially 10 cm apart. An object is located 20 cm beyond one of the lenses. What is the magnification of the image and the focal length of the equivalent lens of the combination? [$\times\frac{1}{2}$; 10 cm]

**3-3** What must be the separation of two thin lenses of power +10 D and −4 D respectively, if the power of the combination is to be +10 D? Locate the principal focal points and planes for the combination. What would the power be if the lenses were in contact?
[10 cm; $FH_1 = 14$; $F'H_2 = 0$; $HH_1 = 4$; $H'H_2 = 10$; +6 D]

**3-4** A negative and a positive lens, of focal length 15 and 20 cm respectively, are in contact. What is the focal length of the combination? What is the focal length if the lenses are separated by 30 cm? With the separated combination, an object is situated in turn at (a) 18 cm (b) 42 cm beyond the concave lens and at (c) 18 cm (d) 42 cm beyond the convex lens. Calculate the distances of the respective images from the corresponding second lenses.
[−60 cm; +12 cm; (a) +42 cm; (b) +39 cm; (c) −14 cm; (d) +18 cm]

**3-5** A positive meniscus lens with radii of curvature 10 and 15 cm is made of glass of refractive index 1·5. Calculate its focal length. Two such lenses are arranged coaxially a distance 30 cm apart with their concave surfaces facing each other. Show that the principal planes are crossed and symmetrically placed 10 cm apart. [+60 cm]

**3-6** In a Huygenian eyepiece the separation of the two thin convex lenses is equal to half the sum of their focal lengths. If the focal length of the field lens is twice that of the eye lens, show that the power of the eyepiece is $3/4f$ where $f$ is the focal length of the eye lens.

**3-7** A parallel ray close to the optic axis is incident onto a hemispherical lens of radius $a$ and refractive index $n$ and is brought to a focus outside the lens. Show that the principal foci are located at distances $na/(n-1)$ and $a/n(n-1)$ from the plane surface. Hence, show that one principal point is located at the vertex of the curved face and that the other is at a point inside the lens and distance $a(n - 1)/n$ from the vertex.

**3-8** Show that the combined power of two thin lenses of power $P_1$ and $P_2$ respectively is given by $P = P_1 + P_2 - dP_1P_2$, where $d$ is their separation. Hence, show that the power of a single thick lens of thickness $d$, refractive index $n$, and radii of curvature $r_1$ and $r_2$ is given by

$$P = (n - 1)\left(\frac{1}{r_1} - \frac{1}{r_2} + \frac{n - 1}{n} \cdot \frac{d}{r_1 r_2}\right)$$

An equiconvex lens of radii 10 cm and refractive index 1·5 has a thickness of 2 cm, what is its power? [+9·67 D]

**3-9** Show how to determine by a graphical construction the position of the first focal and principal points of a thick diverging lens.

# 4 Aberrations of mirrors and lenses

## 4-1 Introduction

The primary purpose of most optical systems of lenses and mirrors is to produce an image of some object. It is important to know how well this job can be done in practice. In the previous chapter we saw how it is possible to find the exact position of the axial image by tracing the path of a ray from an axial point object. We also saw that primary rays differently inclined to the axis produce images at different points on the axis. Thus the net effect of all the rays is to produce a composite image which is the sum of all the images, each slightly displaced from all the others. The composite image is thus not a perfect representation of the object and is said to be affected by aberration. If, in addition, the object is off-axis, other forms of aberration will result. Before discussing the different types of aberration, let us investigate whether a perfect image can ever be formed by a lens or mirror, even in theory. To do this, it is first necessary to discuss Fermat's principle.

## 4-2 Fermat's principle

*When a ray of light travels from one point to another through any set of optical media, the actual optical path taken is, to the first approximation, equal to other optical paths closely adjacent to the actual path taken.*

This statement, known as Fermat's principle, means that if $ds$ is an element of the path of a ray in a medium of varying refractive index, $n$ (see Fig. 4-1), then the optical path of a ray from A to B, $\int_A^B n\, ds$, will be an extremum, usually a minimum, for the real ray A to B. The *optical path* is the sum of the products of all the geometric distances multiplied by their corresponding refractive indices taken along the ray path. By Fermat's principle, a ray path *closely* adjacent to this, shown dotted in the figure, will at the most be only of second-order difference in length.

Consider, for example, how this principle applies to reflection at a plane mirror, as in Fig. 4-2. A ray from A to B is reflected at O, forming an image of B at B′. If now the point O is displaced to O′ so that the ray path is AO′B, then no matter how it is displaced, the path AOB will always be shorter than the path AO′B. Thus the true optical path is an extremum, in this case a minimum, and is in agreement with Fermat's principle.

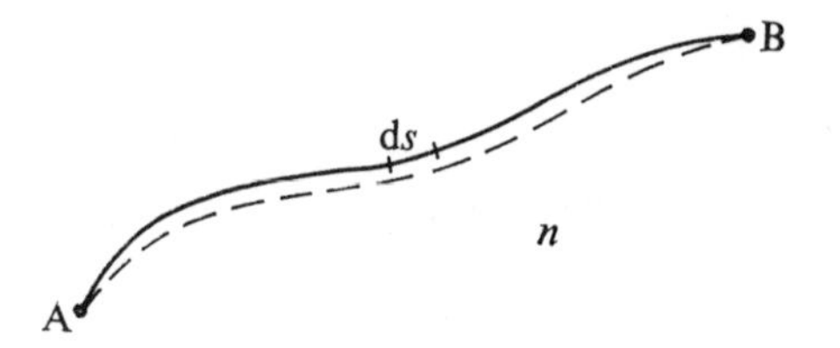


Fig. 4-1 Illustrating Fermat's principle

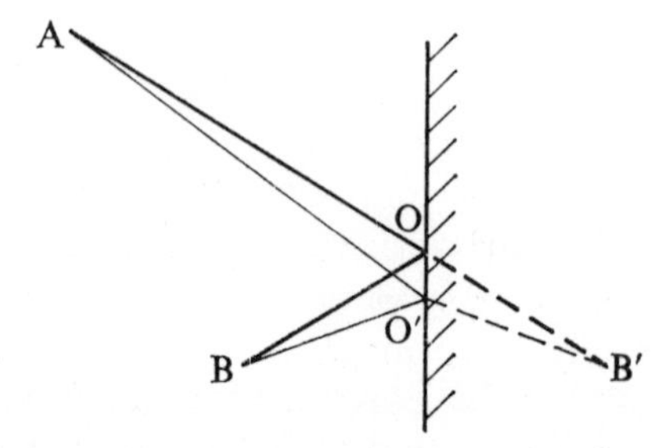


Fig. 4-2 Fermat's principle applied to reflection at a plane mirror

A focusing system is formed if many rays can be drawn from a point A to a point B, each taking a different route (Fig. 4-3) but all having exactly the same optical path length—that is, $(AA')n + (A'B')n' + (B'B)n''$ is constant for each ray.

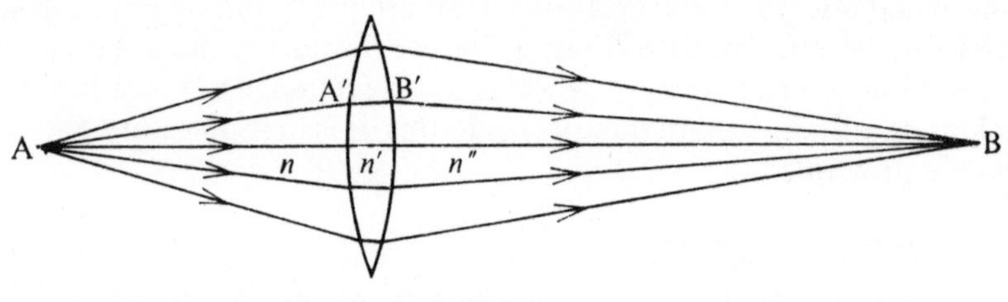


Fig. 4-3

## 4-3 Sine condition for perfect image formation

Consider a perfect image A′B′ (Fig. 4-4) formed by an optical system of an object AB. Aaa′A′ and Bbb′B′ represent conjugate ray paths. Then,

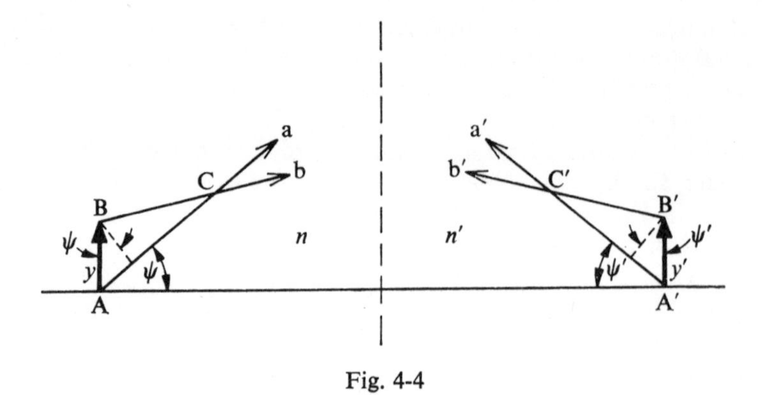


Fig. 4-4

by Fermat's principle, the optical path of any ray, Aaa′A′, is a constant and similarly so is the optical path Bbb′B′. Thus their difference,

$$\text{Aaa}'\text{A}' - \text{Bbb}'\text{B}' = \text{const.}$$

i.e., $$(\text{AC} + \text{Caa}'\text{C}' + \text{C}'\text{A}') - (\text{BC} + \text{Cbb}'\text{C}' + \text{C}'\text{B}') = \text{const.} \quad (4\text{-}1)$$

Also, by Fermat's principle

$$\text{Caa}'\text{C}' = \text{Cbb}'\text{C}' \quad (4\text{-}2)$$

Therefore, by subtraction

$$(\text{AC} + \text{C}'\text{A}') - (\text{BC} + \text{C}'\text{B}') = \text{const.}$$

i.e., $$(\text{AC} - \text{BC}) + (\text{C}'\text{A}' - \text{C}'\text{B}') = \text{const.} \quad (4\text{-}3)$$

i.e., $$ny \sin \psi + n'y' \sin \psi' = \text{const.} \quad (4\text{-}4)$$

providing $y$ and $y'$ are small, and where $n$ and $n'$ are the refractive indices of the object and image spaces respectively.

For perfect image formation Eqn (4-4) must be true for all ray paths, including that for which $\psi = \psi' = 0$. For this to be true, the constant of Eqn (4-4) must be zero. Also, by the sign convention of the previous chapter, $\psi$ is regarded as a negative angle and therefore

$$ny \sin \psi = n'y' \sin \psi' \quad (4\text{-}5)$$

This is known as *Abbe's Sine Condition* and is the condition for the formation of a perfect image of a small lateral object centred on the axis. Note that no restriction has been placed on the size of the angles involved. Equation (4-5) implies that the ratio of the sines of the angles made with the axis by each pair of corresponding incident and refracted rays is constant, since the lateral magnification $y'/y$ must be constant for perfect image formation. If the lens obeys this condition, a perfect image will be formed on axis. So long as $y$ and $y'$ are small, as specified for Eqn (4-4), then a lens will also give a near perfect image just off the axis. This will be considered further when coma is discussed in Section 4-7.

## 4-4 Herschel's relation

Consider Fig. 4-5, then, in the same way as for the sine condition of the previous section and as in Eqn (4-3)

$$(\text{AC} - \text{BC}) + (\text{C}'\text{A}' - \text{C}'\text{B}') = \text{const.} \quad (4\text{-}6)$$

that is, providing $x$ and $x'$ are small

$$nx \cos \psi + n'x' \cos \psi' = \text{const.} \quad (4\text{-}7)$$

Then, as before, this must be true for all angles, including $\psi = \psi' = 0$, i.e.,

$$nx + n'x' = \text{const.} \quad (4\text{-}8)$$

Therefore, combining Eqns (4-7) and (4-8),

$$nx(1 - \cos \psi) = n'x'(1 - \cos \psi') \quad (4\text{-}9)$$

since $x$ and $x'$ are in opposite directions and may be regarded as having opposite signs. Applying the sign convention to the angles does not affect the equation, since $\cos(-\psi) = \cos\psi$. Rewriting Eqn (4-9) in the form

$$nx \sin^2 \frac{\psi}{2} = n'x' \sin^2 \frac{\psi'}{2} \tag{4-10}$$

gives one form of *Herschel's relation* for the perfect imaging of a longitudinal small object. Since the longitudinal magnification $x'/x$ must be constant, it will be seen that Abbe's sine condition and Herschel's relation [Eqns (4-5) and (4-10)] are not generally compatible. Thus, an optical system can be corrected to give a perfect image of an object centred on the axis if the object is small and is either perpendicular or parallel to the axis but not both. Thus it is usually impossible to form a perfect image of a three-dimensional object even if it is small.

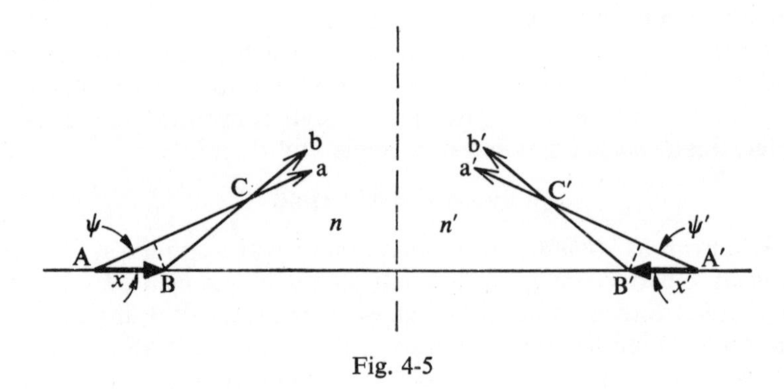


Fig. 4-5

The only situation in which both conditions could hold simultaneously is when $\psi = \psi'$. Then Eqn (4-5) becomes

$$ny = n'y' \tag{4-11}$$

and Eqn (4-10) becomes

$$nx = n'x' \tag{4-12}$$

$$\therefore \quad \frac{y'}{y} = \frac{x'}{x} = \frac{n}{n'} \tag{4-13}$$

that is, the lateral magnification, $y'/y$, and the longitudinal magnification, $x'/x$, are both equal to the ratio of the refractive indices of the object and image spaces.

Although the two conditions may be satisfied separately for arbitrarily large angles, they do not hold if the object and image are large or are off-axis. In practice, the majority of instruments effectively involve two-dimensional objects and images—for example, a camera which is focused on an object at a particular distance and forms an image on a flat film. Thus the Abbe sine condition is of great importance and in fact may be considered as one of the most important principles of geometric optics.

However, there are many cases in practice where the sine condition is either not satisfied due to poor optical design or cannot hold because the object and image are large or off-axis. The resulting image is poor due to aberrations. These aberrations will now be discussed individually in detail.

## 4-5 Third-order theory

In the previous chapters, formulae have been derived that involve the paraxial rays only, that is, all the angles involved are small and therefore $\sin\psi = \psi$. These formulae were mainly originally derived by Gauss and are therefore termed 'Gaussian'. In practice, the apertures of optical systems are comparatively large so that the paraxial rays are only a small fraction of the rays accepted. The Gauss theory was therefore extended by von Seidel to include the next term in the expansion of sine, that is, he used

$$\sin\psi = \psi - (\psi^3/3!) \tag{4-14}$$

Since this term in the expansion involves the angle raised to the third power, the resulting theory is called *third-order theory*, whereas the Gauss theory is called *first-order theory*. Seidal extended the theory as a series of five correction terms to the Gauss theory. When these corrections, called the *five Seidel sums*, are individually all equal to zero in a particular optical system then the paraxial, or Gaussian, formulae hold. When any one or more of the sums is not zero the paraxial formulae are not exact and aberrations result. Each non-zero sum leads to a particular type of aberration. The Seidel sums are such that they must be taken in order, so that a particular sum being zero can only lead to the absence of a particular aberration if the proceeding sums are also zero. The five non-zero Seidel sums lead to the following five aberrations, in order, spherical aberration, coma, astigmatism, curvature of field, and distortion. These aberrations are also called the five *monochromatic* aberrations since light of only one wavelength is considered. If more than one wavelength is used, little variation results in the Seidel aberrations but chromatic aberration has to be considered, a discussion of which will be deferred until Chapter 6. In practice, there are insufficient design parameters to allow all the Seidel sums to be made individually and simultaneously zero and thus a perfect system cannot be made. The normal practice is to examine each aberration separately and to make zero only those aberrations that are of the greatest importance in the particular optical system. The less important aberrations are neglected. For large aperture systems the fifth-order terms in the expansion of sine become important and in turn higher order aberrations may result, but these become very difficult to handle theoretically.

## 4-6 Spherical aberration

This occurs when the first of the Seidel sums is not zero. As the previous chapter showed, paraxial and marginal rays from a point object on the axis do not generally come to the same point focus. The distance between the two point foci measured along the axis—that is, the difference between the image distances—is the *longitudinal spherical aberration* whereas the radius of cross-section of the pencil of rays at the paraxial focus is the *lateral spherical aberration*. Spherical aberration is generally considered as positive when the marginal ray comes to a focus inside the paraxial focus. The blurring produced by spherical aberration is the result of separate overlapping images of slightly different magnification and focus.

Spherical aberration can be controlled to some extent by 'bending' the lens—that is, by varying the curvature of the refracting faces. In this way lenses can be produced which have different shapes but the same diameter and power. The amount of bending is measured by a term called the *shape factor*, $q$, defined as

$$q = \frac{r_2 + r_1}{r_2 - r_1} \tag{4-15}$$

where $r_1$ and $r_2$ are the two radii of curvature. For a thin lens the spherical aberration can be minimized by bending the lens but it cannot be completely removed. As would reasonably be expected, the least amount of spherical aberration occurs when the work of refraction is equally divided between the two surfaces. Thus for a converging lens used such that the object and image are at equal distances from the lens, the minimum spherical aberration occurs with an equi-convex form—that is, the shape factor, $q$, equal to zero. For light parallel to the axis the optimum lens shape depends on lens thickness and occurs when $q$ is near to unity. The parallel light is incident onto the side with the shorter radius of curvature. When the lens is reversed, the spherical aberration is increased. Thus a lens must be designed to give the least spherical aberration for the conditions under which it is to be used.

Although the spherical aberration produced by a single lens can be minimized, it cannot normally be eliminated. It can, however, be eliminated if a system of two or more lenses is used. For example, in a positive system, a lens of positive focal length and with a small amount of positive spherical aberration may be used in conjunction with a low power lens of negative focal length having a correspondingly higher negative spherical aberration, to produce a system with an overall positive focal length and in which the positive and negative spherical aberrations cancel each other. By a suitable choice of different glasses and by designing the two lenses to have surfaces with a common radius of curvature, the two lenses may be joined at their matching surfaces with Canada balsam* to produce a cemented *doublet*.

By combining lenses with spherical aberrations of opposite signs, the aberrations may thus be made to cancel out, but only for a particular zone. For example, if there is no spherical aberration between paraxial rays and marginal rays, it does not follow that there is no spherical aberration due to paraxial rays and rays other than marginal, although such aberration will generally be small. In many cases it is therefore best to

* Canada balsam is made from the sap of the North American balsam fir and is normally dissolved in xylene. For cementing purposes, it is baked to harden it by driving off the solvents. When dry it has a refractive index of about 1·54.

balance out aberrations due to rays passing through the lens at different heights by reducing to zero the spherical aberration between paraxial rays and rays passing through some arbitrary zone, such as a zone corresponding to 0·7 of the aperture, rather than for the marginal rays.

The only way in which the spherical aberration may be reduced to zero for a single lens or mirror is to aspherize a surface—that is, to make a surface non-spherical. Thus, for example, a parabolic mirror will focus parallel light to a point focus free of spherical aberration. Rays from a point source on axis are similarly focused by an ellipsoidal mirror. Lenses are more difficult to aspherize because the required refracting surface is not generally of simple geometric form. Aspherizing is an expensive process and it is therefore usual to keep to spherical surfaces and reduce the spherical aberration by extra surfaces and by a suitable choice of radii.

## 4-7 Coma

If the second of the Seidel sums is not zero, then *coma* results. When spherical aberration is zero, all rays from an axial point object are focused to an axial point image; however, the rays from a point object off the axis may not be focused to one point. In Fig. 4-6(a), the lens is free from spherical aberration, so that all rays from an axial point A are imaged at a point A′, whereas the rays from a point B, just off-axis, are not simply imaged to one point. The lateral magnification is generally different for rays passing through different zones of a lens. Thus, if the lens is considered as divided into circular zones, numbered 0 to 5, with zone 0 at the centre of the lens and zone 5 the outer zone, then rays from point B passing through the central zone will be imaged at $B_0'$, while those passing say

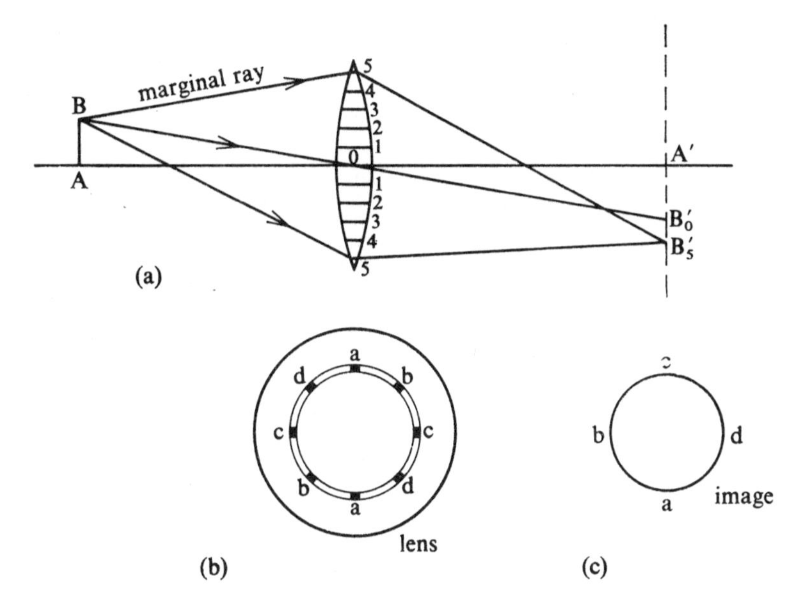


Fig. 4-6 Formation of coma

through zone 5 will be imaged at $B'_5$, both points lying on the plane through A′. Rays from B passing through other zones will be focused at points between $B'_0$ and $B'_5$. Thus, if only rays in the plane of the figure were concerned, point B would be imaged as a line. But this is not so, since the zones are circular and skew rays—that is, rays not in the plane of the figure—are involved.

Figure 4-6(b) shows one particular zone in the plane of the lens. Two rays from point B passing through the diametrical points 'aa' in this zone are brought to a focus at a point 'a' in the image plane [Figure 4-6(b)]. Similarly, rays through the points 'bb' are brought to a focus at 'b', and so on for the rays through other diametrical points of the zone. The resultant image is a circle, called the *comatic circle.*

Rays passing through other circular zones of the lens produce similar circular images. These circles increase in diameter and become further from the axis as the square of the diameter of the corresponding zone increases. Thus the total image due to all the rays passing through the lens from point B will consist of a series of overlapping circles, as in Fig. 4-7.

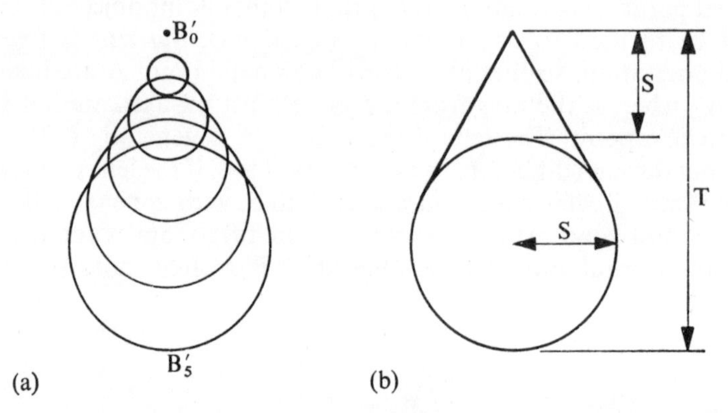


Fig. 4-7 Coma

Most of the light goes into the region of the smallest circles so that the image has a comet-like appearance, and hence the name coma for this type of aberration. The magnitude of the coma is given in two ways, as the *sagittal coma*, $S$, and as the *tangential coma*, $T$, where $S$ and $T$ are the lengths as indicated in Fig. 4-7(b) and where $T = 3S$. The coma in Fig. 4-6 is said to be positive, since the lateral magnification due to the marginal rays is greater than that due to the central rays. For the converse, the coma is said to be negative.

The fact that the magnification is different for rays through different zones means that the focal lengths as measured for these rays are different. If a lens is free from both spherical aberration and coma, these focal lengths are all the same. In the previous chapter, the focal length was defined as the distance between the principal focus and the corresponding principal plane. For the focal length to be a constant for all rays the principal plane of first-order theory must become a spherical surface, termed the principal surface, in third-order theory. As shown in Fig. 4-8 the centre of curvature of the principal surface, H′, is at the principal focus. This shows that, whereas spherical aberration is concerned with the position of the focal point, coma is concerned with the shape of the principal surface.

Since coma is positive or negative, it may, like spherical aberration, be eliminated from a system by the use of two lenses of opposite sign. It is of great practical importance that a doublet can be corrected for spherical aberration and coma at the same time. Unlike spherical aberration, coma may be completely eliminated in a single lens by 'bending',—that is, varying the shape factor. The spherical aberration of a lens shaped to eliminate coma will be minimal or near minimal.

As for spherical aberration, coma can usually be eliminated only for a particular zone and then only when the system is used in the manner for which it was designed. If an image free from both coma and spherical aberration for all zones can be formed for one particular object point, the system is said to be *aplanatic*. In practice, however, this term is often applied if these aberrations are eliminated for just one zone.

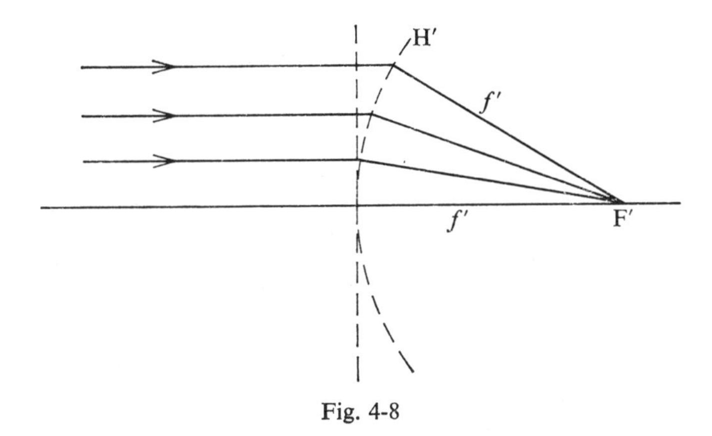


Fig. 4-8

As we have said before, for a lens to be free from spherical aberration and coma the lateral magnification for all zones of the lens must be constant—that is,

$$\frac{\text{height of image}}{\text{height of object}} = \frac{y'}{y} = \text{const.} \tag{4-16}$$

Thus, the sine condition [Eqn (4-5)]

$$\frac{\sin \psi}{\sin \psi'} = \frac{y'}{y} \cdot \frac{n'}{n} = \text{const.} \tag{4-17}$$

which was derived assuming no spherical aberration, also becomes the condition for the absence of coma due to an off-axis object. Departure from the sine condition is thus a measure of the coma, always providing that all other aberrations are absent. It will also be noted that if Herschel's relation [Eqn (4-10)] is satisfied for a system for all $\psi$—that is, a system corrected for spherical aberration for more than one object distance—then it cannot be corrected for coma, since the sine condition cannot be satisfied at the same time.

Various systems have been devised to be truly aplanatic, but only one is of particular interest. This is a refracting sphere which is aplanatic for a particular pair of conjugate object and image points, Fig. 4-9. This sphere has a radius OC $= r$ and refractive index $n$. The object point A is such that AO $= r/n$ and it will be shown that the image will be at B such that BO $= nr$. Then

$$\frac{\mathrm{BO}}{\mathrm{OC}} = n = \frac{\mathrm{OC}}{\mathrm{AO}} \tag{4-18}$$

and since $\angle$BOC $= \angle$AOC the triangles BOC and COA are similar. Therefore

$$\psi' = \theta' \quad \text{and} \quad \psi = \theta \tag{4-19}$$

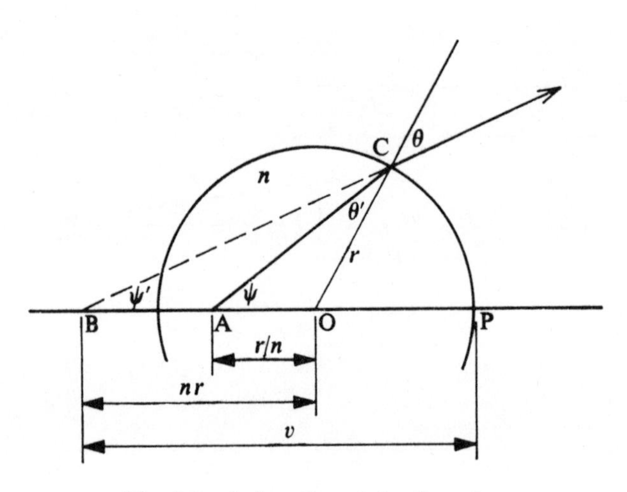


Fig. 4-9 Aplanatic points of a sphere

where the angles are as shown in the figure. Also in the triangle COA, by the sine rule

$$n = \frac{\mathrm{OC}}{\mathrm{AO}} = \frac{\sin \psi}{\sin \theta'} \tag{4-20}$$

and therefore, since $\sin \psi' = \sin \theta'$ and $\sin \psi = \sin \theta$ by Eqn (4-19)

$$\frac{\sin \theta}{\sin \theta'} = n = \frac{\sin \psi}{\sin \psi'} \tag{4-21}$$

Thus the original proposal of the image being at B is in fact correct since Snell's law is satisfied.

Also, applying the sine rule to triangle BOC, where $v$ is the image distance BP, and by Eqn (4-21)

$$\frac{v - r}{r} = \frac{\sin \theta}{\sin \psi'} \tag{4-22}$$

$$\therefore\ v = r + r\frac{\sin \theta}{\sin \psi'}$$

$$= r + r\frac{\sin \theta}{\sin \theta'}$$

$$= r + nr \tag{4-23}$$

The image distance is independent of the angle $\psi$ made by the incident ray with the axis. That is, there is no spherical aberration, and, since the sine condition is satisfied, according to Eqn (4-17), there is no coma. The sphere is therefore aplanatic for the two points A and B.

A sphere of this kind is used for the front lens of high-power microscope objectives. The object cannot of course be embedded in the glass sphere, but by grinding part of the sphere away and using a drop of oil of the same refractive index as the glass, between the glass and the specimen, the required optical condition can be achieved.

If the rays diverging from the image at B are in turn incident on a meniscus lens, as in Fig. 4-10, in which the centre of curvature of the first

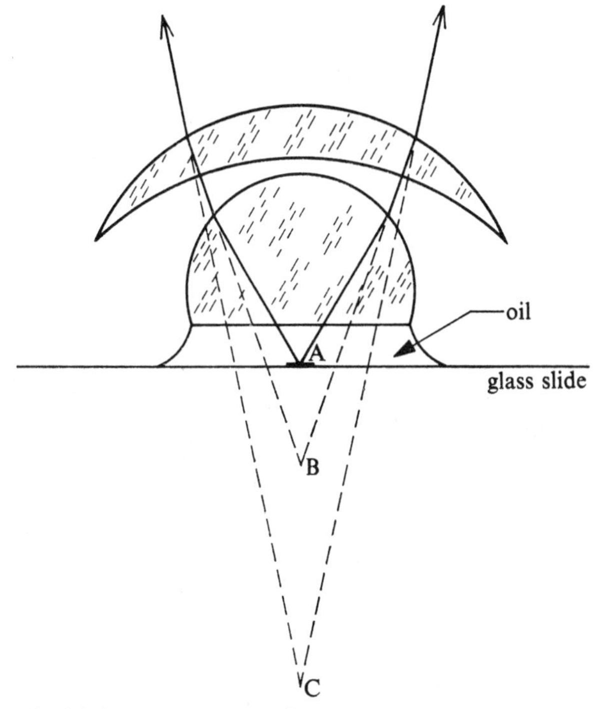


Fig. 4-10 An Amici front component for an oil immersion microscope objective

surface is at B, these rays will not be deviated, since they are incident normally to the surface. If also the second surface of the meniscus lens is of such a radius of curvature that B is at its first aplanatic point, these rays will be refracted again without spherical aberration or coma, and appear to come from the second aplanatic point, C, for this surface. In such an arrangement, which follows the design first suggested by Amici, magnification is thus achieved by both lenses aplanatically. Unfortunately, this process cannot practically be carried on further due to the increasing amount of chromatic aberration.

## 4-8 Astigmatism

If the first two Seidel sums are both independently zero, a small point object either on-axis or close to it will be imaged without spherical aberration or coma. If the third sum is also zero, a point object some distance from the axis will also be correctly imaged as a point and the image is then said to be *stigmatic*. If the third sum is not zero, astigmatism occurs

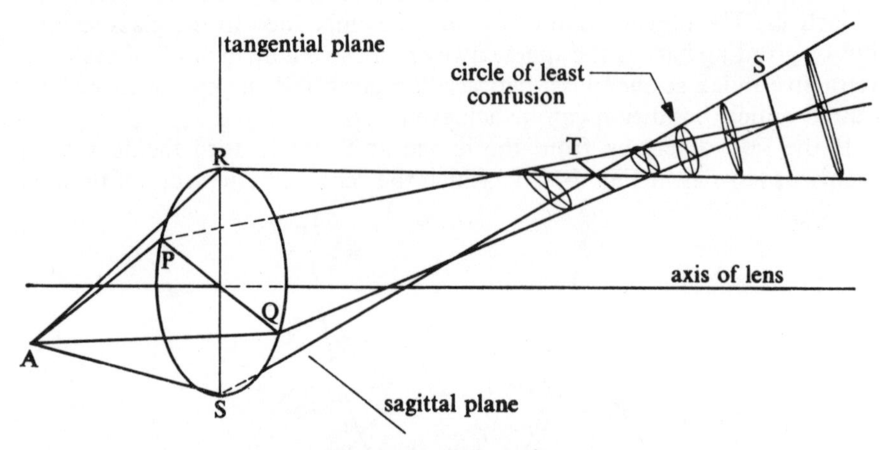


Fig. 4-11 Astigmatism

and the image is said to be *astigmatic*. As the distance of the object from the axis increases, the skew rays become less important, leading to astigmatism rather than coma.

Consider the astigmatic lens shown in perspective in Fig. 4-11; various rays from an off-axis point object at A are shown refracted. Rays in the tangential plane, defined by RSA, are brought to a focus at T, whilst rays through the sagittal plane, defined by PQA, are brought to a focus at S. If a screen is moved through the focusing region, the images seen are an ellipse, a line, an ellipse, a circle, an ellipse, a line, and an ellipse; a point image is never formed. The best representation of the object that can be achieved is a circular disc, termed the *circle of least confusion.*

If the positions of the images T and S are determined for a large number of extra-axial objects such as A—that is, in effect for a large plane object normal to the axis—it is found that the loci of T and S are two ellipsoidal surfaces, as shown in section in Fig. 4-12. The amount of astigmatism, the *astigmatic difference*, is the distance between the two surfaces as measured along the chief ray—that is, the ray drawn through the object point in question and the centre of the entrance pupil, which in this case is the centre of the lens. On the axis, the difference is zero but it increases approximately as the square of the image height. It is also partly dependent on the object distance. The difference is termed positive when, as in the figure, the tangential image surface is formed nearer to the lens than the sagittal image surface and negative when the converse is true.

To correct astigmatism, it is necessary to make the two surfaces coincident, that is, to make the astigmatic difference zero. Little can be achieved by bending a lens. However, the shapes of the image surfaces and their difference, can be controlled by varying the separation of two or more lenses or by the introduction of suitably placed stops, which determine the zones of a lens most effective for refraction of different rays. When the

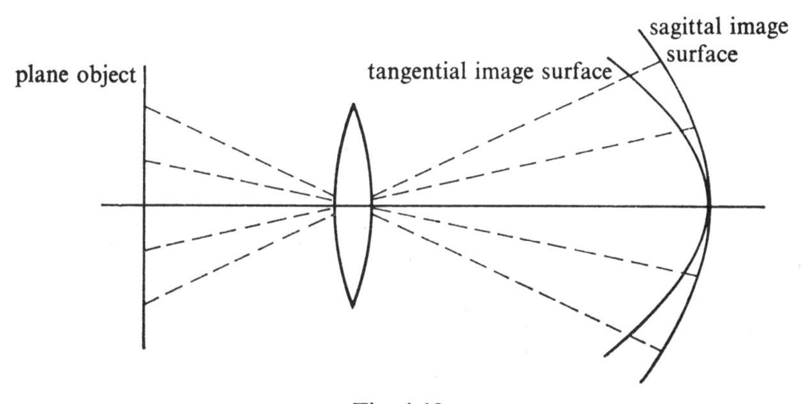


Fig. 4-12

two image surfaces coincide, astigmatism is eliminated and point images may be formed. The single tangential-sagittal image surface is termed the *Petzval surface*. The fact that this surface is curved gives rise to the next type of aberration, curvature of field.

It should be noted that the astigmatism discussed in this chapter is not the same as the defect of the eye referred to as astigmatism. Astigmatism in the eye is due to the eye lens having what is effectively a cylindrical component. This results in different magnifications in different radial directions so that a point object will be imaged as a line even when it is on the axis of the eye lens.

## 4-9 Curvature of field

When the first three Seidel sums are individually and simultaneously zero and the fourth sum is not zero, curvature of field results. The Petzval surface discussed in the previous section is curved, thus a flat screen cannot be used to form a uniformly in-focus image. This is of great significance in photography where, due to manufacturing techniques, flat plates and film are used.

The image field can be flattened by the use of a stop (Fig. 4-13); this causes the tangential and sagittal surfaces to be curved in opposite directions. The circle of least confusion is formed approximately midway between the two surfaces, giving what is effectively a flat field. Although

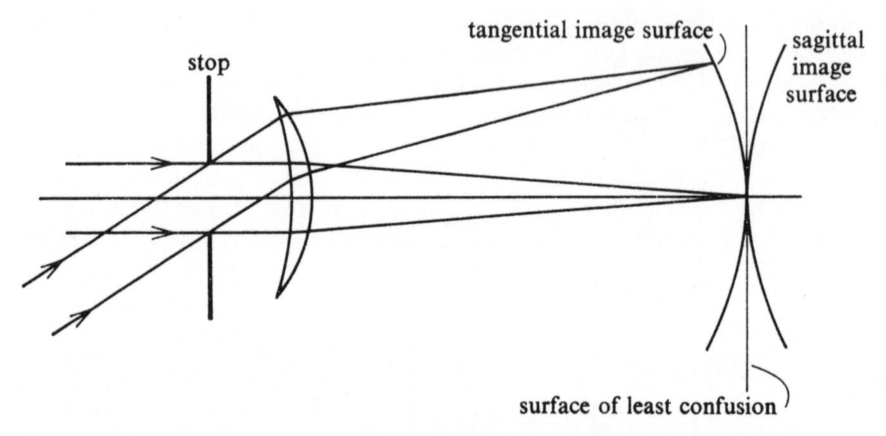


Fig. 4-13

there is an astigmatic difference between the surfaces, this is less objectionable in practice than a curved focal surface with no astigmatism. This design is used extensively in cheap cameras; it gives a fairly good overall image with some blurring at the edge of the field due to the astigmatism.

For better definition at the edge of the field, it is not sufficient to flatten the field in this way. The development of new types of glass makes better corrected systems possible. In these systems the two image surfaces cross over making the astigmatic difference equal to zero for points near to the edge of the field (Fig. 4-14); an approximately flat field is obtained with a minimum of astigmatism. Such a lens is termed an *anastigmat*.

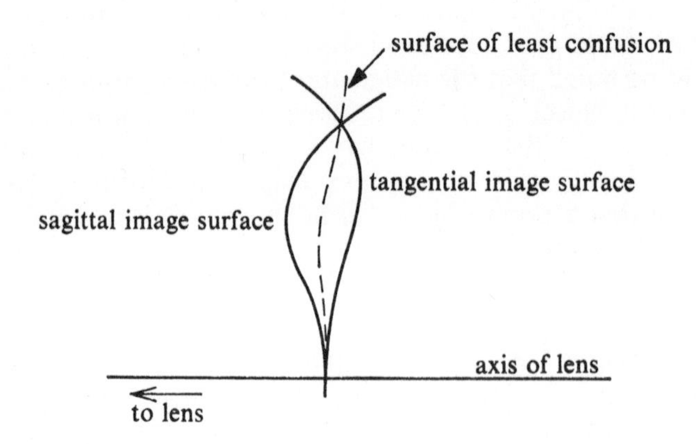


Fig. 4-14

## 4-10 Distortion

This is the fifth of the Seidel aberrations and results if the fifth sum is not zero while the other four sums are zero both separately and simultaneously. If a lens has this type of aberration its image of a plane object normal to the axis does not have a geometrically similar form. Thus, the lateral magnification may not be constant over the image plane causing *distortion* as in Fig. 4-15 where the image should correctly be a square. If the magnification increases with distance from the axis, the distortion is said to be positive or 'pincushion', as in Fig. 4-15(b). Conversely, with a decreasing magnification, the distortion is negative or 'barrel', as in Fig. 4-15(c). In a pinhole camera there is always a direct relation between every object point and its image point, so that the image is completely free from distortion.

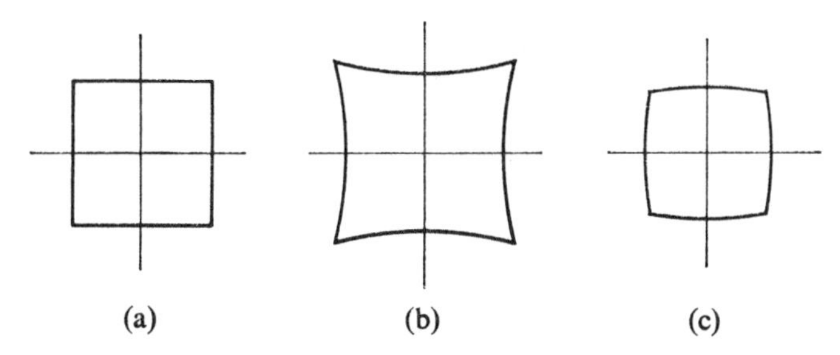


Fig. 4-15 Illustrating (a) no distortion, (b) pincushion distortion, and (c) barrel distortion

Stops can be used to control distortion in the following way. In Fig. 4-16(a), an object plane normal to the axis is imaged as a plane through A′, which is also normal to the axis, assuming the first four of the Seidel sums are each zero. (The lens will therefore be at least a doublet but is drawn as a simple lens for clarity.) Thus all the points of the object plane will be sharply focused on the image plane but not in their correct spatial relationship if distortion is present. If a stop is placed in front of the lens as in Fig. 4-16(a), rays from an axial point A on the object will pass through the centre of the lens and be focused at A′. Oblique rays from off-axis points such as B and C will pass through the upper part of the lens and come to a focus at B′ and C′ respectively. In the figure, the stop also acts as the entrance pupil*, and the rays drawn are chief rays since they are rays from the object passing through the centre of the entrance pupil. After passing through the system, these rays will pass through the centres of the appropriate exit pupils at $E'_B$ and $E'_C$.

The principal planes for the chief rays from the object points B and C are not identical, nor are the positions of their exit pupils $E'_B$ and $E'_C$.

* The entrance pupil is defined as the image, formed by all the lenses preceding it, of the aperture stop that limits the extreme marginal rays through an optical system. The exit pupil is the image of this aperture formed by all the lenses following it. Thus the exit pupil is also the image of the entrance pupil, whilst, for example, the objective lens of a telescope is usually both the aperture stop and entrance pupil. For a single lens on its own, its aperture acts as both the entrance and exit pupils.

The separation $E'_B E'_C$ is in fact spherical aberration of the exit pupil. If the principal planes and the exit pupils were respectively coincident, the lateral magnification would be constant for any rays between the object and image and there would be no distortion. As it is, however, in Fig. 4-16(a)

$$\frac{A'C'}{AC} < \frac{A'B'}{AB} \tag{4-24}$$

and thus the lateral magnification decreases with distance from the axis, resulting in negative or barrel distortion.

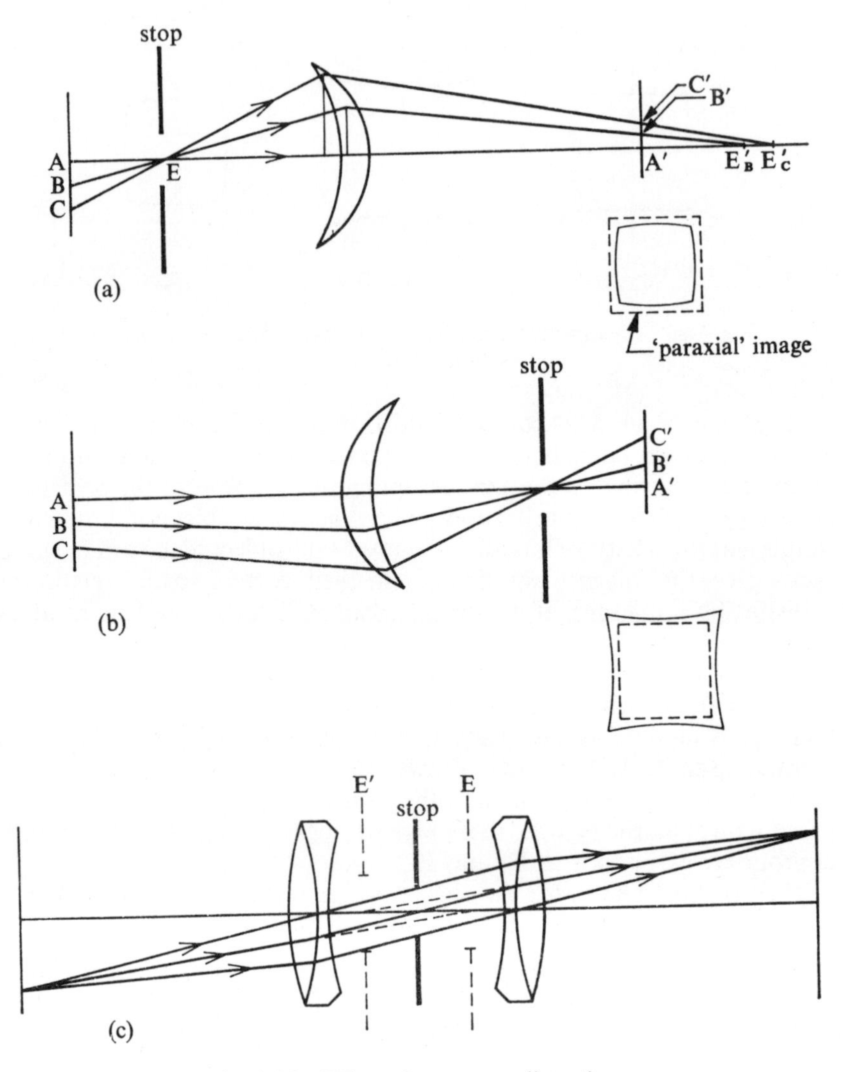


Fig. 4-16 Effect of a stop on distortion

If the stop is on the other side of the lens as in Fig. 4-16(b), the opposite effect is produced, that is

$$\frac{A'C'}{AC} > \frac{A'B'}{AB} \tag{4-25}$$

The lateral magnification thus increases with distance from the axis, resulting in positive or pincushion distortion.

By combining the two situations to form a symmetrical system, with the stop between the lenses, as in Fig. 4-16(c), the equal but opposite sense distortions are made to cancel out. A symmetrical system in which the object and image distances are the same may thus be made completely free from distortion. For other than unit magnification, however, it becomes necessary to correct the system for spherical aberration with respect to the entrance and exit pupils, E and E′, as well as for the object and image planes. Such a condition is impossible to achieve in practice, but nevertheless by balancing the aberrations a virtually distortion-free lens with only small amounts of spherical aberration and astigmatism may be made. A lens system such as this, in which the spherical aberration has been corrected for the exit and entrance pupils, is used for photographic purposes and is called a rapid rectilinear lens or orthoscopic doublet.

The simplest optical system which is completely free from distortion is the plane mirror, which, of course, has unit magnification.

## 4-11 Applications

Spherical aberration and coma can be controlled by a suitable choice of lens shape. Of the aberrations due to the object being well off-axis, astigmatism and curvature of field need separated lenses for their control, while distortion is controlled by the proper use of stops. Since all the aberrations are interdependent, and any change in design to affect one aberration will also have an effect on all the other aberrations, it is impossible to make all the aberrations simultaneously zero. So for every optical system we must decide which are the most important aberrations to be corrected. Then, by reducing these aberrations to small amounts and balancing others, a good workable lens may be obtained.

A camera lens generally has a large aperture. This allows a large amount of light to enter the camera in the short exposure time that is required for cameras held by hand and for photography of moving objects. The optical system of a camera must also produce a true image of a wide-angle field on a flat plate. The aberrations of these systems have been satisfied by various designers in various ways. In the symmetrical lens arrangement the field is flattened by allowing some astigmatism and spherical aberration to remain, as in the rapid rectilinear lens. By extending the number of individual lenses and keeping the symmetrical arrangement, each half may be corrected for spherical aberration, astigmatism, and lateral chromatic aberration, and when combined, coma, curvature of field, distortion and lateral chromatic aberration can also be reduced to negligible amounts, as in the Goertz 'Dagor' or Zeiss 'Triple Protar'. Using an unsymmetrical arrangement, as in the Zeiss 'Tessar', spherical aberration, astigmatism, curvature of field, and chromatic aberration may be made negligible. In modern photographic objectives it is usual to allow some spherical aberration to remain, leaving more freedom to control distortion.

Microscopes are generally used to look at small objects which are brought to the centre of the field; these are near or on-axis point objects. Thus spherical aberration and coma are the important aberrations. Chromatic aberration is of less importance here since monochromatic light may be used, or since the instrument may be treated as a whole, chromatic aberration from the objective may be compensated in a suitable eyepiece. Microscopes used for photomicrography must have a flat field with a minimum of astigmatism. For colour photomicrography, good colour correction becomes of importance. Similar corrections must be made in telescopes, but here, since object fields subtend smaller angles at the objective than in the microscope and thus the focal length of the objective is large compared to its diameter, it is somewhat easier to deal with aberrations. Also, if the telescope is not going to be used for photography a flat field is not so important. Since therefore the only important aberrations in a telescope are spherical aberration, coma, and chromatic aberration, a very high order of correction may be achieved.

## PROBLEMS

**4-1** Which aberrations should be minimized in (a) simple magnifiers, (b) microscopes, (c) telescopes, (d) cameras? Give your reasons and discuss how and to what extent the corrections are achieved in practice.

**4-2** Differentiate clearly and discuss the differences between astigmatism and coma.

**4-3** Calculate the radii of curvature of a 1-cm thick meniscus lens of refractive index 1·5 if it is to be aplanatic for two points 5 cm apart. [9 cm; 6 cm]

**4-4** If an object is to be located 6 cm from the concave side of a meniscus lens of thickness 1 cm and refractive index 1·60, find the radii of curvature of the two lens surfaces and the distance of the image from the concave surface, if the image is to be aplanatic. [6 cm; 4·31 cm; 10·21 cm]

# 5 Optical instruments

## 5-1 Cameras

A camera consists basically of a lens that focuses an image of an object onto a photographic plate or film and a shutter that controls the time for which the image is allowed to fall onto the film. Providing that the object is stationary and a long exposure is practicable, a perfectly good photograph may be obtained using only a simple lens system and a small entrance pupil. In an extreme case, an image may be obtained using a pinhole camera without a lens. Cameras that require long exposures and stationary objects, however, are not much use for practical photography. Most modern cameras have lenses corrected to give a sufficiently good image with a large entrance pupil, thus allowing sufficient light to enter in a short exposure time. With a sufficiently short exposure time, the photography of rapidly moving objects is no problem. We have already discussed the requirements and design of camera lenses (Section 4-11) and little more will be said here.

The lens, which may be a simple or a compound system, is mounted as a unit and its position is adjustable, usually by means of a bellows or screw arrangement, so that the image may be focused on the plane of the film [Fig. 5-1(a)]. In other cameras the separation of the components may be adjusted to focus the image on the film; this has the disadvantage that the aberrations may increase. The shutter allows light to reach the film and is calibrated for various opening times—perhaps, for example, $\frac{1}{30}$, $\frac{1}{60}$, $\frac{1}{125}$, and $\frac{1}{250}$ s. It may be located close to the lens or, alternatively, may be a focal plane shutter in the form of a roller blind placed immediately in front of the film.

The camera is sighted on the object to be photographed by a variety of methods. For example, Fig. 5-1(b) shows a viewfinder in which the object is imaged on a small ground-glass screen built into the camera body. Anything seen in the viewfinder is imaged on the film, allowing for the slightly different viewing positions of the viewfinder and taking lens. Figure 5-2 shows single- and double-lens reflex cameras. In the single-lens reflex, the action of pressing the exposure button first moves the front silvered mirror out of the way and then exposes the film. This arrangement has the advantage that the object is imaged on the glass screen exactly as it is on the film. The double-lens reflex has the added advantage that the object remains in view while the exposure is made.

In all cameras the focal length of the lens should be short, to give a compact camera unit, but should be slightly greater than the length of the diagonal of the film negative being exposed. This is a photographer's 'rule of thumb' for satisfactory illumination and sufficient freedom from aberration. Thus a 35-mm camera—that is, one using strip film 35 mm wide and exposing a picture area of 24 × 36 mm—should have a lens with a focal length a little greater than about 43 mm.

The stop referred to in Section 4-10 and Fig. 4-16 is a variable aperture diaphragm, termed the *aperture stop*. It is located either in front of the lens or between the lens components in more complicated systems and controls the amount of light entering the system, acting as the entrance pupil. A further stop, the *field stop*, is located immediately in front of the

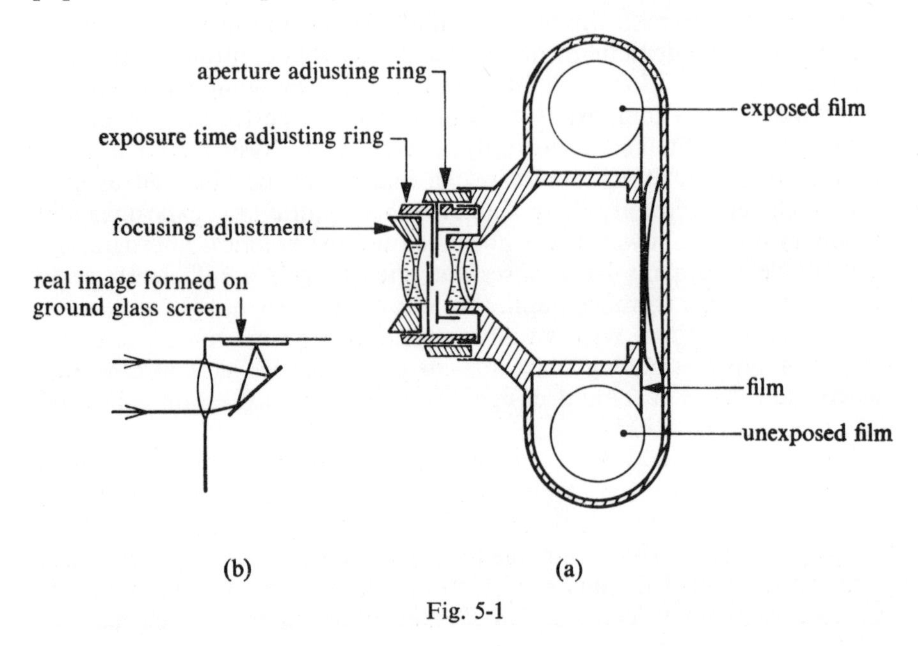


Fig. 5-1

film and serves to limit the field of view. For cameras with the aperture stop in front of the lens, the entrance pupil diameter may be measured directly, but, for an aperture stop situated inside the lens system, the entrance pupil is the image of the stop as seen from the front of the lens and not the actual stop diameter.

The time taken for the photographic image to be formed is a function of the amount of light falling on the photographic film; the measure of the ability of the lens to collect light is its *speed*. The amount of light entering the lens is proportional to the luminance of the source, in this case the object (see also Chapter 7), and to the solid angle subtended by the aperture stop at the source. *Luminance* is the number of lumens emitted per steradian (unit solid angle) by unit area of the source or object and is constant under specific conditions. The solid angle is the angle subtended by an area $A$ on the surface of a sphere of radius $r$, and is by definition equal to $A/r^2$ steradians. Thus the solid angle subtended at the source or object by the aperture stop is $\frac{1}{4}\pi a^2/x^2$, where $a$ is the diameter of the aperture stop and $x$ is the distance to the object, and, therefore, the amount of light entering the lens from the object is proportional to $a^2/x^2$. How much of this light falls on unit area of the actual image formed at the film varies inversely as its area, which varies inversely as the square of the lateral magnification—that is, as $(x/x')^2$, where $x'$ is the image distance.

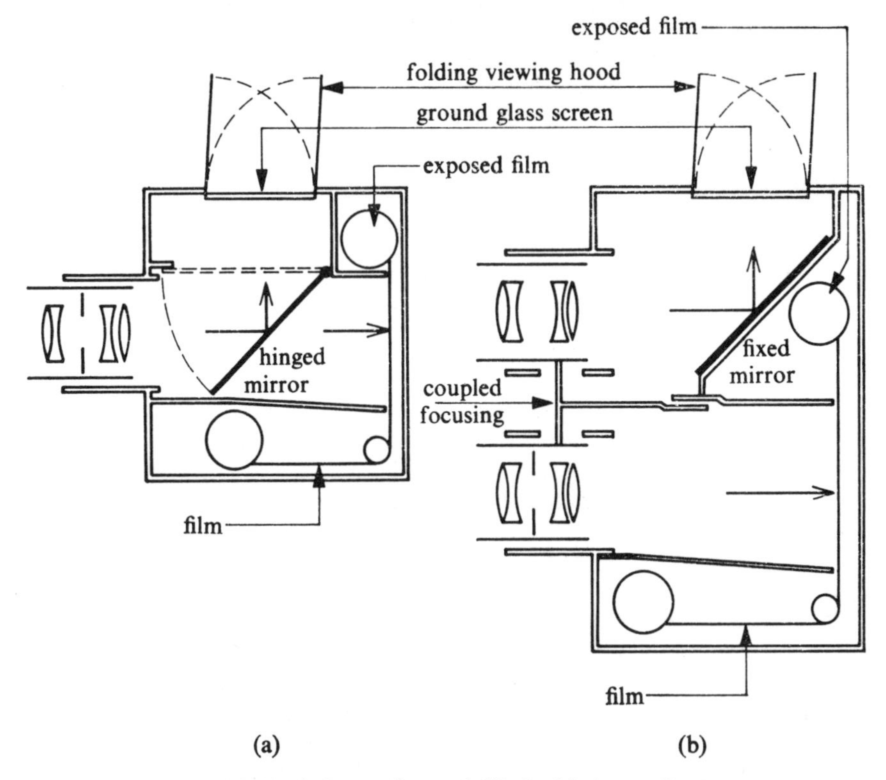


Fig. 5-2 (a) Single-lens reflex and (b) double-lens reflex cameras

Thus the illumination of the image, that is the number of lumens falling on unit area is proportional to

$$\left(\frac{a}{x}\right)^2\left(\frac{x}{x'}\right)^2 = \left(\frac{a}{x'}\right)^2 \tag{5-1}$$

For a distant object, the image distance becomes approximately equal to the focal length of the lens and the illumination then becomes proportional to $(a/f)^2$.

The number $a/f$, therefore, is useful in determining the necessary exposure time to build up the correct density photographic image. However, since $a/f$ is usually a fractional number, the reciprocal is normally quoted; this is called the *focal ratio* or *f-number* and is defined as the ratio of the focal length of the lens, $f$, to the diameter of the entrance pupil, $a$, i.e.,

$$f\text{-number} = \frac{f}{a}$$

The illumination of the image is thus proportional to $1/(f\text{-number})^2$. For example, a lens of focal length 45 mm and linear aperture 15 mm is an $f/3$ lens. This means that an $f/3$ lens is 'faster' than an $f/3$ lens stopped down to $f/4$ in the ratio $4^2 : 3^2$. The $f/3$ lens is thus $\frac{16}{9} = 1{\cdot}78$ times faster than the $f/4$ lens; putting it another way, the exposure time required to photograph an object would have to be increased 1·78 times if an $f/4$ lens were used instead of the $f/3$. The adjustable aperture stop is normally calibrated to give known $f$-values. For example, a typical series of stop numbers would be

$$f\text{-}\quad 22 \quad 16 \quad 11 \quad 8 \quad 5{\cdot}6 \quad 4 \quad 2{\cdot}8$$

that is, the corresponding illuminations of the image would be as

$$\frac{1}{22^2} : \frac{1}{16^2} : \frac{1}{11^2} : \frac{1}{8^2} : \frac{1}{5{\cdot}6^2} : \frac{1}{4^2} : \frac{1}{2{\cdot}8^2}$$

i.e.,

$$\frac{1}{484} : \frac{1}{256} : \frac{1}{121} : \frac{1}{64} : \frac{1}{31{\cdot}36} : \frac{1}{16} : \frac{1}{7{\cdot}84}$$

The corresponding illuminations are, therefore, approximately in the ratio

$$1 : 2 : 4 : 8 : 16 : 32 : 64$$

Thus increasing the aperture by one stop—for example, from $f/8$ to $f/5{\cdot}6$—halves the required exposure time. This means that the exposure time can readily be reduced for the photography of quickly moving objects or the amount of light entering increased for poor lighting conditions.

Increasing the aperture not only lets more light in, it also increases the lens aberrations. With commercially available cameras aberrations are normally kept within satisfactory limits at the full aperture available with the particular camera. With large apertures, the focusing also becomes much more critical. Only rays for a particular object distance are in focus at one time. Thus in photographing a scene, only part of it will be sharply focused on the film and objects nearer or further than a particular distance become progressively more out of focus. However, a certain amount out of focus is not readily noticed in the final photographic image and can therefore be tolerated. The distance between near and far objects for which a satisfactorily sharp image is formed is called the *depth of field*. With large apertures, rays cut the axis more steeply, thus making the focusing more critical and hence the depth of field less. Thus a good camera, with, say, a lens of 45-mm focal length set to give correct focus for an object at 3 m (approx. 10 ft), will give a sharp focus for objects from about 2·5 m (8 ft 3 in) to 3·9 m (12 ft 9 in) away for an aperture setting of $f/2{\cdot}8$. For an aperture of $f/16$ everything beyond about 1·4 m (4 ft 8 in) is in focus. Thus a small aperture (large $f$-number) gives a large depth of field but involves a long exposure time, whilst a large aperture gives a short depth of field with consequent critical focusing but a short exposure time. The most satisfactory balance for each photograph is a matter of compromise.

## 5-2 Projectors

Basically a projector is a camera used in reverse with the film image forming the object to be projected. A camera lens used in reverse is thus a very good projection lens but, since this is costly and since larger aberrations may be tolerated in a projector lens, it is normal to use a lens combination of somewhat simpler construction. The focal length of the lens depends on the magnification required and the convenient distance to the screen. The optimum magnification for correct perspective gives an image on the

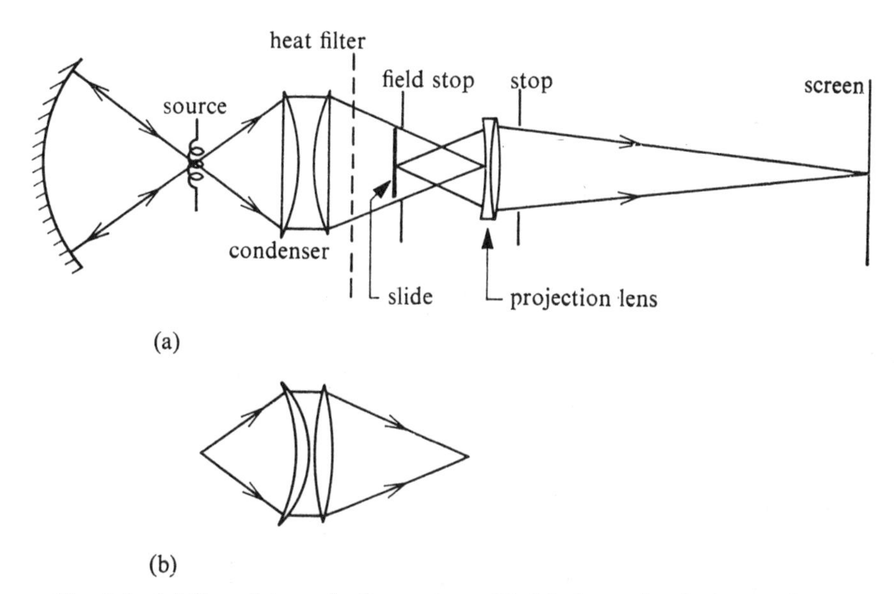


Fig. 5-3 (a) Complete projection system with (b) alternative design condenser

screen which subtends the same angle at the viewer's eye as did the original scene at the camera. Thus for a projector at the rear of a hall and with the screen image viewed from a point approximately midway between the projector and screen, the focal length of the projector lens should be about twice that of the taking camera lens.

The main problem with projection systems is to obtain sufficient illumination of the slide so as to achieve, in turn, sufficient illumination at the screen. Condenser lenses are used to collect light from what is normally a comparatively small source (such as an arc or a tungsten filament lamp) and focus it to fill the entrance pupil of the projection lens [see Fig. 5-3(a)] thus giving uniform illumination at the screen. The condenser lens should have a diameter greater than the length of the diagonal of the slide so that the cone of light it produces covers the whole area of the slide, as in the figure; in this way the edge of the slide becomes the field stop for the whole system.

To promote running economy and heat dissipation, the source should be small. This, however, means that the magnification produced by the condensing lens must be greater in order to fill the entrance pupil of the projector lens. This in turn means an increase in the cost of the condenser lens, since spherical aberration and coma should be minimal, and the necessity for a short source to condenser-lens distance with the consequent possibility of over-heating and damage. In this last respect, the lens design of Fig. 5-3(b) is better since the clearance is increased. However, to reduce these difficulties the maximum magnification that is produced by the condenser is kept as low as possible. Condensers, however, focus heat rays, as well as light rays, and it is therefore necessary to use a filter of heat-absorbing glass to protect the slide from over-heating. It is also usual to incorporate a fan in the projector housing for general cooling of the system.

The area and uniformity of the source may be improved by placing a concave mirror behind it, at a distance equal to the radius of curvature of the mirror. This mirror then forms an image of the source immediately beside the original. Alternatively, the back half of the projection lamp envelope may be silvered.

## 5-3 Simple magnifiers

As an object is brought closer to the eye it appears larger, since it subtends a larger angle at the eye. It cannot, however, be brought too close, since the eye is unable to focus on very near objects. A magnifying lens is used to produce a virtual magnified image of a real object (Fig. 5-4). Such a lens should be held close to the eye so as to give a large field of view. At the same time, its position relative to the object should be adjusted to form the image at the *least distance of distinct vision* (*near point*) so as to subtend the largest angle compatible with comfortable viewing. This distance varies slightly with the individual but by convention is taken as 25 cm, or 10 in.

Up to now, we have used the term *magnification* in the sense of *lateral magnification*: it was defined as the ratio of the image to object size and is the same as the ratio of the image to object distance. When dealing with optical instruments it is often more convenient to define magnification in terms of the angles subtended at the eye by the object and image. The actual forms of the definitions vary, but this is of little importance since the results obtained are for all practical purposes equivalent. Thus *angular magnification* may be defined as

$$\text{angular magnification} = \frac{\text{angle subtended at eye by image}}{\text{angle subtended at eye by object}}$$

$$= \frac{\theta'}{\theta} \tag{5-2}$$

The actual value varies with the position of the object and its corresponding image position. In many instances, the angles involved are small and angular magnification can be written as

$$\frac{\theta'}{\theta} = \frac{\tan \theta'}{\tan \theta} = \frac{\sin \theta'}{\sin \theta} \tag{5-3}$$

*Magnifying power* may be considered as a measure of how effective an optical system is when used in conjunction with the eye. It may be defined as

magnifying power (or apparent magnification)

$$= \frac{\text{angle subtended at eye by image when formed at the near point}}{\text{angle subtended at eye by object when at the near point}} \tag{5-4}$$

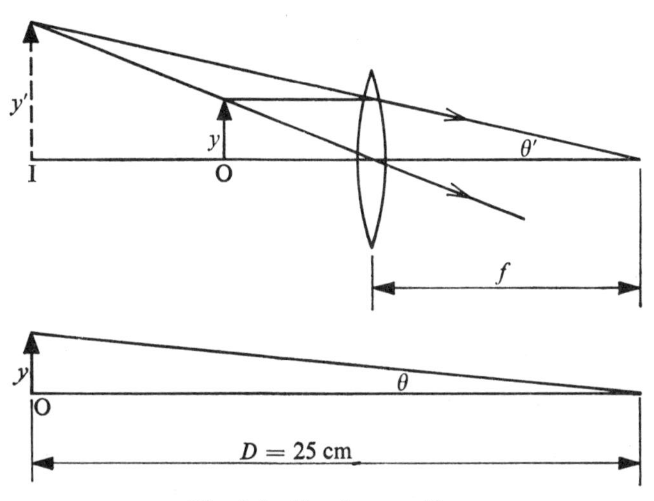


Fig. 5-4 Simple magnifier

The magnifying power will thus be a fixed value for a given instrument. In some instruments the object and image are both located effectively at infinity and it is common to refer to this case as magnifying power also.

The magnifying power of a simple magnifier, like the one shown in Fig. 5-4, is thus equal to

$$\frac{\theta'}{\theta} \simeq \frac{\tan \theta'}{\tan \theta} = \frac{y/f}{y/D}$$

$$= \frac{D}{f} \tag{5-5}$$

If $f$ is measured in metres, then since the least distance of distinct vision, $D = 25$ cm or $\frac{1}{4}$ metre

$$\text{magnifying power} = \frac{D}{f} = PD = \frac{P}{4} \tag{5-6}$$

where $P$ is the dioptric power of the lens.

**Example 5-1** Calculate the magnification produced by a simple magnifying lens of focal length 2·5 cm, when used by a person whose least distance of distinct vision is 25 cm.

The image distance is required to be 25 cm. Applying the thin-lens formula:

$$\frac{1}{v} - \frac{1}{u} = \frac{1}{f}$$

$$\frac{1}{-25} - \frac{1}{u} = \frac{1}{+2{\cdot}5}$$

i.e.,

$$\frac{1}{u} = -\left(\frac{1}{2{\cdot}5} + \frac{1}{25}\right) = -\frac{11}{25}$$

$$\therefore\ u = -\frac{25}{11}$$

Lateral magnification,

$$\frac{v}{u} = \frac{-25}{-25/11}$$

$$= \times 11,\ \text{erect}$$

whereas, magnifying power,

$$\frac{D}{f} = \frac{25}{2{\cdot}5}$$

$$= \times 10$$

Thus, depending on how the magnification is defined, slightly different results are obtained. In practice this difference is of little significance since the given position of the near point is only an approximation anyway. If greater accuracy were required, it would be necessary to measure the position of the near point for the particular conditions. Power thus has the obvious advantage of ease of calculation.

Ideally a magnifier lens should be free from aberrations and should have a large working distance—that is, lens to object distance—to allow freedom of manipulation of the object and adequate illumination. The simplest design of magnifier is a double-convex lens (Fig. 5-5). This design is perfectly adequate when used for low magnification but as the magnification is increased aberrations become more obtrusive. Distortion, curvature of field, and the lack of chromatic correction are especially troublesome.

The early high-magnification lenses were simple spheres of glass and they produced a very poor image. Wollaston obtained a better image by using two hemispheres of glass separated by a small stop. This system was modified by both Brewster and Coddington who used complete spheres with a groove cut around the equators to act as a stop. The difference in their designs was in the shape of the groove. By limiting the angles of incidence of rays, this stop brings about a considerable improvement in image quality. However, this design still has the disadvantages of a short working distance and pronounced curvature of field. Modern simple magnifiers are made with powers up to ×20 and usually consist of symmetrical triplets (Fig. 5-5). The lenses are cemented together with Canada balsam, resulting in little light loss by internal reflections and, due to the symmetrical arrangement, may be used either way up. This type of design has the advantages of good aberration correction, in particular, excellent colour correction, also a large field of view, and a comparatively long working distance.

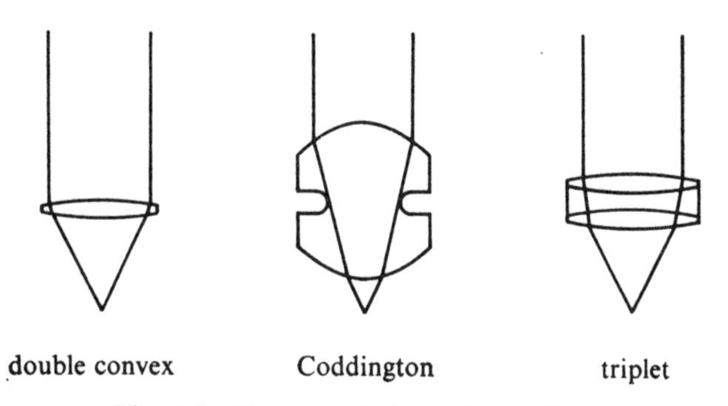


Fig. 5-5 Common designs of magnifiers

## 5-4 Galilean telescope

It is probable that Hans Lippershey constructed the first telescope in Holland in 1608, but its optical arrangement is not known. It is possible that the invention may be of earlier date although Lippershey was probably the person who developed it. It is known that he in fact constructed binocular telescopes. In the following year, Galileo heard of Lippershey's telescope and, without knowing any details, himself designed and constructed a telescope.

Essentially a telescope consists of two lenses mounted coaxially: a long focus lens, or objective, and a short focus lens, or eyepiece. In practice these are both compound lenses or systems. Galileo used a negative eyepiece in his telescope (Fig. 5-6). Rays from a distant object, making a small angle $\theta$ with the optic axis, are effectively parallel and are therefore brought to a focus at A′ by the objective lens at O, where OA is its focal length. Before coming to a focus, however, the rays are further deviated by the eyepiece lens at E. This is located at a distance equal to its focal length from A so that the incident rays are deviated as parallel rays, making an angle $\theta'$ with the optic axis. The object is effectively at infinity and so also is its image. By slightly changing the separation of the lenses, the image could be formed at a nearer position but, in practice, since the eye is relaxed to focus on the distant object, it is natural to use the telescope to form its image at a distance to obviate eye strain. Thus, the angle subtended at the eye by a distant object has been effectively increased, causing the object to appear larger. It is immaterial whether the angles are considered to be subtended at the eye or at the telescope objective, since the length of the telescope is negligible compared with the object and image distances. Since these distances may be considered infinite, we cannot use them to calculate the magnification. The only magnification with meaning involves angles. Thus the angular magnification

$$\frac{\theta'}{\theta} = \frac{\tan\theta'}{\tan\theta} = \frac{AA'/f_E}{AA'/f_O} = \frac{f_O}{f_E} \tag{5-7}$$

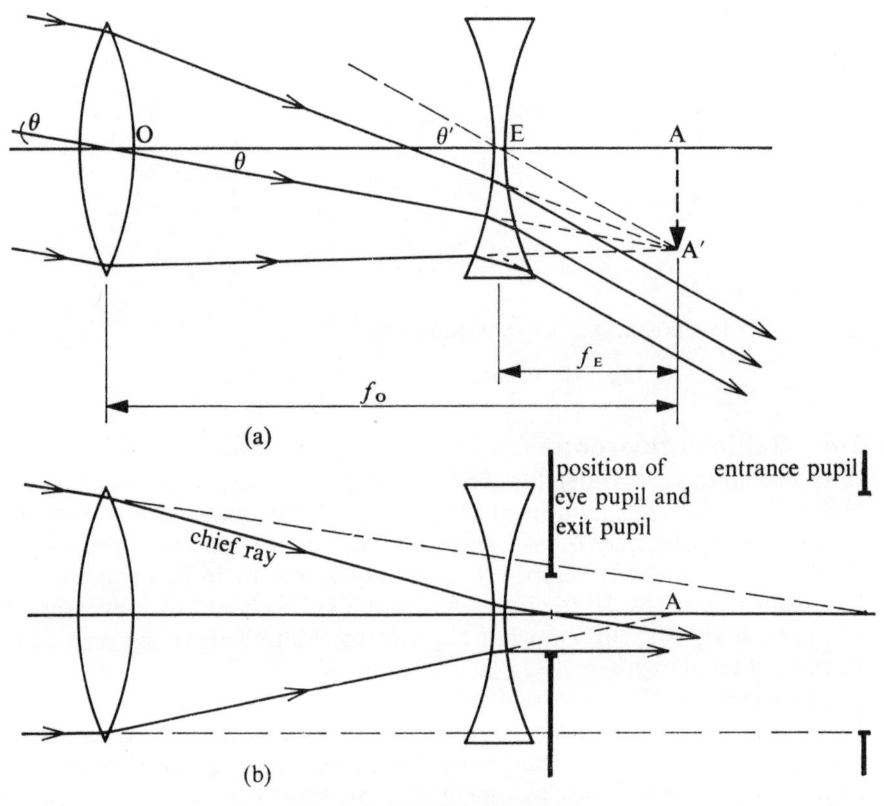


Fig. 5-6 Galilean telescope

where $f_O$ and $f_E$ are the focal lengths of the objective and eyepiece respectively. Since the object and image are effectively at the same distance, it is common practice to call this angular magnification 'magnifying power', or simply 'magnification'. The image is erect and virtual.

If it is assumed that the edge of the objective lens limits the cone of rays entering the telescope, then the objective may be regarded as the aperture stop. The exit pupil then becomes the image formed by the eyepiece of this aperture. With a negative eyepiece lens, this image is formed between the eyepiece and the objective. In theory the eye pupil should be placed at this point, but since this is obviously impossible the eye must simply be positioned as close to the eyepiece lens as possible. This means that the eye pupil now becomes the exit pupil and, since it also limits the cone of useful rays, it is also the aperture stop. The image of the aperture stop produced by all the lenses to its left, in this case both the objective and the eyepiece, is the entrance pupil which is thus formed behind the observer [Fig. 5-6(b)]. The path of a chief ray is also shown. This is a ray that passes, or appears to pass, through the centre of the entrance pupil and through the centre of both the aperture stop and exit pupil.

As the magnification is increased, by increasing the focal length of the objective, the angular field of view decreases. This means that in practice a magnification of two to three times is the maximum. The great advantages of this type of telescope—its short overall length and erect image—make it an ideal lightweight telescope for use in opera glasses. Because overall magnification is low, aberration is not a serious problem and, providing the telescope is corrected for chromatic aberration, a satisfactory image is produced with a comparatively simple lens design.

## 5-5 Astronomical telescope

This is basically similar to the Galilean telescope except that a positive eyepiece is used. Parallel rays incident from a distant object at an angle $\theta$ to the optic axis [Fig. 5-7(a)] are brought to a focus at A′, where OA is equal to the focal length $f_O$ of the objective as before. It is common to position a stop at A to obtain a clean edge to the field. E, the positive eyepiece lens, is situated at a distance such that AE is equal to $f_E$, its focal length. Thus the rays are deviated by the eyepiece to form parallel rays at an angle $\theta'$ to the axis. As before, the angular magnification

$$\frac{\theta'}{\theta} = \frac{\tan\theta'}{\tan\theta} = \frac{AA'/f_E}{AA'/f_O} = \frac{f_O}{f_E} \tag{5-8}$$

This time, however, the object and image rays are inclined in opposite senses to the axis and the image is therefore inverted. This is the reason why the telescope is called ‘astronomical’, since for astronomical use it does not matter whether the image is inverted.

The diameter of the objective lens limits the size of the cone of rays to be admitted and is therefore the aperture stop. In this telescope it is also the entrance pupil, and the exit pupil is its image formed by all the lenses to the right of the aperture stop, which in this case is only the eyepiece [Fig. 5-7(b)]. Figure 5-7(b) also shows a chief ray and a ray parallel to the axis, which together define the diameter and position of the exit pupil. By a comparison of the similar triangles, the magnification

$$\frac{f_O}{f_E} = \frac{E_n}{E_x} \tag{5-9}$$

where $E_n$ and $E_x$ are the diameters, respectively, of the entrance and exit pupils, and where $E_n$ is the diameter of the clear aperture of the objective.

This gives a quick way of establishing the magnification of the telescope. By focusing the telescope for infinity and pointing it at the sky,* the exit pupil may be located by moving a screen back and forth behind the eyepiece until a position is found where the emergent beam is at its smallest and most sharply defined. This then gives the diameter and position of the exit pupil and hence the magnification may be calculated.

The exit pupil is also called the *eye circle* or *Ramsden circle* and is the position at which the eye pupil should be placed for correct viewing. If the diameter of the exit pupil is greater than that of the eye pupil, not all of the light will get into the eye. The maximum illumination of the retinal image occurs when the pupils are the same size, but in practice it is found

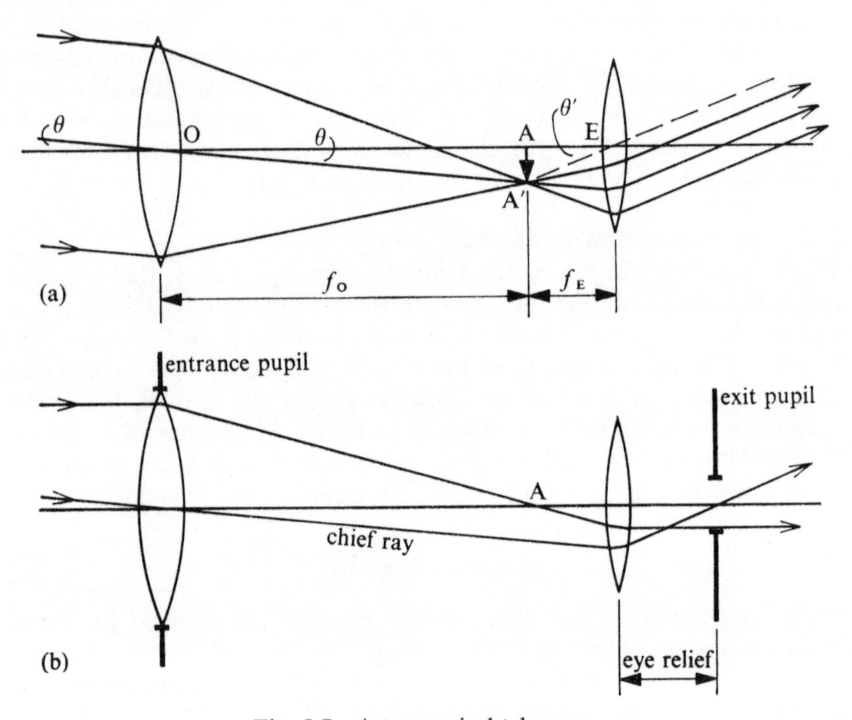


Fig. 5-7 Astronomical telescope

to be better to have the exit pupil slightly smaller than the eye pupil for ease and comfort of viewing. For daylight viewing, the overall magnification and the diameter of the objective are arranged to give an exit pupil of 1·5–2-mm diameter. The distance between the eye and the eyepiece—that is, the distance of the exit pupil from the eyepiece—is the *eye relief*. The optimum eye relief depends on the use of the telescope. Where comfort is the only consideration, a distance of about 8 mm is usually satisfactory but, if, for example, the telescope is to be used for gun sighting,

* On no account should a telescope be used to look at the sun, as serious damage may be done to the eye.

the eye relief is greater to allow for recoil. The actual amount of eye relief may be calculated as follows. The exit pupil is the image of the entrance pupil as formed by the eyepiece. Therefore, with respect to the eyepiece, the object distance is $f_O+f_E$, as in Fig. 5-7(b), and the eye relief is the image distance $v$. Applying the thin-lens formula [Eqn (2-18)]

$$\frac{1}{v}-\frac{1}{u}=\frac{1}{f}$$

$$\frac{1}{v}-\frac{1}{-(f_O+f_E)}=\frac{1}{+f_E}$$

$$\therefore \frac{1}{v}=\frac{1}{f_E}-\frac{1}{f_O+f_E}$$

$$=\frac{(f_O+f_E)-f_E}{f_E(f_O+f_E)}$$

i.e.,

$$\text{eye relief} = v = \frac{f_E(f_O+f_E)}{f_O} \tag{5-10}$$

To increase the magnification of a telescope it is necessary to have an eyepiece of short focal length or an objective of long focal length. An eyepiece of very short focal length presents difficulties—for example, the eye relief distance becomes short and the positioning of the eye pupil becomes

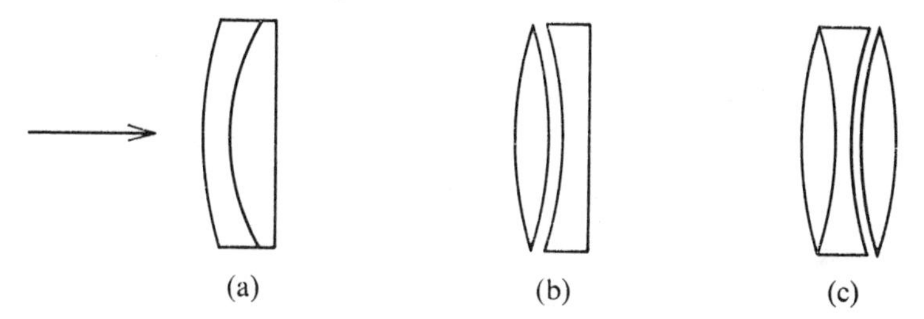


Fig. 5.8 Typical telescope objectives

critical—whereas an objective of long focal length involves small angles of refraction and consequently a high degree of aberration correction can be achieved.

The problems of eyepiece design are common to other instruments, such as microscopes, and will therefore be discussed separately. A common design of telescope objective, the cemented doublet, is shown in Fig. 5-8(a). Such an arrangement can be corrected for spherical and chromatic aberration but not usually for coma, although this can be quite small. The elimination of coma is not, however, always essential, since for many telescopes only good central definition is important. For large aperture objectives, two slightly separated lenses or a combination of cemented and separated lenses are often used to give greater freedom in design.

## 5-6 Terrestrial telescopes

A terrestrial telescope is an astronomical telescope with an additional inverting element to give an erect image. This element consists of a system of lenses or prisms.

In Fig. 5-9(a) and by comparison with Fig. 5-7(b), it will be seen that by the introduction of two similar lenses $L_1$ and $L_2$ the image at A′ is inverted and reformed at A″. This real image is then magnified in the usual way by

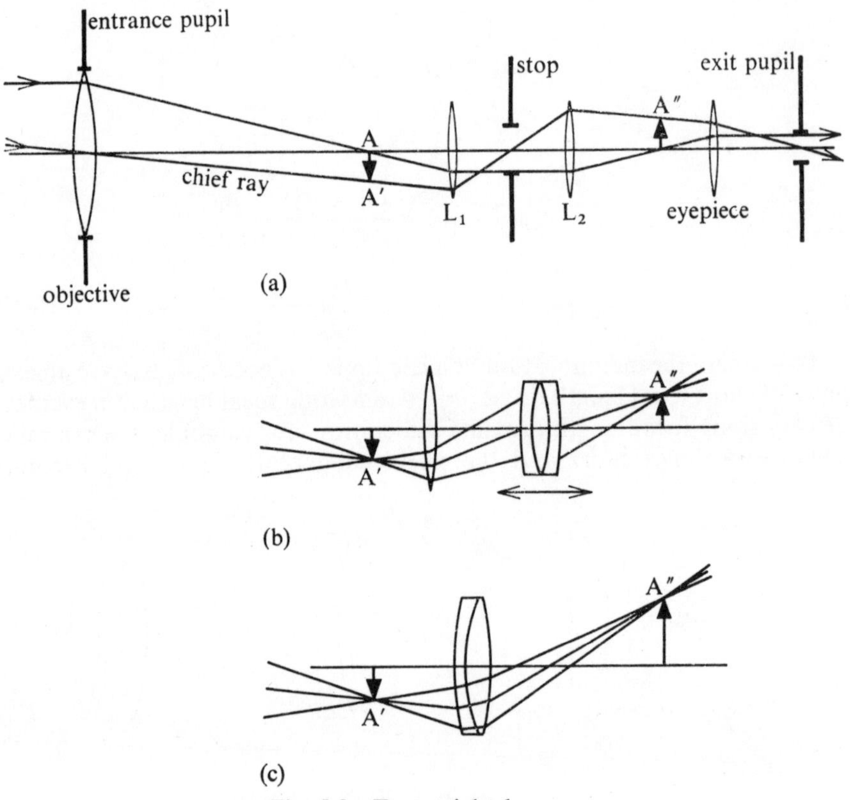


Fig. 5-9 Terrestrial telescope

the eyepiece. The erecting system and the eyepiece are then moved as one unit for focusing. Although drawn as single lenses for clarity, the objective is normally at least a doublet and the eyepiece a two-component combination.

By adjusting the separation of the erecting lenses, the magnification may be changed. The actual erecting lens is usually a cemented triplet [Fig. 5-9(b)]. If the erecting lens is moved towards the first image at A′ it produces a magnified image at A″, now further to the right. The eyepiece is then moved to focus on the image at the new position of A″. For objects not at infinity, the objective forms the image A′ at a different position. For focusing, the complete erecting system and eyepiece have to be slid back as a unit.

By mounting the triplet erecting lens and the eyepiece in telescopic tubes connected by a suitable scroll thread, the magnification of the telescope may be changed whilst preserving the focus. The variable magnification with such telescopes is usually from about five to twenty times.

A single corrected lens may also be used as an erector [Fig. 5-9(c)]. However, this is more suitable as a fixed magnification erector, as the lens can only be at its best aberration correction at a particular magnification.

The magnification of such telescopes is a combination of the magnification of the erector system and of the basic astronomical telescope. The magnification of the complete telescope is thus given by $M f_O/f_E$, where $M$ is the magnification produced by the erector. To obtain a high magnification with this type of telescope an objective of long focal length is required; added to the length of the erector system this produces a somewhat unwieldy instrument. An erector system made up of prisms can produce a short instrument—for example, binoculars.

## 5-7 Binoculars

Figure 5-10 shows two of a number of alternative arrangements of prisms to reverse a pencil of rays both left to right and top to bottom. The

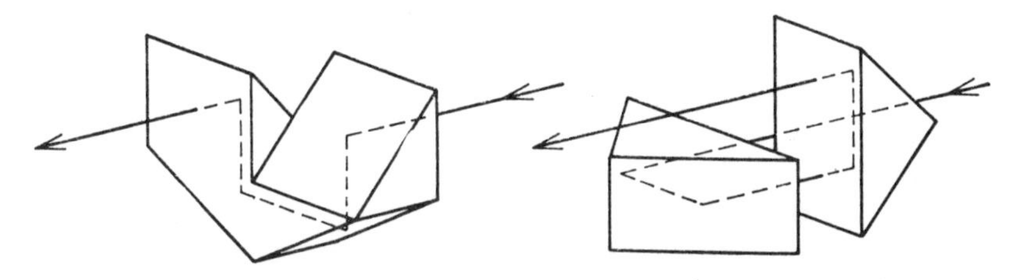

Fig. 5-10 Prism erecting systems

insertion of such an arrangement between the objective and eyepiece of an astronomical telescope converts it for terrestrial use and considerably shortens its length. Two such identical telescope arrangements constitute a pair of binoculars. It is important to ensure that the emergent ray pencil is parallel to the incident rays, otherwise it is not possible to superimpose the two images from the separate telescopes without considerable eye strain. Generally this means that the 90° prism angles should be accurate to $\pm 3'$ and the 90° edge must be parallel to the hypotenuse face—that is, there must be no pyramidal error. As well as accurately made prisms, good aberration correction is required and therefore the objective is normally a doublet or triplet and the eyepiece a combination involving separated components. Since the path through the glass prisms may amount to several centimetres, small amounts of spherical and chromatic aberration, astigmatism, and coma are introduced. Chromatic aberration is corrected in the eyepiece and the others are dealt with by leaving corresponding amounts of residual aberrations of opposite sign in the objective.

Both telescopes are focused together by a single central control. In addition, one or both of the eyepieces may be focused independently, over a calibrated range of about $-5$ to $+5$ diopters, to allow for differences in the observer's eyes.

Because binoculars are usually held in the hand, they should be light in weight and compact. The 'folded' design is useful from this point of view and also brings their centre of gravity within the user's hands. It is difficult to hold binoculars with more than $\times 10$ magnification sufficiently steady without a stand and so most hand-held instruments have a magnification between $\times 5$ and $\times 8$.

Binoculars should also have a large field of view and good light-gathering power. The field is controlled by the eyepiece aperture. A 5° object field corresponds to a $5 \times 10 = 50°$ image field for a $\times 10$ magnification, which is quite large enough for normal requirements. The light-gathering power depends on the clear aperture of the objective. Providing the exit pupil almost fills the eye pupil and the glass surfaces are 'bloomed' to reduce the amount of light lost by reflection, the image should not appear noticeably darker than the object viewed direct.

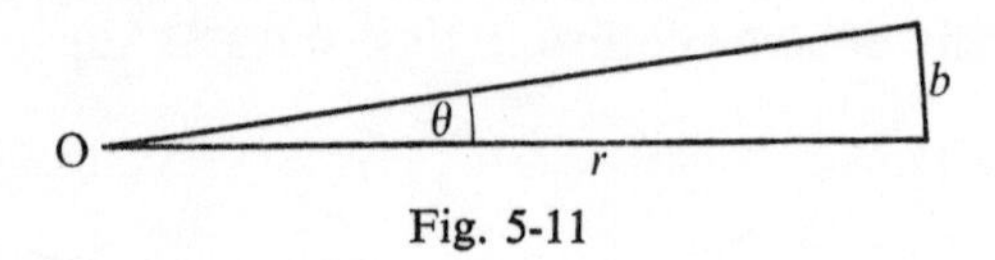


Fig. 5-11

It is usual to describe a pair of binoculars as, for example, 8 × 30. This means that they have a magnification of 8 times and an objective with a clear aperture of 30 mm. This gives an exit pupil of 3·75 mm. In twilight, the pupil of the eye has a diameter of about 7 mm and therefore 'night glasses' are scaled up in size to give an exit pupil of approximately the same size as the eye pupil. These binoculars can be used in daylight, but they are somewhat cumbersome.

The two halves of a pair of binoculars are hinged to adjust the interocular distance to suit the observer. Due to the arrangement of the erecting prisms, the separation of the objectives is about twice this distance. This means that the radius of stereoscopic vision is double that due to the magnification alone. Consider Fig. 5-11, where O is a distant object viewed by a person whose interocular distance is $b$ and $b \ll r$. Then

$$r = \frac{b}{\theta} \tag{5-11}$$

If $r$ is a great distance, then $\theta$ becomes equal to $\alpha$, the limiting angle of resolution of the eye. Thus, stereoscopic vision just ceases when $r = r'$ such that

$$r' = \frac{b}{\alpha} \tag{5-12}$$

$r'$, the *radius of stereoscopic vision*, is approximately 200 m (700 ft) with an interocular distance, $b$, of about 60 mm and a limiting angle of resolution,

$\alpha = 1$ min of arc $= 0{\cdot}00029$ rad. This means that points which are more than about 200 m from the eyes cannot be distinguished from points which are an infinite distance away, except by comparison of their size with known objects at known distances. With binoculars of magnification $M$, the limiting angle of resolution $\alpha$ is reduced to $\alpha/M$ and the interocular distance $b$ is increased to $b'$, the separation of the objectives. Thus the stereoscopic radius of the binoculars becomes $S$, where

$$S = \frac{b'}{\alpha/M} = \frac{Mb'}{\alpha}$$

$$= r' \frac{Mb'}{b} \qquad (5\text{-}13)$$

For binoculars giving a magnification of $\times 10$ and with the separation $b'$ of the objectives twice that of the interocular distance $b$, the stereoscopic radius is increased 20 times over that of the eyes alone, that is, from about 200 m to about 4 km (approx. $2\frac{1}{2}$ miles). Separating the objectives also has the advantage that it increases the *penetration effect*. If, for example, the view is restricted by bushes or a snow storm, the object may appear momentarily obscured from one eye but not from the other, and so the object is more continuously in view than it would be to the naked eye. Obviously there is less chance of both lines of sight being simultaneously obscured if the lines are well separated.

## 5-8 Theodolites

Telescopes with special properties are required for sighting and the accurate determination of angles—for example, in surveying. They must be able to stand up to use in wet weather and, since they are normally mounted on tripods, they must be compact and remain balanced as the focus is changed. These initial requirements can be met by an internal focusing arrangement (Fig. 5-12). Since for measurement purposes it is of little importance if the image is inverted, an astronomical telescope with its fewer number of lenses can be used as the basic instrument. Instead of altering the distance between the objective and the eyepiece, the instrument is focused by adjusting the position of a negative lens between the objective and the eyepiece. This has the advantages that the centre of gravity changes only slightly and that the objective and eyepiece can be sealed into the telescope to keep out the weather. The internal focusing lens inside the main telescope barrel moves coaxially as its position is adjusted by means of a rack and pinion. The pinion shaft is fitted with a stuffing box where it projects through the main telescope tube, again to keep out the weather. Another advantage of the negative focusing lens is that it gives a compact telescope, since in effect it increases the focal length of the objective, thus giving high magnification without the disadvantage of a long telescope tube or an eyepiece of inconveniently short focal length. Since high definition is required with such a telescope, the complete instrument, including the focusing lens, should be well corrected

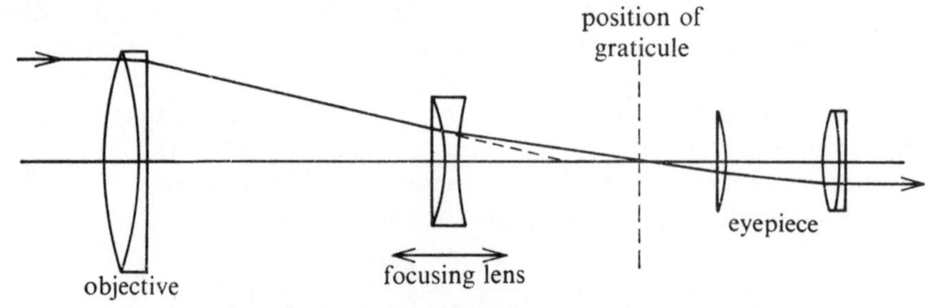


Fig. 5-12 Internal focusing of a telescope

for spherical and chromatic aberration and coma. Some astigmatism is allowable at the edge of the field, since the edge is only used as a finder.

For sighting purposes it is necessary to have some form of cross line. This is either ruled or formed by a photographic process onto a glass plate, and is called a *graticule*. It is located at the focal plane of the eyepiece. There are two parallel lines on the graticule which subtend an angle, usually one in a hundred, at the second nodal point of the objective. This angle is called the *stadia interval* and is equivalent to 34′ 22·6″ (2,062·6 seconds of arc). To measure a distance the telescope is focused on a distant graduated staff; the length of staff seen between the two stadia lines is approximately one hundredth part of the distance of the staff from the telescope. If the staff to telescope distance is changed, the telescope must be refocused. This means that the distance between the objective and its rear image plane is changed and the stadia interval no longer subtends the correct angle at the telescope. However, Fig. 5-13(a)

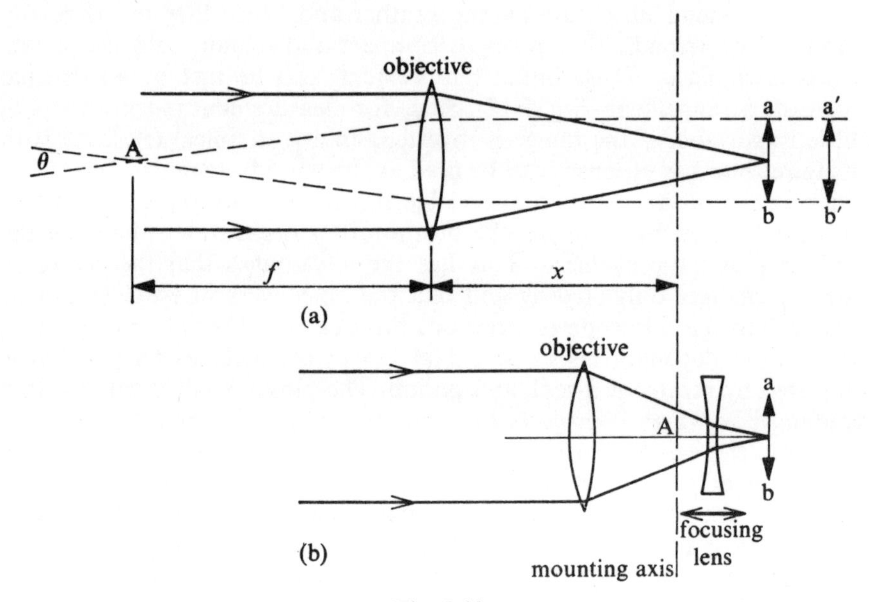


Fig. 5-13

shows that for the external focusing telescope—that is, one in which the eyepiece and graticule are moved together for focusing—there is a point in front of the object to which distances may be measured and agree correctly with the measurement using the stadia marks. The distance of this point from the centre line of the telescope, which normally coincides with the axis of the tripod, is termed the 'constant'. If ab is the stadia interval corresponding to angle $\theta$, then as the telescope is refocused the image is formed at some other plane, say, a′b′, but, as is seen from the figure, the same stadia marks correspond to the same angle $\theta$ no matter in which plane the image is formed. Thus all distances must be referred to the point A, termed the *anallatic point*, with the distance $(f + x)$ being added to give the true distance of the staff from the centre line of the telescope and tripod. Providing the instrument is focused by adjusting the position of the eyepiece and graticule and not by moving the objective, this distance $(f + x)$ is a constant. However, with the internal focusing telescope—that is, one incorporating a movable negative lens between the fixed objective and eyepiece—it is possible to choose the telescope length and the powers of the objective and negative focusing lens such that compensating effects are introduced. These keep the angle subtended by the staff at the centre of the instrument, the tripod mounting axis, the same as the angle subtended by the stadia marks, irrespective of the distance of the staff. That is, the anallatic point A now lies on the central vertical axis of the instrument [Fig. 5-13(b)]. This figure also shows the comparative shortening in length of an internal focusing telescope over an external one of the same power. In practice the compensation is not exactly correct but the error introduced is generally less than one in a thousand, so the 'constant' may normally be disregarded. Thus, if the staff intercept is 2·50 m, as measured by a 1-in-100 stadia interval, the distance of the staff from the centre of the theodolite, neglecting the 'constant' is $250 \pm 0{\cdot}25$ m.

Besides measuring distances by means of the stadia intercept, theodolites are used for measuring the angles subtended by two distant objects by sighting on the objects in turn and measuring the angle subtended on a graduated circle. This means that it should be possible to sight the telescope with the greatest accuracy possible. In Chapter 9 we shall show that the minimum angle which can be resolved by a telescope is equal to $1{\cdot}22\lambda/a$, where $\lambda$ is the wavelength of the light and $a$ is the diameter of the circular aperture which limits the beam, in this case the clear aperture of the objective. With an objective aperture of, say, 4 cm and a mean wavelength of 6000 Å, the minimum angle of resolution with this telescope will be $1{\cdot}83 \times 10^{-5}$ rad (3·8 seconds of arc). Since the unaided eye can just resolve two points separated by about 1′, the eyepiece must have a magnification of at least ×16, so that the eye can see all that is resolved by the objective. A higher magnification than this does not provide any more information and gives a dimmer image.

The resolution of the eye is at its maximum when the pupil diameter is about 2 mm. This means that a telescope with a 4-cm diameter objective—

that is, one with a 4-cm diameter entrance pupil—should have an overall magnification of ×20 to give a 2-mm exit pupil and hence fill the eye pupil. Together with the magnification required by the eyepiece, this is a satisfactory overall magnifying power, since it makes full use of the resolutions of both the objective and of the eye and gives an eyepiece with a convenient focal length. Thus there is an optimum magnification depending on the clear diameter of the objective lens.

## 5-9 Reflecting telescopes

To increase the resolution of a telescope we can increase the aperture of the objective up to a certain practical limit. The Yerkes Observatory refracting telescope with its 40-in diameter and 62·5-ft focal length objective is the largest in existence. The limit is set firstly by the problem of casting a slab of glass sufficiently large which is free from bubbles, striations, or other internal defects, and secondly by weight. The lens must be supported by its rim which causes distortion by the flexure of the glass under its own weight. There is also the problem of supporting a large mass high up in the observatory in such a way that it can be moved freely and precisely. The refracting telescope does have the advantage that it can be corrected for coma and made to have a larger field of definition than a comparable reflecting telescope, but for astronomical use this is of limited importance, since the object appears comparatively small and is arranged to be on axis.

To obtain substantially better resolution it is necessary to use a reflecting telescope, of which the largest is at Mount Palomar Observatory and has a 200-in diameter mirror with a focal length of 55 ft. Basically, the reflecting telescope is a concave mirror accepting virtually parallel light from a distant object and bringing it to a focus, where the image formed is examined by a magnifying eyepiece. A spherical mirror produces spherical aberration and in order to bring parallel rays to a single focus it is necessary to use a paraboloid mirror. The 200-in Mount Palomar mirror was first polished to a spherical surface with a radius of curvature of 110 ft, corresponding to a sag at the centre of the mirror of approximately $3\frac{3}{4}$ in. A change in this sag of approximately 0·005 in was then required to make the spherical mirror surface paraboloid and so correct for spherical aberration. Since there is no refraction, there is no chromatic aberration. The reflecting telescope also has the advantage that the heavy weight of the mirror is low down, where it may more easily be supported. This is important, because the mirror should have a thickness equal to about one sixth of its diameter, and a large mirror will therefore be very heavy and need a considerable amount of metal support to prevent it distorting under its own weight.

Owing to the comparatively large diameters of these mirrors, localized heating can cause distortions of the mirror surface. It is therefore necessary to use a material with a low coefficient of thermal expansion, such as fused quartz.

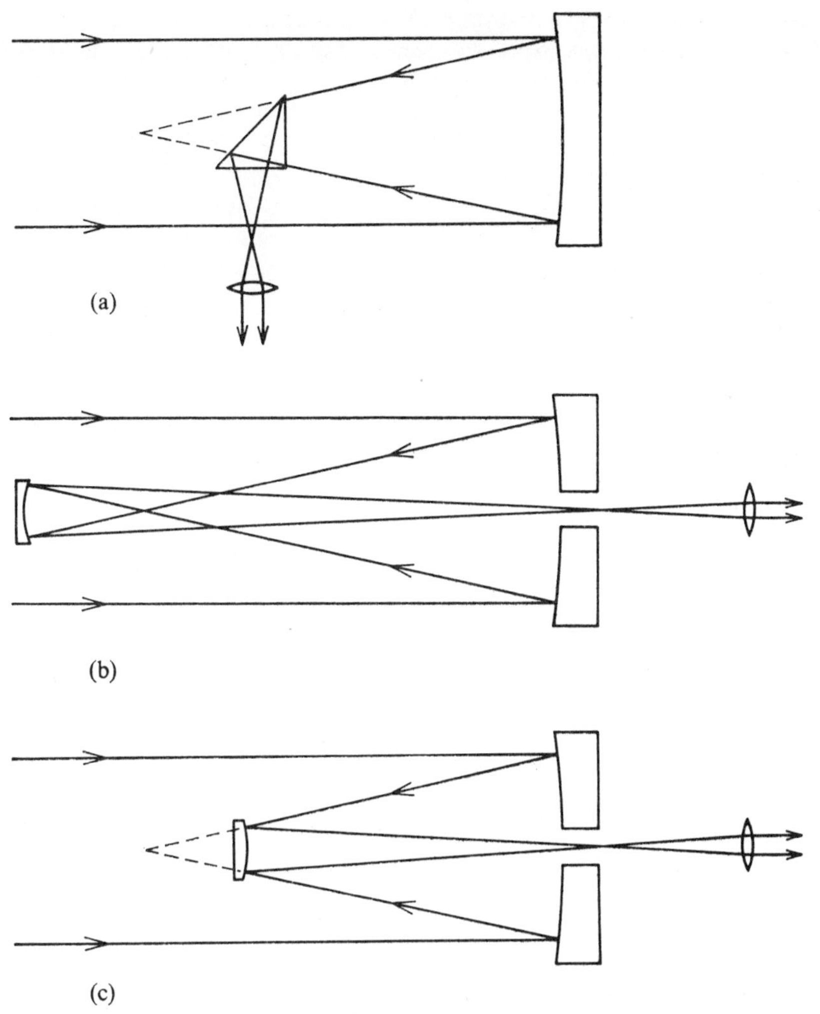


Fig. 5-14 Optical systems of (a) Newtonian, (b) Gregorian, and (c) Cassegrain reflecting telescopes

Having used the mirror to form an image, this image has then to be viewed. Obviously it cannot be simply viewed using an eyepiece, otherwise the observer's head would get in the way of the beam. Therefore the image has to be formed in a more convenient position. Figure 5-14 shows the usual systems for doing this—the Newtonian, Gregorian, and Cassegrain. In the Newtonian system a flat mirror or totally reflecting prism reflects the rays to one side. In the Gregorian arrangement a small concave mirror placed beyond the focus of the main mirror reflects the rays back through a hole in the main mirror to form the image behind it. Correctly, the auxiliary mirror should have an ellipsoidal form. In the Cassegrain system a small convex mirror, which should correctly be of hyperbolic section, reflects the rays back through a hole in the main mirror. Both the

Gregorian and Cassegrain systems have the advantage that the direction of viewing is in the direction of the incoming rays and both systems effectively increase the focal length of the main mirror and hence increase the magnification. The Cassegrain arrangement, however, leads to a flatter field than the Gregorian and is therefore the more popular. It is the Cassegrain arrangement that is used in the 200-in Mount Palomar telescope. In practice, there is little purpose in trying to build larger telescopes, since their resolution is limited by the non-homogeneity of the earth's atmosphere. This, of course, is not limiting if the telescope is mounted on an earth satellite.

## 5-10 Eyepieces

Many eyepieces have been designed to serve many different purposes. For example, a paper by E. Wilfred Taylor* describes 27 types. However, in most cases their use is very specialized and only a small number of types are in common use.

The purpose of an eyepiece is to magnify the image produced by an objective without causing a deterioration in the quality of the image and, if necessary, to correct any residual aberrations. It should also give an eye relief distance—that is, distance between the last lens of the system and the observer's eye—of about 8 mm. If the eye relief distance is less than this, eyelashes get in the way and, if it is more, there may be difficulty in positioning the eye pupil correctly which would cause cut-off in part of the angular field. The image field produced by the eyepiece should also be flat to prevent undue eye strain and so that, for example, a graticule scale is sharply in focus over its whole length. It is normal to attempt to satisfy these requirements with a two-lens system. Eyepieces are commonly made with magnifying powers of $\times 5$, $\times 10$, $\times 15$, and occasionally higher powers for astronomical telescopes.

The *Huygenian eyepiece* consists of two plano-convex lenses made from glass of the same refractive index and mounted with the plane surfaces towards the eye. The ratio of the focal lengths of the field lens to eye lens (Fig. 5-15), $f_F:f_E$, varies from 3:1 to 2:1 depending on the manufacturer. The lenses are separated by a distance equal to half the sum of the individual focal lengths:

$$\text{separation, } s = \frac{f_F + f_E}{2} \tag{5-14}$$

It may also be shown that the spherical aberration will be a minimum when the separation is equal to $(f_F - f_E)$; this condition is not usually satisfied except when the focal lengths are in the ratio 3:1. The focal length of the complete eyepiece is given by Eqn (3-10):

$$f = \frac{f_F f_E}{f_F + f_E - s} \tag{5-15}$$

* Journal of Scientific Instruments, 1945, **22**, 43–48.

Thus if $f_F = 2f_E$, $f$ will be equal to $\frac{4}{3}f_E$. Figure 5-15 shows the relative positions of the principal and focal points for an eyepiece in which the field lens has a focal length three times that of the eye lens, that is, $f_F = 3f'$, say, and $f_E = f'$, so that $f = \frac{3}{2}f'$.

Rays from the objective are brought to a focus at a point in the plane through the first principal focus $F_1$ of the eyepiece. With the eyepiece in position, a ray parallel to the axis is deviated by the field lens towards its focus at $F_F$. On reaching the eye lens this ray is further deviated to pass through the second principal focus $F_2$ of the eyepiece. Similarly, since the object as produced by the objective is located at the first focal plane of the eyepiece, any other ray is deviated to emerge parallel to the first ray. Its actual path may be determined graphically by use of the principal planes,

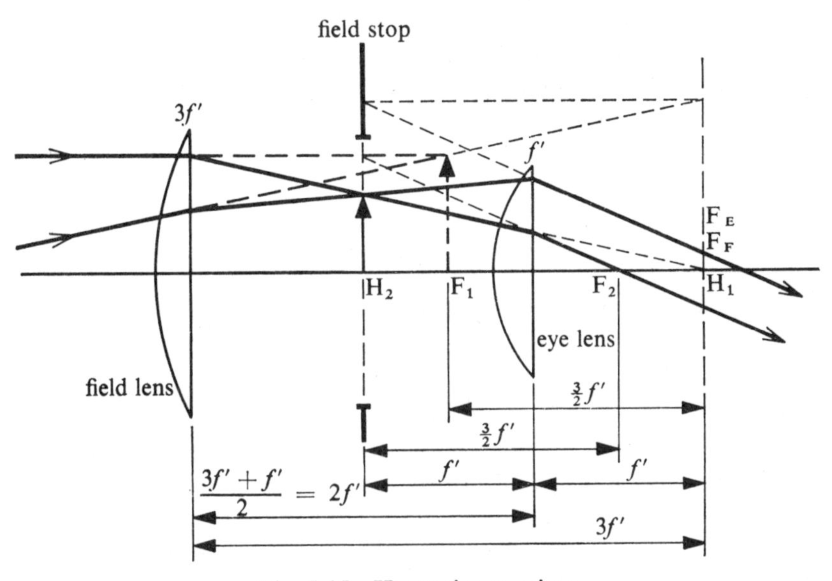


Fig. 5-15 Huygenian eyepiece

as in the figure. The exit pupil, which is the image of the entrance pupil of the complete optical system, is located by the point in which a chief ray, drawn through the centre of the entrance pupil, is deviated to cut the axis.

An important factor with this type of two-lens design is its colour correction. For example, the power of the combination is $P$, where

$$P = P_F + P_E - sP_FP_E \tag{5-16}$$

from Eqn (3-12), where $P_F$ and $P_E$ are the powers of the field and eye lens respectively. Therefore, since plano-convex lenses are used

$$P = \frac{n-1}{r_F} + \frac{n-1}{r_E} - \frac{s(n-1)^2}{r_Fr_E} \tag{5-17}$$

from Eqn (2-36), where $r_F$ and $r_E$ are the radii of the field- and eye-lens convex surfaces, and $n$ is the refractive index of the glass for a particular wavelength. For a different wavelength the refractive index is $n + \delta n$ and the corresponding total power

$$P' = \frac{n + \delta n - 1}{r_F} + \frac{n + \delta n - 1}{r_E} - \frac{s(n + \delta n - 1)^2}{r_F r_E} \tag{5-18}$$

Therefore the difference in power due to light of different wavelengths is

$$P' - P = \frac{\delta n}{r_F} + \frac{\delta n}{r_E} - \frac{s[2\delta n(n - 1) + (\delta n)^2]}{r_F r_E}$$

$$= \delta n \left[ \frac{1}{r_F} + \frac{1}{r_E} - \frac{2s(n - 1)}{r_F r_E} \right] \tag{5-19}$$

neglecting $(\delta n)^2$ as small. Therefore, the condition that the power shall be unchanged for light of different wavelengths (that is, $P' - P = 0$) is that

$$\frac{1}{r_F} + \frac{1}{r_E} - \frac{2s(n-1)}{r_F r_E} = 0 \tag{5-20}$$

$$\therefore\ s = \frac{r_E r_F}{2(n - 1)} \left( \frac{1}{r_F} + \frac{1}{r_E} \right) \tag{5-21}$$

$$= \frac{1}{2} \left( \frac{r_E}{n - 1} + \frac{r_F}{n - 1} \right)$$

$$= \frac{f_E + f_F}{2} \tag{5-22}$$

which is in agreement with Eqn (5-14). This means that the eyepiece is virtually free from chromatic error, unless it is to be used with a graticule. To be in focus, the graticule has to be placed in the plane of the field stop (Fig. 5-15) and, since this is only viewed by the eye lens, it shows some colouration. One objection to this type of eyepiece, apart from its use with graticules, is that the eye relief is rather short for comfortable viewing. There is also some spherical aberration, astigmatism, and rather pronounced curvature of field.

The *Ramsden eyepiece* also consists of two plano-convex lenses but these have the same focal length and their convex faces towards each other. To satisfy the condition for achromatism, the lenses should be separated by a distance equal to half the sum of their focal lengths, which, in this case, is equal to either of their focal lengths. This has the disadvantages that any dust on the surface of the field lens is sharply focused by the eye lens and that the eye relief is very small when used with a telescope. To overcome these difficulties the separation is usually made two thirds of either of the focal lengths and this, in fact, introduces only a small amount of chromatic error. Such an eyepiece is shown in Fig. 5-16. The image produced by the objective, which now becomes the object for the eyepiece, is arranged to be in the plane through the first focal point $F_1$ of the eyepiece. As for the Huygenian eyepiece, an incident ray parallel to the axis is deviated first towards the focus of the field lens $F_F$ and then by the eye lens through $F_2$, the second principal focus of the eyepiece. Any other ray is deviated to emerge parallel to this first ray, as in the figure.

As the object is outside the system, a Ramsden eyepiece is used for all measuring purposes, since a graticule can be located at the first image position—that is, at the position of the field stop—and magnified with the object without any additional chromatic or other aberration. Since the object is outside the system, this type of eyepiece is sometimes termed positive, whereas the Huygenian is termed negative.

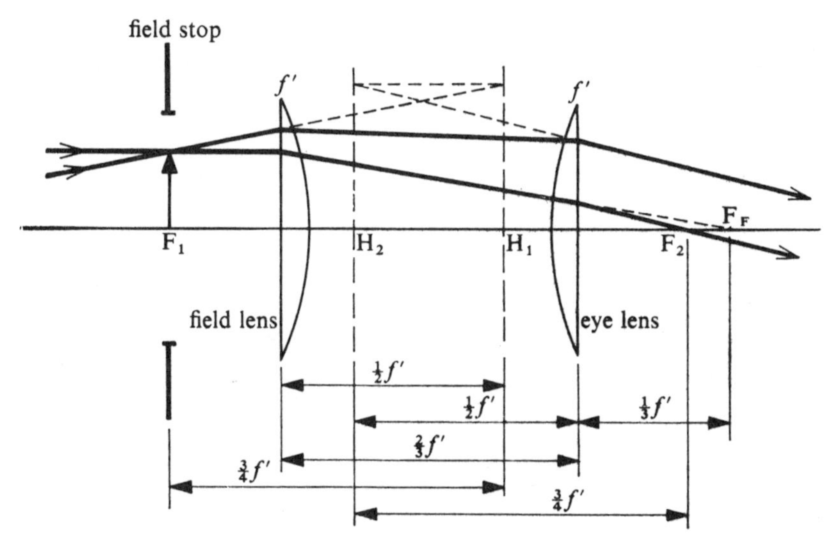


Fig. 5-16 Ramsden eyepiece

If the focal length of each lens is $f'$ and their separation $\frac{2}{3}f'$, the total focal length of the eyepiece is $f$ where

$$f = \frac{f'^2}{2f' - \frac{2}{3}f'} \tag{5-23}$$

i.e.,

$$= \tfrac{3}{4}f'$$

by Eqn. (3-10).

The Ramsden and the Huygenian eyepieces have approximately the same angular field of view, which is about 40°–50° for a magnifying power of ×10. The Ramsden, however, has somewhat less spherical aberration and distortion but, due to the incorrect separation, has more chromatic aberration in its basic form.

The Kellner eyepiece is similar to the Ramsden but has a double-convex field lens and a cemented doublet as the eye lens. This form of eyepiece is corrected for the chromatic error introduced by the incorrect separation of the lenses in the Ramsden and is therefore also known as the 'achromatized Ramsden eyepiece'. There are other eyepiece designs for special purposes. These usually have rather more lenses and are used to give greater angular fields, in the region of 70° for a ×10 eyepiece, or to give greater eye-relief distances, which may be up to the focal length of the eyepiece. These long eye-relief eyepieces are usually meant for use in gun sights or to be used by spectacle wearers.

To make all eyepieces readily interchangeable, their outside diameter has been standardized to 1·25 in (31·8 mm) for telescope use and 0·917 in (23·3 mm) for microscope use.

## 5-11 Microscopes

In the microscope the objective, which is a compound lens system, is used to produce a magnified, real, inverted image. This image is then further magnified to produce a virtual image at the near point—that is, at about 25 cm from the eye (Fig. 5-17). Since both the objective and eyepiece are compound lens systems, the rays are drawn with reference to the principal planes. Thus rays may be drawn from a small object of height $y$ to pass through the first and second principal foci $F_O$ and $F'_O$ of the objective to form a real image of height $y'$. The image $y'$ is then further magnified by the eyepiece, which behaves as a simple magnifier. If the final image is to be formed at infinity, $y'$ will be located at the first principal focus $F_E$ of the eyepiece. In practice, the microscope is normally adjusted to form the final image at the near point—that is, at about 25 cm from the eye. This means that the image $y'$ is located just inside the first principal focus $F_E$ of the eyepiece.

The lateral magnification produced by the objective is thus

$$M_O = \frac{y'}{y} \simeq \frac{l}{f_O} \qquad (5\text{-}24)$$

where $f_O$ is the focal length of the objective and $l$ is the *optical tube length* defined as the distance between the back focal plane of the objective and the first focal plane of the eyepiece, which is approximately equal to the distance to the intermediate image $y'$. The eyepiece behaves as a simple magnifier and hence has a magnifying power equal to

$$M_E = \frac{D}{f_E} \qquad (5\text{-}25)$$

from Eqn (5-6), where $D$ is the least distance of distinct vision and $f_E$ is the focal length of the eyepiece. The total magnification of the microscope is then $M_O . M_E$. Microscope objectives are normally manufactured with powers of ×5, ×10, ×20, ×40, and ×100 and eyepieces with powers of ×5, ×10, and ×15. If the magnification obtained is required with greater accuracy, so that the size of an object may be measured, a divided graticule is located at the plane of $y'$ and an object of known length is used. Hence the divisions of the graticule are calibrated, and the size of an unknown object determined in terms of these divisions.

The *mechanical tube length* is the actual length of the microscope tube as measured from the top, where the shoulder of the eyepiece rests, to the bottom where the shoulder on the screw-in objective locates. This has been standardized at 160 mm. Generally the mechanical and optical tube

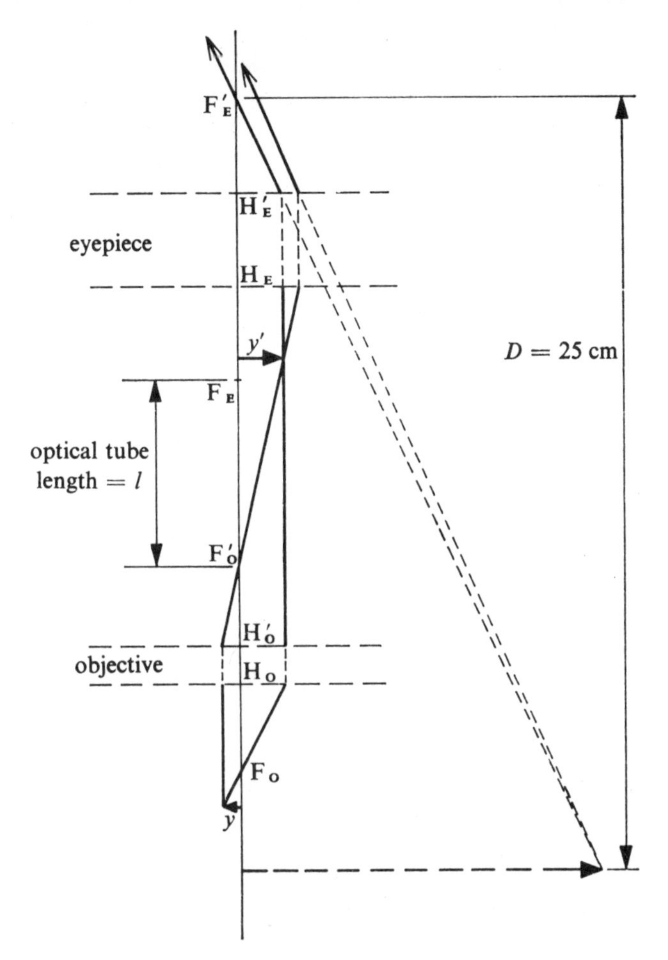


Fig. 5-17 Ray diagram for a microscope

lengths are approximately the same. The thread on the objective has also been standardized by the Royal Microscopical Society. Since the diameter of the eyepieces is also standardized, all eyepieces and objectives should be interchangeable to give an image which is approximately in focus. The *parfocal distance* is the distance from the object to the bottom of the microscope tube where the objective fits. This distance must be constant so that on changing an objective to one of a different power the object should be kept approximately in focus. To get the back focal plane of the objective in the correct position, the lenses are suitably positioned in the objective jacket.

The optical requirements and the design of microscope objectives have already been discussed in Section 4-11 and the use of aplanatic points in Section 4-7. However, as pointed out in Section 1-11, a parallel glass plate introduces spherical aberration and, since an object for microscopy is often mounted on a glass slide with Canada balsam and a cover glass of refractive index about 1·515 (Fig. 5-18), spherical aberration is produced. It is necessary, therefore, at least for the higher powers, to have different objectives corrected for use with covered or uncovered objects. The usual thickness of a cover glass is 0·18 mm. A variation of 0·03 or 0·04 mm in the thickness for which the objective is corrected gives a noticeable deterioration of the image obtained with the highest power objectives. To

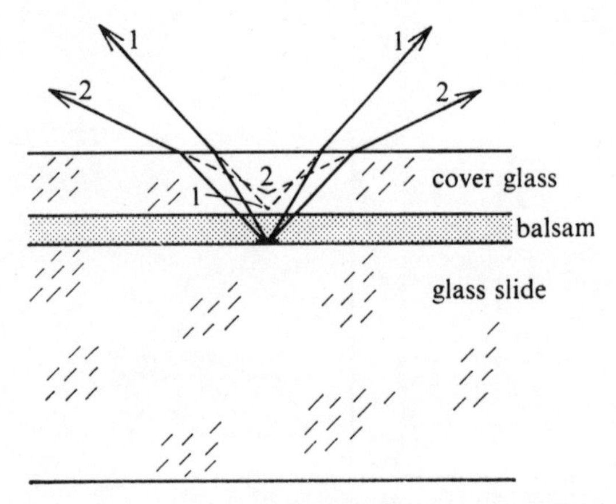


Fig. 5-18 Spherical aberration introduced by cover glass

correct for variations in cover glass and balsam thickness, the draw tube, which is the tube into which the eyepiece drops, is made movable so that the mechanical tube length may be changed from its nominal 160 mm; by shortening the tube length, the spherical aberration resulting from too thick a cover glass can be compensated. For the oil immersion objectives, normally $\times 100$, a drop of oil, usually cedar oil of refractive index 1·517, is placed between the front lens of the objective and the object (Section 4-7 and Fig. 4-10). This oil is of the same refractive index as that of the front lens and hence no spherical aberration is introduced. However, the dispersions of the glass and oil are different and hence some chromatic aberration is introduced and this must be compensated for in the rest of the objective.

It is important to the microscopist to know the highest resolution that can be obtained. Suppose the smallest distance that two point objects can be apart and just be resolved with the microscope as two objects is $y$. On being magnified they have an apparent separation $y'$. Assuming the smallest angle that can be resolved by the eye under these illumination conditions is 2 minutes of arc (0·00058 radians) then $y' = 25 \times 0{\cdot}00058$ cm (Fig. 5-19). Chapter 9 shows that, due to diffraction, the smallest separation, $y$, that two point objects can have and still be resolved as two objects, termed the *resolving power*, is such that

$$\text{resolving power} = \frac{0{\cdot}61\lambda}{NA} \tag{5-26}$$

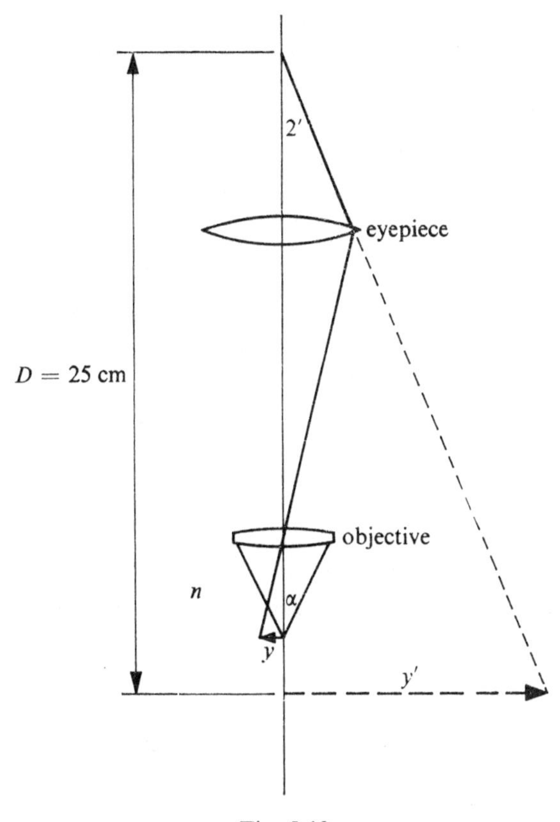


Fig. 5-19

where $NA$ stands for *numerical aperture* and is equal to the product of the refractive index of the medium of the object space and the sine of the semi-aperture of the cone of rays entering through the entrance pupil of the objective (Fig. 5-19)—that is, $NA = n \sin \alpha$. The total magnification produced by the microscope is equal to

$$M = y'/y = (25 . 0{\cdot}00058) \Big/ \left(\frac{0{\cdot}61 . 5 . 10^{-5}}{NA}\right)$$

$$= 475\ NA$$

$$= \text{say, } 600\ NA$$

to allow for some spare magnification, and where the mean wavelength of light is taken as $5 \times 10^{-5}$ cm. Sin $\alpha$ cannot be greater than unity and with oil immersion—that is, $n = 1{\cdot}5$—the largest $NA$ normally available is about 1·3, using a ×100 objective. This gives the total magnification required to just resolve the two point objects as very approximately 1000 times, which is obtainable with a ×100 objective and a ×10 eyepiece. If the power of the eyepiece is increased, no more information is obtained and the image is less bright. The extra magnification is then termed 'empty magnification'.

The expression of Eqn (5-26) is only true if the two point objects are each self-luminous. For a microscope this is not true since the objects are illuminated from an external source; Abbe has shown that for a microscope it is more correct to replace the factor of 0·61 by 0·5. He also showed that the resolution depends on the $NA$ of the condenser used to illuminate

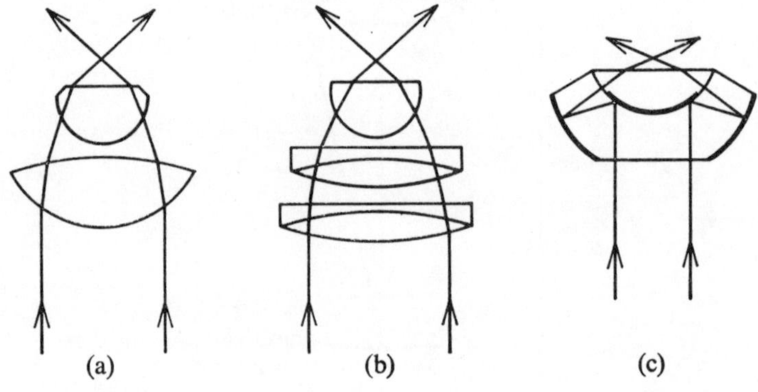


Fig. 5-20 (a) Abbe, (b) achromatic, and (c) dark-field condensers

the object. The best resolution is obtained for *critical illumination*, that is, when the cone of light focused onto the object by the condenser fills two-thirds of the aperture of the objective. Then the

$$\text{resolving power} = \frac{\lambda}{NA_{\text{obj}} + NA_{\text{cond}}} \tag{5-27}$$

For critical illumination the numerical apertures of the condenser and objective are approximately equal, giving

$$\text{resolving power} = \frac{0{\cdot}5\lambda}{NA_{\text{obj}}} \tag{5-28}$$

The design of the substage condenser system thus becomes important. The system should be capable of focusing light from an external source to a point focus so that a cone of rays diverges to fill almost the whole aperture of the objective uniformly with light. Figure 5-20(a) shows the arrangement of the Abbe condenser, which has a maximum $NA$ of about 1·2 using oil immersion. This design, however, has considerable spherical and chromatic aberration. These aberrations may be corrected in an achromatic condenser [Fig. 5-20(b)]. The design requirements are similar to those of the objective but the manufacturing tolerances are not so stringent. An iris diaphragm fitted below the condenser allows the numerical aperture to be adjusted to suit the objective being used.

If a condenser of greater *NA* than the objective is used, then by stopping out a central cone of rays equal in *NA* to that of the objective, direct rays can be prevented from entering the microscope. However, if the field to be magnified is made up of a number of small objects capable of scattering, diffracting, or refracting the light in any way, then some of this light may enter the objective. These small objects are seen as bright objects on a dark background, instead of a bright background. The objects are made far more visible but they may be difficult to identify due to diffraction effects. Besides a simple stop on the axis of the condenser, various systems have been designed to give oblique illumination without the wastage of light caused by a blank stop—see, for example, the condenser in Fig. 5-20(c).

## 5-12 Autocollimators

An autocollimator consists simply of a well-corrected lens with a pinhole aperture accurately positioned at a principal focus of the lens (Fig. 5-21).

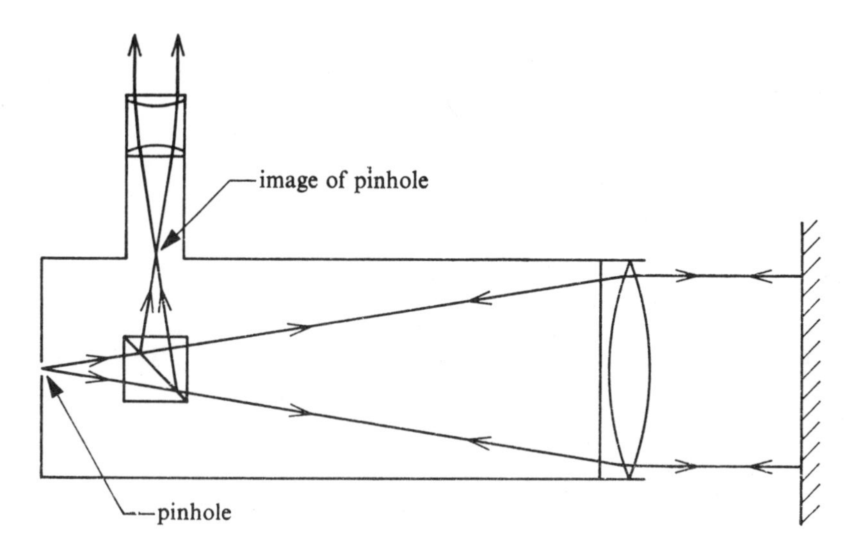


Fig. 5-21 Arrangement of an autocollimator

On illuminating this pinhole, which then approximates to a point source, the lens forms a parallel beam of light. If these parallel rays are incident normally onto a reflecting surface, they return along their own paths. By means of a suitable half-silvered reflector these rays are now reflected to one side of the collimator axis to form an image of the pinhole source. This image is in turn viewed by a Ramsden eyepiece. The half-silvered reflector is in the form of a cube cut diagonally, polished, half-silvered, and cemented together again. This has an advantage over a plain glass plate half-silvered on one face, since there is no ghost reflection from the other surface.

The autocollimator is important in the measurement of small angles. If the reflecting surface used to return the parallel beam is tilted, the position of the image of the pinhole as seen through the eyepiece shifts to one side of its original position. Suppose, for example, that the reflecting surface is tilted through an angle of 1′, then the reflected ray is shifted through 2′. Using a collimating lens of, say, 40 cm focal length, the image is shifted $400 \tan 2' = 400 \times 0{\cdot}0006 = 0{\cdot}24$ mm. This distance is readily measured by a magnifying eyepiece with a calibrated graticule. If, for example, the diameter of the pinhole is 0·20 mm, this is the size of the image. Thus, for a 1′ tilt, the image is shifted by $1\frac{1}{5}$ times its own diameter, so that the graticule can be calibrated. When, for example, a plate of glass is held in front of the autocollimator, two reflections are obtained, one from each side. If the two images formed are superimposed, the plate has parallel faces; if not, the angle between them can readily be measured.

## PROBLEMS

**5-1** A 24 mm × 36 mm transparency is projected onto a screen 3 metres from the projector. If the image size is required to be 1 m × 1·5 m, calculate the focal length of the projector lens. How far should the transparency be from the lens? [7 cm, 7·2 cm]

**5-2** What are the optical qualities required in a good camera? Discuss how and to what extent these are achieved in practice. Discuss the conditions which decide the best choice of aperture and shutter speed.

**5-3** A person whose least distance of distinct vision is 20 cm uses a 5-cm focal length lens to magnify a small object. How far away must the lens be held from the object, and what is the magnification produced? [4 cm, ×5]

**5-4** An astronomical telescope is 2 m long. Calculate the focal length of an eyepiece that would give a magnification of 200 diameters. Explain the advantage of a telescope over a pair of simple sights for purposes of angular measurement. [1 cm]

**5-5** An astronomical telescope of known magnifying power $M$ is focused on a very distant object, and the distance between the lenses is found to be $a$ cm. On focusing the telescope on a nearer object the eyepiece has to be moved out a distance $b$ cm. Show that the distance to the object is given by

$$\frac{aM}{b(M+1)}\left(\frac{aM}{M+1}+b\right)$$

**5-6** A telescope consisting of an objective lens of focal length 50 cm and an eyepiece lens of focal length 2 cm is used to view a distant object. What is the magnification and distance between the lenses if the final image is seen (a) at a great distance (b) at a distance of 25 cm?

[(a) ×25, 52 cm; (b) ×27, 51·85 cm]

**5-7** The objectives of a pair of 10 × 50 binoculars have focal lengths of 30 cm. What is (a) the focal length of the eyepieces, (b) the diameter of the exit pupils, (c) the angular field of the objectives and (d) the eyepieces, and

(e) what is the eye relief? Comment on these figures with respect to the performance of the binoculars.

[(a) 3 cm; (b) 5 mm; (c) 8·7°; (d) 87°; (e) 3·3 cm]

**5-8** A microscope objective of 4-mm focal length forms an image 160 mm from its second focal point. If the focal length of the eyepiece is 50 mm, what is the magnification of the microscope? [×200]

**5-9** Describe and point out the respective merits of Ramsden's and Huygens' eyepieces.

**5-10** A Huygenian eyepiece has lenses of focal length 3 cm and 1 cm, separated by a distance of 2 cm. Determine the positions of the cardinal points. What is its magnifying power when used by a person whose least distance of distinct vision is 25 cm?

[$H_2F_2 = F_1H_1 = 1{\cdot}5$ cm; $F_1F_2 = 1{\cdot}0$ cm; $H_2H_1 = 2{\cdot}0$ cm;
Field lens to $H_2 = 1$ cm; ×16·7]

# 6 Prisms, spectra, and colour

## 6-1 Introduction

The refractive index of a medium depends on the wavelength of the light used (Chapter 1). Consequently when light of various wavelengths passes through a refracting medium, the rays are refracted by different amounts and take different optical paths. Large differences in deviation are obtained if the refracting medium is in the form of a prism. We shall begin this chapter by determining the deviations of incident rays of different wavelengths. This information is important because some substances may be caused to emit light of characteristic discrete wavelengths. If this light is passed through a prism, the rays of different wavelengths are refracted along different paths and the deviations produced may be measured, the wavelengths may be determined, and the substance identified.

Lenses, too, can cause a separation of white light into its constituent colours. The rays refracted along different paths and hence to different axial focal points result in chromatic aberration. By a suitable combination of two lenses this aberration may be corrected (Section 6-6).

## 6-2 Refraction by a prism

The deviation produced by a prism is related to the refractive index of the prism and its refracting angle. Consider a monochromatic ray incident on the face of a prism of refracting angle $A$ and refractive index $n$ (Fig. 6-1), making angles with the normals to the surfaces as shown. Then the deviation of the ray at the first surface is

$$D_1 = \theta_1 - \theta_1' \tag{6-1}$$

This ray is then further deviated at the second surface by an amount

$$D_2 = \theta_2 - \theta_2' \tag{6-2}$$

Hence the total deviation is

$$D = (\theta_1 - \theta_1') + (\theta_2 - \theta_2') \tag{6-3}$$

Also, from the figure

$$A + \alpha = 180° \tag{6-4}$$

as also

$$\theta_1' + \theta_2' + \alpha = 180° \tag{6-5}$$

Therefore

$$\theta_1' + \theta_2' = A \tag{6-6}$$

and the total deviation $D$ is given by

$$\theta_1 + \theta_2 = A + D \tag{6-7}$$

When the angle of incidence $\theta_1$ is changed by a small amount $\mathrm{d}\theta_1$, then, since for the first surface

$$\sin \theta_1 = n \sin \theta_1' \tag{6-8}$$

$$\cos \theta_1 \, \mathrm{d}\theta_1 = n \cos \theta_1' \, \mathrm{d}\theta_1' \tag{6-9}$$

Similarly for the second surface

$$\sin \theta_2 = n \sin \theta_2' \tag{6-10}$$

$$\cos \theta_2 \, \mathrm{d}\theta_2 = n \cos \theta_2' \, \mathrm{d}\theta_2' \tag{6-11}$$

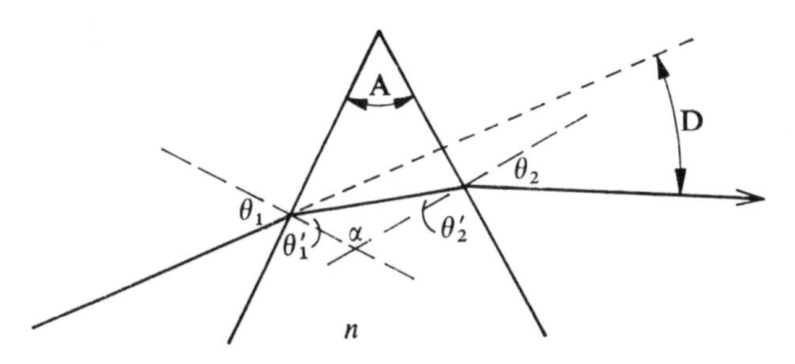


Fig. 6-1 Refraction by a prism

Therefore, from Eqns (6-9) and (6-11)

$$\frac{\mathrm{d}\theta_2}{\mathrm{d}\theta_1} = \frac{\cos \theta_1 \cos \theta_2' \, \mathrm{d}\theta_2'}{\cos \theta_2 \cos \theta_1' \, \mathrm{d}\theta_1'} \tag{6-12}$$

Also, by differentiating Eqn (6-6)

$$\mathrm{d}\theta_1' = -\mathrm{d}\theta_2' \tag{6-13}$$

and hence

$$\frac{\mathrm{d}\theta_2}{\mathrm{d}\theta_1} = -\frac{\cos \theta_1 \cos \theta_2'}{\cos \theta_2 \cos \theta_1'} \tag{6-14}$$

which shows the corresponding change $\mathrm{d}\theta_2$ in the emergent ray due to a small change $\mathrm{d}\theta_1$ in the angle of incidence of the incident ray. If, however, the angle of incidence is decreased sufficiently, then $\theta_2'$ becomes equal to the critical angle for the second face and further decrease in $\theta_1$ results in total internal reflection.

The total deviation of the ray is [Eqn (6-7)]

$$D = \theta_1 + \theta_2 - A$$

For this deviation to be an extremum, its first derivative must be zero

i.e.,
$$\frac{dD}{d\theta_1} = 1 + \frac{d\theta_2}{d\theta_1} = 0 \tag{6-15}$$

and hence, from Eqns (6-14) and (6-15), for an extremum

$$\frac{\cos\theta_1 \cos\theta_2'}{\cos\theta_2 \cos\theta_1'} = 1 \tag{6-16}$$

Squaring

$$\frac{\cos^2\theta_1}{\cos^2\theta_1'} = \frac{\cos^2\theta_2}{\cos^2\theta_2'} \tag{6-17}$$

Therefore, using Eqns (6-8) and (6-10)

$$\frac{1 - \sin^2\theta_1}{n^2 - \sin^2\theta_1} = \frac{1 - \sin^2\theta_2}{n^2 - \sin^2\theta_2} \tag{6-18}$$

For this equation to be satisfied, that is for an extremum

$$\theta_1 = \theta_2 \tag{6-19}$$

and therefore

$$\theta_1' = \theta_2' \tag{6-20}$$

It may be shown that the second derivative $d^2D/d\theta_1^2$ is positive and therefore the extremum is a minimum. However, the proof is somewhat tedious and, since by tracing the paths of several rays of different angles of incidence it becomes obvious that the extremum is not a maximum, the result may be accepted. Thus the condition for *minimum deviation* is that the angle of incidence shall equal the angle of emergence—that is, $\theta_1 = \theta_2$.

Alternatively, the fact that minimum deviation occurs when the ray passes symmetrically through the prism—that is, $\theta_1 = \theta_2$—may be argued as follows. Consider Fig. 6-2 in which an incident ray follows the path PQRS. A ray incident at the same angle but on the other face of the prism follows a corresponding path P′Q′R′S′. The rays are deviated by the same amount. Also, a ray incident along S′R′ follows the path S′R′Q′P′ and is also deviated by the same amount. Thus there are two rays, PQ and S′R′, which are deviated the same amount by the prism, so that this deviation is not unique and therefore cannot be an extremum, such as a

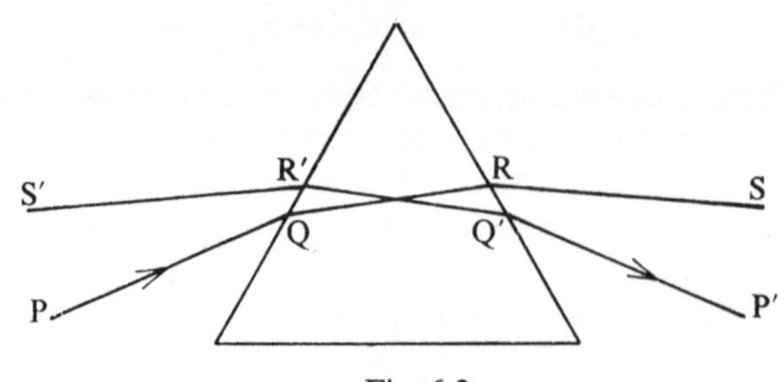


Fig. 6-2

minimum. The deviation is unique and an extremum only when the two rays PQRS and S′R′Q′P′ coincide and this only occurs when the rays pass symmetrically through the prism—that is, when the angle of incidence is equal to the angle of emergence, $\theta_1 = \theta_2$.

If now the ray path is adjusted for minimum deviation so that

$$\theta_1 = \theta_2 = \tfrac{1}{2}(A + D) \tag{6-21}$$

and

$$\theta_1' = \theta_2' = \tfrac{1}{2}A \tag{6-22}$$

from Eqns (6-7) and (6-6), and where $D$ is now the angle of minimum deviation, then the refractive index $n$ of the prism is given by

$$n = \frac{\sin\theta_1}{\sin\theta_1'} = \frac{\sin\frac{1}{2}(A + D)}{\sin\frac{1}{2}A} \tag{6-23}$$

This is an accurate method of determining refractive index, since the angles $A$ and $D$ can be determined with a spectrometer.

## 6-3 Dispersion by a prism

The refractive index of the prism material depends on the wavelength of the light used: red light is deviated least and the violet the most (Fig. 6-3).

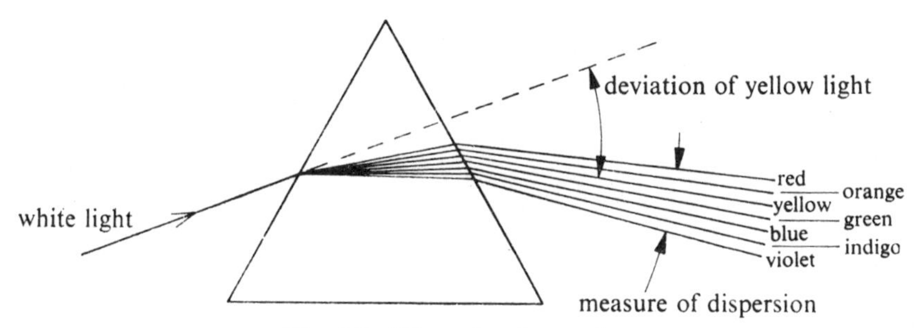


Fig. 6-3 Dispersion by a prism

Incident white light is thus *dispersed* into the colours of the *spectrum*. The deviation of any ray depends on the refractive index $n$ which is in turn dependent on the wavelength $\lambda$.

The quantity

$$\frac{\mathrm{d}D}{\mathrm{d}\lambda} = \frac{\mathrm{d}D}{\mathrm{d}n}\cdot\frac{\mathrm{d}n}{\mathrm{d}\lambda} \tag{6-24}$$

is termed the *angular dispersion* of the prism for a particular angle of incidence $\theta_1$. The term $\mathrm{d}n/\mathrm{d}\lambda$ is determined by the glass used to manufacture the prism and thus the term $\mathrm{d}D/\mathrm{d}n$, which depends on the geometry of the system, becomes of interest since this varies with the angle of incidence. Then, for a particular angle of incidence, that is, $\theta_1 = \text{const.}$,

$$\frac{\mathrm{d}\theta_1'}{\mathrm{d}n} + \frac{\mathrm{d}\theta_2'}{\mathrm{d}n} = 0 \tag{6-25}$$

from Eqn (6-6), and from Eqn (6-7)

$$\frac{d\theta_2}{dn} = \frac{dD}{dn} \tag{6-26}$$

Similarly, differentiating Eqns (6-8) and (6-10) with respect to $n$

$$0 = \sin\theta_1' + n\cos\theta_1' \frac{d\theta_1'}{dn} \tag{6-27}$$

$$\cos\theta_2 \frac{d\theta_2}{dn} = \sin\theta_2' + n\cos\theta_2' \frac{d\theta_2'}{dn} \tag{6-28}$$

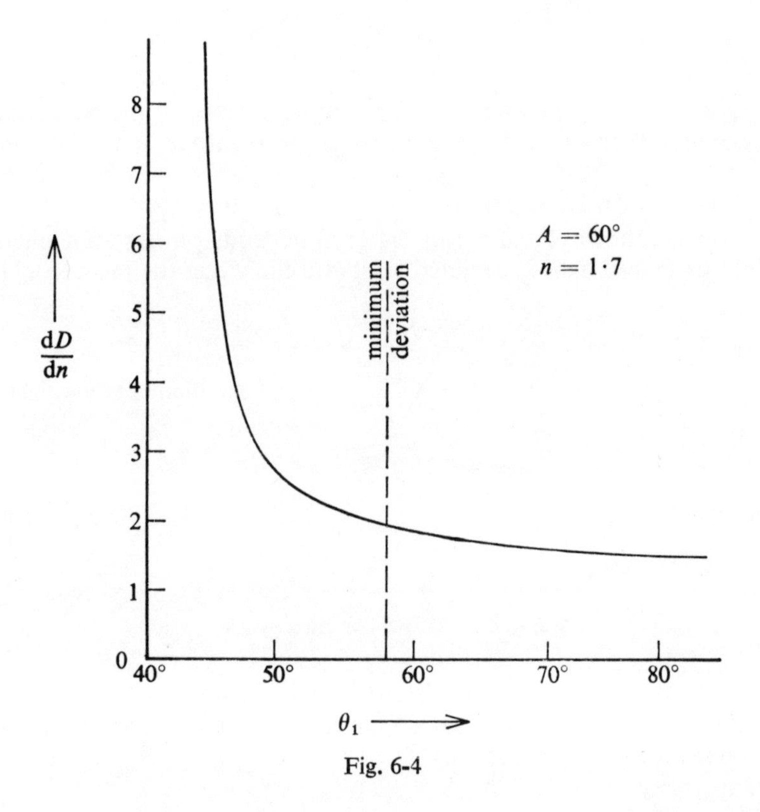


Fig. 6-4

Hence, substituting for $d\theta_1'/dn$ in Eqn (6-27) from Eqn (6-25) and combining with Eqn (6-28)

$$\frac{d\theta_2}{dn} = \frac{\sin(\theta_1' + \theta_2')}{\cos\theta_2 \cos\theta_1'}$$

$$= \frac{\sin A}{\cos\theta_2\cos\theta_1'} = \frac{dD}{dn} \tag{6-29}$$

by Eqns (6-6) and (6-26). Figure 6-4 shows how this varies with the angle of incidence $\theta_1$ for a prism of refracting angle 60° and refractive index 1·7.

In particular, at minimum deviation $\theta_1 = \theta_2$ and $\theta_1' = \theta_2' = \frac{1}{2}A$, then from Eqn (6-29)

$$\frac{dD}{dn} = \frac{2 \sin \frac{1}{2}A \cos \frac{1}{2}A}{\cos \theta_1 \cos \frac{1}{2}A}$$

$$= \frac{2 \sin \frac{1}{2}A}{\cos \theta_1} = \frac{2 \sin \frac{1}{2}A}{\cos \frac{1}{2}(A + D)} \tag{6-30}$$

## 6-4 Thin prisms

When the refracting angle $A$ of the prism and the deviation produced are only of the order of a few degrees, the refraction at minimum deviation [Eqn (6-23)] can be expressed in a much simpler form, since the angles in radians are equal to the sines of the angles. Thus

$$n = \frac{\sin \frac{1}{2}(A + D)}{\sin \frac{1}{2}A}$$

$$= \frac{A + D}{A}$$

and

$$D = (n - 1)A \tag{6-31}$$

If this prism has a refractive index $n_F$ for blue light, then the deviation produced is

$$D_F = (n_F - 1)A \tag{6-32}$$

and similarly for red light

$$D_C = (n_C - 1)A \tag{6-33}$$

Therefore the dispersion—that is, the angle between the red and the blue rays—is

$$D_F - D_C = (n_F - n_C)A \tag{6-34}$$

Also the deviation of yellow light is

$$D_d = (n_d - 1)A \tag{6-35}$$

Hence

$$D_F - D_C = (n_F - n_C)\frac{(n_d - 1)}{(n_d - 1)}A$$

$$= \frac{n_F - n_C}{n_d - 1} D_d$$

that is

$$D_F - D_C = \frac{D_d}{\nu} \tag{6-36}$$

where

$$\frac{1}{\nu} = \frac{n_F - n_C}{n_d - 1} = \text{dispersive power} \qquad \text{[Eqn (1-2)]}$$

## 6-5 Combination of prisms

Prisms made from different types of glass may be combined to produce special effects. Figure 6-5(a) shows an *achromatic prism*; this has the property of producing deviation without dispersion—that is, light of all wavelengths is equally refracted. Crown and flint glasses are used in the prism. A crown glass produces less dispersion than a flint glass for glasses of approximately the same refractive index. It is thus possible to construct two prisms with different refracting angles so that each produces the same amount of dispersion as the other. When combined, one prism will counteract the dispersion but not the whole of the deviation produced by the other prism. Therefore deviation is produced without dispersion. In practice dispersion is not completely eliminated as the cancellation may be effective for the F and C wavelengths but not for the intermediate wavelengths. It is therefore better to obtain exact cancellation for, say, the green and the orange light since these tend to be brighter than the extreme blue and red.

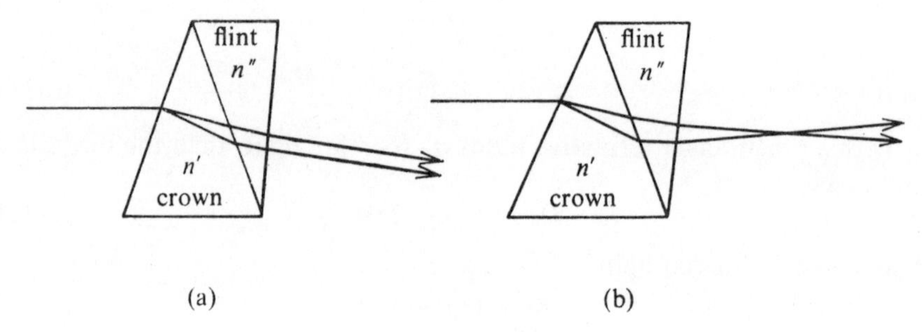


Fig. 6-5 (a) Achromatic and (b) direct-vision prisms

For thin prisms the condition for achromatism follows simply from Eqn (6-34). Then the dispersion introduced by the first prism must be equal to that of the second, that is,

$$D_F - D_C = (n'_F - n'_C)A' = (n''_F - n''_C)A'' \tag{6-37}$$

where the single and double superscripts refer to the first and second prisms respectively. Hence the condition for achromatic *thin* prisms is that

$$\frac{A'}{A''} = \frac{n''_F - n''_C}{n'_F - n'_C} \tag{6-38}$$

From Eqn (6-35), the deviation is given by

$$D'_d - D''_d = (n'_d - 1)A' - (n''_d - 1)A'' \tag{6-39}$$

Similarly, by a suitable choice of glasses and refracting angles, a *direct vision prism* may be produced [Fig. 6–5(b)]. This disperses white light into the colours of the spectrum but does not deviate the yellow light.

By Eqn (6-35), the deviation of the yellow light produced by the first prism is

$$D'_d = (n'_d - 1)A' \tag{6-40}$$

and by the second prism

$$D''_d = (n''_d - 1)A'' \tag{6-41}$$

Since the total deviation produced is to be zero

$$D'_d - D''_d = (n'_d - 1)A' - (n''_d - 1)A'' = 0$$

i.e.,

$$\frac{A'}{A''} = \frac{n''_d - 1}{n'_d - 1} \tag{6-42}$$

From Eqn (6-34) the dispersion, blue to red, is

$$\begin{aligned} D_F - D_C &= (D'_F - D''_F) - (D'_C - D''_C) \\ &= (n'_F - n'_C)A' - (n''_F - n''_C)A'' \end{aligned} \tag{6-43}$$

If, however, the refracting angles are not small, then the above formulae are no longer applicable and no useful simple equations can be derived.

## 6-6 Achromatic thin lenses

A pair of thin lenses made of crown and flint glass respectively may be considered as a combination of two thin prisms whose refracting angles

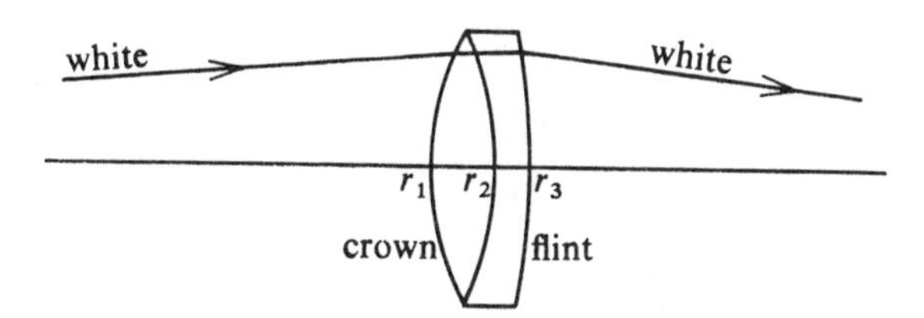


Fig. 6-6 An achromatic doublet

vary with the distance from the axis. In the same way as for an achromatic prism, the shapes and refractive indices of the lenses may be chosen such that the dispersion introduced by the first lens is cancelled by the dispersion of the second but still leaves a net deviation. Figure 6-6 shows a typical lens combination. One face of each lens has a common radius of curvature with one face of the other lens so that they may be cemented together with Canada balsam. If the lenses are thin, simple conditions for achromatism apply. If they are not thin, so that the principal planes for the red light and for the blue light do not coincide, then, although the red and blue may be made to come to the same point on the axis so that *longitudinal* chromatic aberration is absent, the focal lengths and hence the magnification is different for the red and the blue. The images formed have different sizes depending on their colour. This is known as *lateral* chromatic aberration. Except for very thick lenses, the lateral chromatic aberration is small if the combination is corrected longitudinally.

Thin-lens combinations may be dealt with in the following way. Let $f_d$ be the required focal length of an achromatic doublet, then

$$\frac{1}{f_d} = \frac{1}{f'_d} + \frac{1}{f''_d} \tag{6-44}$$

where the single superscript refers to the crown lens and the double to the flint lens. Let the crown lens have refractive indices $n'_F$, $n'_d$, and $n'_C$, and the flint lens refractive indices $n''_F$, $n''_d$, and $n''_C$, for blue, yellow, and red light respectively. Then in terms of the refractive indices and radii of curvature

$$\frac{1}{f_d} = (n'_d - 1)\left(\frac{1}{r_1} - \frac{1}{r_2}\right) + (n''_d - 1)\left(\frac{1}{r_2} - \frac{1}{r_3}\right) \tag{6-45}$$

where the radii are numbered in order, as in Fig. 6-6, and the combination has a common interface, radius $r_2$. Similarly,

$$\frac{1}{f_F} = (n'_F - 1)\left(\frac{1}{r_1} - \frac{1}{r_2}\right) + (n''_F - 1)\left(\frac{1}{r_2} - \frac{1}{r_3}\right) \tag{6-46}$$

$$\frac{1}{f_C} = (n'_C - 1)\left(\frac{1}{r_1} - \frac{1}{r_2}\right) + (n''_C - 1)\left(\frac{1}{r_2} - \frac{1}{r_3}\right) \tag{6-47}$$

Then for achromatism of the blue and red light

$$f_F = f_C$$

and therefore

$$(n'_F - 1)\left(\frac{1}{r_1} - \frac{1}{r_2}\right) + (n''_F - 1)\left(\frac{1}{r_2} - \frac{1}{r_3}\right)$$

$$= (n'_C - 1)\left(\frac{1}{r_1} - \frac{1}{r_2}\right) + (n''_C - 1)\left(\frac{1}{r_2} - \frac{1}{r_3}\right) \tag{6-48}$$

i.e., $$n'_F\left(\frac{1}{r_1} - \frac{1}{r_2}\right) + n''_F\left(\frac{1}{r_2} - \frac{1}{r_3}\right) = n'_C\left(\frac{1}{r_1} - \frac{1}{r_2}\right) + n''_C\left(\frac{1}{r_2} - \frac{1}{r_3}\right) \tag{6-49}$$

Therefore

$$\frac{1/r_1 - 1/r_2}{1/r_2 - 1/r_3} = \frac{r_3}{r_1}\left(\frac{r_2 - r_1}{r_3 - r_2}\right) = -\frac{n''_F - n''_C}{n'_F - n'_C} \tag{6-50}$$

Since $(n''_F - n''_C)$ and $(n'_F - n'_C)$ are both positive, the negative sign implies that one lens must be positive and the other negative, as would be expected from Fig. 6-6. Also, for yellow light, the focal lengths of the individual crown and flint lenses are given by

$$\frac{1}{f'_d} = (n'_d - 1)\left(\frac{1}{r_1} - \frac{1}{r_2}\right) \tag{6-51}$$

$$\frac{1}{f''_d} = (n''_d - 1)\left(\frac{1}{r_2} - \frac{1}{r_3}\right) \tag{6-52}$$

Therefore, dividing Eqn (6-51) by Eqn (6-52) and rearranging

$$\frac{r_3}{r_1}\left(\frac{r_2 - r_1}{r_3 - r_2}\right) = \left(\frac{n''_d - 1}{n'_d - 1}\right)\frac{f''_d}{f'_d} \tag{6-53}$$

Equating this with Eqn (6-50)

$$\frac{f''_d}{f'_d}\left(\frac{n''_d - 1}{n'_d - 1}\right) = -\frac{n''_F - n''_C}{n'_F - n'_C} \tag{6-54}$$

and therefore

$$\frac{f'_d}{f''_d} = -\frac{(n''_d - 1)/(n''_F - n''_C)}{(n'_d - 1)/(n'_F - n'_C)} = -\frac{\nu''}{\nu'} \tag{6-55}$$

Thus for achromatism, it is necessary to satisfy Eqn (6-55)

$$f'_d\nu' + f''_d\nu'' = 0$$

and Eqn (6-44)

$$\frac{1}{f_d} = \frac{1}{f'_d} + \frac{1}{f''_d}$$

where $f_d$ is the required focal length of the combination and $f'_d$ and $f''_d$ are the focal lengths of the individual crown and flint lenses.

Also, from Eqn (6-55)

$$-\frac{\nu''}{f'_d} = \frac{\nu'}{f''_d}$$

Then, adding $\nu'/f'_d$ to both sides gives

$$\frac{\nu'}{f'_d} - \frac{\nu''}{f'_d} = \frac{\nu'}{f'_d} + \frac{\nu'}{f''_d} \tag{6-56}$$

i.e.,

$$\frac{1}{f'_d}(\nu' - \nu'') = \nu'\left(\frac{1}{f'_d} + \frac{1}{f''_d}\right)$$

$$= \nu'\left(\frac{1}{f_d}\right) \tag{6-57}$$

then

$$f'_d = f_d\left(\frac{\nu' - \nu''}{\nu'}\right) \tag{6-58}$$

Similarly, by adding $-\nu''/f''_d$ to both sides

$$f''_d = -f_d\left(\frac{\nu' - \nu''}{\nu''}\right) \tag{6-59}$$

Hence, having determined the required individual focal lengths for the two lenses, the radii may be determined from

$$\frac{1}{f'_d} = (n'_d - 1)\left(\frac{1}{r_1} - \frac{1}{r_2}\right) \tag{6-60}$$

$$\frac{1}{f''_d} = (n''_d - 1)\left(\frac{1}{r_2} - \frac{1}{r_3}\right) \tag{6-61}$$

**Example 6-1** Design a cemented doublet lens to have a focal length of 20 cm, using the following glasses.

| | $n_d$ | $n_F$ | $n_C$ |
|---|---|---|---|
| Borosilicate Crown | 1·5097 | 1·5152 | 1·5073 |
| Dense Flint | 1·6258 | 1·6381 | 1·6206 |

The first calculations to be made are the two $\nu$-values:
*Crown lens*

$$\nu' = \frac{n'_d - 1}{n'_F - n'_C} = \frac{1{\cdot}5097 - 1}{1{\cdot}5152 - 1{\cdot}5073}$$

$$= \frac{0{\cdot}5097}{0{\cdot}0079} = 64{\cdot}5190$$

*Flint lens*

$$\nu'' = \frac{n''_d - 1}{n''_F - n''_C} = \frac{1{\cdot}6258 - 1}{1{\cdot}6381 - 1{\cdot}6206}$$

$$= \frac{0{\cdot}6258}{0{\cdot}0175} = 35{\cdot}7600$$

The two focal lengths must next be calculated:
*Crown lens*

$$f'_d = f_d\left(\frac{\nu' - \nu''}{\nu'}\right) = 20\left(\frac{64{\cdot}5190 - 35{\cdot}7600}{64{\cdot}5190}\right)$$

$$= \frac{20 \times 28{\cdot}7590}{64{\cdot}5190} = 8{\cdot}9149 \text{ cm}$$

*Flint lens*

$$f''_d = -f_d\left(\frac{\nu' - \nu''}{\nu''}\right) = -20\left(\frac{64{\cdot}5190 - 35{\cdot}7600}{35{\cdot}7600}\right)$$

$$= -\frac{20 \times 28{\cdot}7590}{35{\cdot}7600} = -16{\cdot}0845 \text{ cm}$$

As a check

$$\frac{1}{f'_d} + \frac{1}{f''_d} = \frac{1}{8{\cdot}9149} - \frac{1}{16{\cdot}0845}$$

$$= 0{\cdot}1122 - 0{\cdot}0622$$

$$= 0{\cdot}0500$$

that is, in agreement with

$$\frac{1}{f_d} = \frac{1}{20} = 0{\cdot}0500$$

Supposing now that it is required that the crown lens shall be equi-convex (that is, $r_1 = -r_2$) then

*Crown lens*

$$\frac{1}{f'_d} = (n'_d - 1)\left(\frac{1}{r_1} - \frac{1}{r_2}\right) = (n'_d - 1)\frac{2}{r_1}$$

i.e.,

$$r_1 = 2(n'_d - 1)f'_d$$
$$= 2(1{\cdot}5097 - 1)8{\cdot}9149$$
$$= 9{\cdot}0878 \text{ cm} = -r_2$$

*Flint lens*

$$\frac{1}{f''_d} = (n''_d - 1)\left(\frac{1}{-9{\cdot}0878} - \frac{1}{r_3}\right)$$

i.e.,

$$\frac{1}{-16{\cdot}0845} = (1{\cdot}6258 - 1)\left(\frac{1}{-9{\cdot}0878} - \frac{1}{r_3}\right)$$

$$\therefore \frac{1}{r_3} = \frac{1}{16{\cdot}0845 \times 0{\cdot}6258} - \frac{1}{9{\cdot}0878}$$
$$= 0{\cdot}0993 - 0{\cdot}1100 = -0{\cdot}0107$$

i.e.,

$$r_3 = -93{\cdot}4579 \text{ cm}$$

Thus the doublet has the same shape as that shown in Fig. 6-6, where $r_1 = +9{\cdot}09$ cm, $r_2 = -9{\cdot}09$ cm, and $r_3 = -93{\cdot}46$ cm.

## 6-7 Spectrometer

In Section 6-2 we discussed the dispersion of white light by a prism into its constituent colours. If a screen is used to receive these colours, it will be seen that they overlap with each other. No matter how small a source is used there will always be overlap and the spectrum is said to be *impure*. To obtain a *pure* spectrum—that is one in which adjacent colours do not overlap—a spectrometer is used, although the finite width of the spectrometer slit means that there is always some small overlap even with this type of spectrum. In the spectrometer a narrow slit acts as a source. It may be adjusted in width and to a vertical position, and is located in the first focal plane of a converging lens (Fig. 6-7). This lens-slit combination, termed a *collimator*, is used to produce a parallel beam of light. This light is incident onto the prism face. The prism itself is mounted on a rotatable *table*, which may be tilted so that the refracting edge of the prism is vertical and parallel to the source slit. The beam of light is then refracted by the prism as a series of parallel beams each at a slightly different inclination, depending on the colour. These rays are focused by a *telescope*, so that each constituent colour is brought to a separate focus. The telescope can be moved about a vertical axis passing through the centre of the table so that the angle between the axis of the telescope and the axis of the collimator may be read from a divided circle. By centring a particular spectrum colour on the centre line of the graticule of the telescope eyepiece, the deviation for that colour (wavelength) may be determined. By adjusting both the telescope and prism position, the position for minimum deviation may be found for each wavelength. This instrument is called a *spectrometer*. The *spectroscope* is a similar instrument, without the measuring facility, used only for viewing the spectrum. If a photographic record is required, the eyepiece is removed and replaced by a camera. The combined instrument is then known as a *spectrograph.*

Before using the spectrometer it is first necessary to adjust the collimator to give a parallel beam of light and the telescope to receive it—that is, the telescope must be focused for infinity. First, however, it is necessary to adjust the telescope eyepiece to the position where the graticule is sharply in focus when the eye is relaxed and focused at infinity. The rays are then made parallel by *Schuster's method.* To do this the complete spectrometer is set up, with the prism on the spectrometer table and the prism and telescope rotated to the position for minimum deviation.

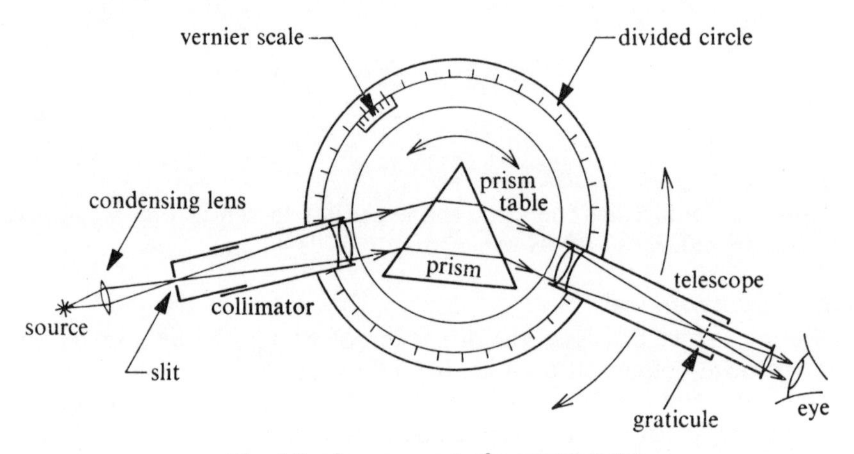


Fig. 6-7 Arrangement of a spectrometer

The source of light should be one that gives a distinctive isolated spectral line rather than a blur of colour. The telescope is then rotated so as to increase the angle made with the 'straight-through' position of the collimator. The spectral line image is thus shifted to the edge of the field as seen through the telescope. This image is then brought back to the centre of the telescope field by rotating the prism. The prism may be rotated in either direction because there are two positions of the prism that give a central image [Fig. 6-8(a)]. When the prism is in position A, that is, the light is falling less obliquely onto the prism face, the slit to lens distance in the collimator is adjusted to obtain the sharpest image as seen through the telescope. The prism is then rotated to position B so that the rays from the collimator fall more obliquely onto the prism face, and now the telescope is focused to obtain the sharpest image. This process is repeated several times—with position A adjust the collimator, with position B adjust the telescope—until the image appears equally sharp and without parallax on the graticule with the prism in either position. The collimator and telescope are then set for parallel rays.

It is also necessary to set the prism table level so that the refracting edge of the prism is parallel to the collimator slit and to the vertical rotation axis of the spectrometer. This may be done simply by arranging the prism as in Fig. 6-8(b), where P, Q, and R are the three table levelling screws and $A$ is the refracting angle of the prism. Face AC is made perpendicular to the line joining screws R and Q. Parallel light from the collimator is reflected from the faces AB and AC as shown. The telescope is first rotated to receive the reflected image of the collimator slit from face AC. Any of the levelling screws may be adjusted to centralize the image vertically in the field of the telescope. The telescope is then rotated to receive the reflection from face AB. Again the image is centralized vertically but this time by the adjustment of screw P only. This has the effect of tilting the reflecting face AB but only shifts face AC in its own plane and does not therefore upset the first setting. At this time the orientation of the collimator slit should also be adjusted so that it appears vertical with either reflection position.

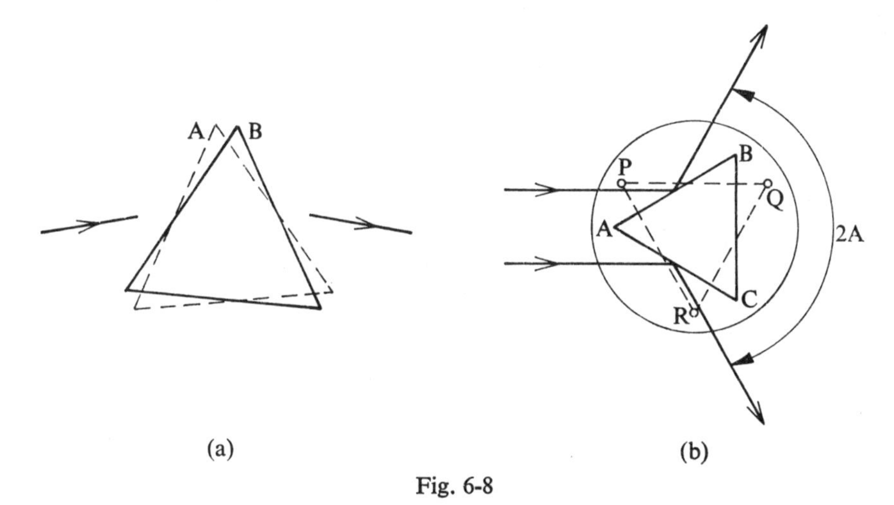


Fig. 6-8

Figure 6-8(b) also shows that the angle between the reflected rays, as measured by the two angular positions of the telescope, is equal to twice the refracting angle $A$ of the prism. Thus both angle $A$ and the angle of minimum deviation for a given wavelength may be measured. The angle of minimum deviation may be determined either by measuring the angle between the minimum deviation position of the telescope and the 'straight-through' position without the prism, or, preferably, by rotating the prism and telescope to obtain the minimum deviation position on the other side of the collimator axis and halving the angle between the two minimum deviation positions of the telescope.

Using the refracting angle $A$ and the angle of minimum deviation $D$, we can find the refractive index of the prism material [Eqn (6-23)] for given wavelengths of light, and hence determine its dispersive power. Alternatively, a calibration graph of angles of deviation against known wavelengths may be made and used to determine unknown wavelengths. The actual resolving power of a prism spectrometer is discussed in Chapter 9.

## 6-8 Spectra

So far we have discussed the dispersion of white light by a prism into its constituent colours. The spectrum formed is called a *continuous spectrum*, since it is a continuous band of colour spreading from red, through orange, yellow, green, blue, and indigo, to violet—that is, through the full range of the visible spectrum without any clear demarcations. In fact it may spread further, into the infra-red, where wavelengths are greater than the red of the visible spectrum, and into the ultra-violet, where the wavelengths are shorter than those of the blue end. Such radiations are invisible to the human eye, but they may be detected and measured by suitable instruments. This type of continuous spectrum is emitted by all incandescent solids and some dense gases. As a solid is heated it first emits heat in the form of infra-red radiation. As its temperature rises it becomes 'red-hot', when it emits wavelengths extending into the red end of the visible spectrum. At higher temperatures it becomes 'white-hot', when it then emits wavelengths extending into the blue end of the visible spectrum, which with the mixture of other wavelengths present make up 'white' light. Such emission of the various wavelengths is due to the thermal vibration of the molecules within the solid. The wavelength of light emitted is related to the frequency of vibration of the molecules in the hot solid. Due to collision with neighbouring molecules there can be no specific frequency of vibration and so white light is emitted.

If the spectrum caused by burning organic compounds, or, generally, the spectrum obtained by suitably excited freely moving molecules, as opposed to atoms, is examined by a spectrometer, a *band spectrum* is seen. This appears as broad bands which are more intense at one edge and fade away towards the other edge. On closer examination it is found that these bands have a very fine line structure.

Whereas band spectra are due to freely vibrating molecules—that is, as gases or vapours—*line spectra* are due to vibrations of free atoms of the elements. The line spectrum of an element is a characteristic of the particular element and is normally only produced by the element in its gaseous state. A line spectrum is formed when the atomic structure of an atom has been disturbed and is due to the radiant energy emitted as the electrons return to their stable energy levels. If two carbon electrodes are brought together and separated to form an arc, the glowing electrodes themselves emit a continuous spectrum, whereas the arc, which contains vaporized electrode material as well as the atmospheric gases, emits a mixture of line and band spectra. An arc may be struck in a vacuum to exclude the atmospheric gases. If one of the electrodes is a metal, or contains some metal, the arc contains its vapour and the line spectrum of the metal is emitted. Similarly, if an electrode is sparked onto a metal surface, the vapour emits the spectrum of the metal, although in some cases this spectrum may be different from that due to the arc. Thus it is usual to refer to both *arc* and *spark* spectra as appropriate.

The spectra referred to above are all examples of *emission* spectra. If white light—that is, a continuous spectrum as emitted by an incandescent solid—is passed first through a vapour and then into the spectrometer, it is found that the continuous spectrum is crossed by a series of dark lines or bands. These lines form the *absorption spectrum* of the vapour and are in the same positions as the lines or bands that are emitted by the vapour if heated. Thc cold vapour absorbs those wavelengths that it emits if hot. The wavelengths of a considerable number of such lines occurring in the sun's spectrum have been measured by Fraunhofer who labelled the more prominent lines with letters of the alphabet (Section 1-2). The sun is surrounded by a relatively cool layer of vapour which absorbs light of these wavelengths emitted by the central hot globe of the sun. Like emission spectra, absorption spectra may be in the form of lines or bands. The line absorption spectra are obtained only with monatomic gases and vapours; band absorption spectra are formed by molecules. The low pressure and the high temperature of the sun's vapour cause gases, that under normal terrestrial conditions exist in molecular form, to exist as separate atoms and hence form line spectra. Absorption spectra may also be obtained by passing light through solids and liquids.

Absorption and emission spectra may be used, not only in the visible range but also in the infra-red and ultra-violet for qualitative and quantitative analysis. For quantitative analysis the intensity of lines produced on a photographic plate is measured and referred to a calibrated standard.

## 6-9 Colour

Colour is a difficult thing to define; it is not only a question of physics but depends very much on the physiological and psychological condition of the observer. It was Newton who first realized that white light is a mixture of colours and that bodies appear coloured because they reflect certain wavelengths more strongly than others. The problem becomes one of how to describe a particular colour precisely so that all observers know and agree the exact colour.

Since the colour of a body depends on how it reflects light of different wavelengths, this property may be used to measure colour. The instrument used is a *spectrophotometer*. In this a source of white light is dispersed by a prism into the colours of the spectrum and then a narrow range of wavelengths is isolated by a slit. The coloured light from this slit is then split into two beams, one of which falls onto the specimen under test and the other onto a standard white reflecting surface which is capable of reflecting virtually all the light incident upon it. The ratio of the luminance of the surfaces is measured by means of photocells for the particular wavelength. This is repeated for other wavelengths by traversing the spectrum across the isolating slit. The results are plotted graphically, as in Fig. 6-9, which shows that if this particular specimen were illuminated by white light it would appear predominantly yellow.

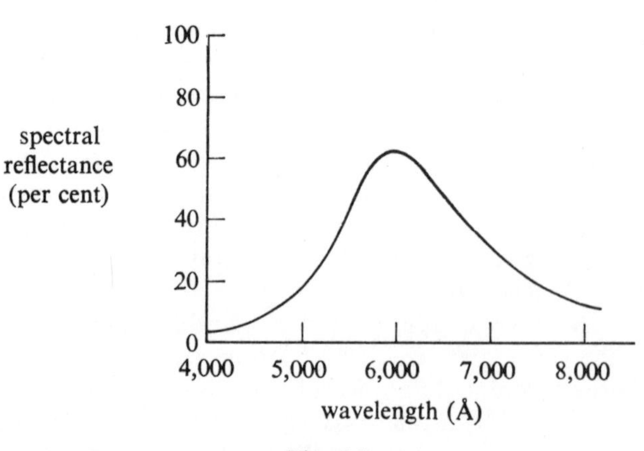


Fig. 6-9

The spectrophotometer can thus be used to define a particular colour when illuminated by monochromatic light of various wavelengths, but it cannot describe the colour when the surface is illuminated by light having a mixture of wavelengths. For this reason *trichromatic colorimeters* have been devised. Basically these are instruments in which the field of view is split so that the specimen under examination may be illuminated by light having the required mixture of wavelengths, whilst the other half of the field acts as a comparison field. Here there is a mixture of the three primary colours, red, green, and blue, produced by specified sources, such that the relative proportions of each may be varied, the quantities of each being read on some arbitrary scale. The three primary colours are also called the basic stimuli and hence the colour under examination is said to be described by its *tristimulus values*. The actual colour being measured may thus be plotted on a colour triangle, as shown in Fig. 6-10. This is an equilateral triangle in which the vertices B, G, R represent the three primary stimuli blue, green, and red. All colours made by mixing blue and

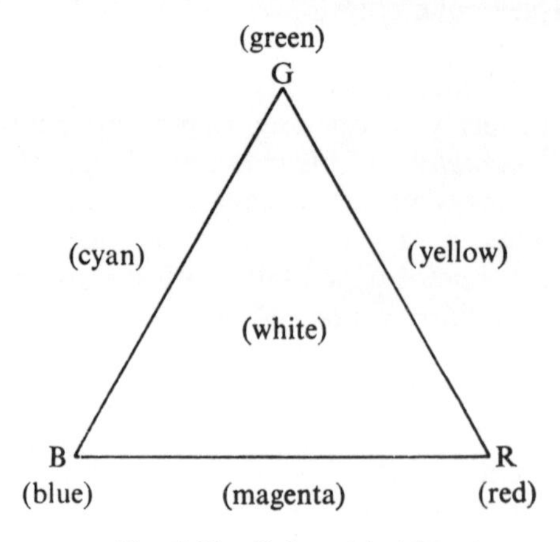


Fig. 6-10 Colour triangle

green lie along the line BG at points depending on the ratio of the amounts of the blue and green stimuli. Colours, including white, formed by mixing various proportions of the three primaries lie on points within the triangle.

Unfortunately, different observers tend to arrive at slightly different tristimulus values for the same colour. This problem has not been overcome completely, even by the use of photocells. If, however, a number of observers are used, mean tristimulus values may be determined. This was done by W. D. Wright (1928) at the University of London with 10 observers and by J. Guild (1931) at the National Physical Laboratory, Teddington, on 7 observers, for illumination by specified wavelengths. Thus, for example, they specified that a unit of radiant energy in the form of green light of wavelength 5,450 Å was equivalent to a mixture of 0·3597 units of red primary, 0·9803 units of green primary, and 0·0134 units of blue primary. Their tristimulus values for the various wavelengths were adopted by the Commission Internationale de l'Eclairage (C.I.E.) as standard (1931). In this way a standard observer is defined, and all colorimetric results can be related to those of the standard observer. However, the weakness in this system is that there are no three monochromatic lights whose mixture can duplicate all the spectrum colours. In fact, if the three primaries are all monochromatic, all spectral colours are represented by points lying just *outside* the triangle. Thus, to match a spectral yellow lying just outside the triangle, a certain amount of blue primary must be added to it to bring it into the line GR so that it can be matched by a mixture of green and red. That is, a negative amount of blue is required to specify the match as a mixture of red, green, and blue.

By convention, since white is a neutral colour, it may be matched by equal amounts of each of the three primaries so that white lies at the centre of the triangle. The amounts of each primary required are not measured in lumens but in units of each stimuli. The size of each unit is such that one unit of red plus one unit of green plus one unit of blue produces a sensation in the observer equivalent to that of mean noon sunlight, that is,

$$1{\cdot}0(\mathrm{R}) + 1{\cdot}0(\mathrm{G}) + 1{\cdot}0(\mathrm{B}) = w(\mathrm{W}) \qquad (6\text{-}62)$$

where (R) represents one unit of the red primary, and so on. If quantities of each unit of primary colour are added to give a certain colour C, it is convenient to define the amount of colour, not in lumens, but in units of that colour, for example,

$$1{\cdot}0(\mathrm{C}) = r(\mathrm{R}) + g(\mathrm{G}) + b(\mathrm{B}) \qquad (6\text{-}63)$$

where one unit of C is termed a *trichromatic unit* or T-unit and is defined by the trichromatic equation [Eqn (6-63)] where

$$r + g + b = 1 \qquad (6\text{-}64)$$

Equation (6-63) is then termed a *unit trichromatic equation* and $r$, $g$, $b$ are termed the *trichromatic coefficients*. Thus writing Eqn (6-62) in the form of a unit trichromatic equation

$$1{\cdot}0(\mathrm{W}) = 0{\cdot}33(\mathrm{R}) + 0{\cdot}33(\mathrm{G}) + 0{\cdot}33(\mathrm{B}) \qquad (6\text{-}65)$$

If Eqn (6-63) is written as

$$c(C) = r'(R) + g'(G) + b'(B) \tag{6-66}$$

where

$$c = r' + g' + b' \tag{6-67}$$

and where (C) is still measured in trichromatic units, then Eqn (6-66) is called the *tristimulus equation* and $r'$, $g'$, $b'$ the tristimulus values. Also

$$\left.\begin{aligned} r &= \frac{r'}{r' + g' + b'} \\ g &= \frac{g'}{r' + g' + b'} \\ b &= \frac{b'}{r' + g' + b'} \end{aligned}\right\} \tag{6-68}$$

for conversion to the unit trichromatic equation.

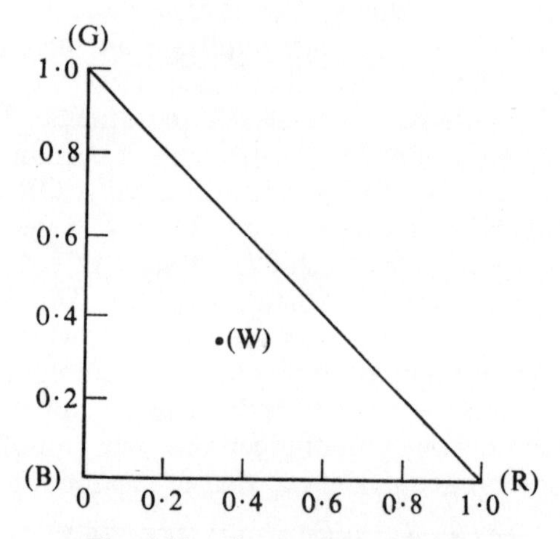


Fig. 6-11 Chromaticity diagram

Since the sum of the coefficients in the unit trichromatic equation is unity, it is in fact only necessary to specify two of the coefficients. By convention those of the red and green primary stimuli are usually given. Thus the plotting of a particular colour on a diagram is made simpler, since only two co-ordinates are required. Such a diagram, called a *chromaticity chart* (Fig. 6-11), is the colour chart of the unit trichromatic system and the colour thus defined is called the *chromaticity*.

As we have said, use of such tristimulus equations may involve negative tristimulus values for some colours, that is, for those lying outside the triangle. Therefore it would be better to use three different primaries so that all colours would be described by positive tristimulus coefficients. For this reason the C.I.E. system makes use of three new primaries, (X), (Y), and (Z). In fact these new primaries are imaginary colours but are related to the real primaries (R), (G), (B) by three unit trichromatic equations such that

$$\left.\begin{aligned}(X) &= 1{\cdot}275(R) - 0{\cdot}278(G) + 0{\cdot}003(B)\\(Y) &= -1{\cdot}740(R) + 2{\cdot}768(G) - 0{\cdot}028(B)\\(Z) &= -0{\cdot}743(R) + 0{\cdot}141(G) + 1{\cdot}602(B)\end{aligned}\right\} \quad (6\text{-}69)$$

In practice, the tristimulus values $r'$, $g'$, and $b'$—that is, the number of (R), (G), and (B) units—are converted by a somewhat laborious computation into $x'$, $y'$, and $z'$—that is, the relative amounts of the C.I.E. non-real primaries. These are then normalized such that

$$\left.\begin{aligned}x &= \frac{x'}{x' + y' + z'}\\y &= \frac{y'}{x' + y' + z'}\\z &= \frac{z'}{x' + y' + z'}\end{aligned}\right\} \quad (6\text{-}70)$$

and where

$$x + y + z = 1 \quad (6\text{-}71)$$

It is thus sufficient to plot the values of $x$ and $y$ to give a chromaticity chart. Thus Eqn (6-63) may now be written as

$$1{\cdot}0(C) = x(X) + y(Y) + z(Z) \quad (6\text{-}72)$$

In Fig. 6-12 the chromaticity co-ordinates $(x, y)$ for the monochromatic spectral colours are plotted, and the wavelengths are in angstrom units. The line joining these points on a chromaticity diagram, whether the chromaticity co-ordinates are reduced to the C.I.E. system or not, is termed the *spectrum locus*. The points representing the C.I.E. primaries may also be plotted on Fig. 6-12, since the point $(x, y) = (1, 0)$ represents the X primary (red), (0, 1) the Y primary (green), and (0, 0) the Z primary (blue) by Eqns (6-71) and (6-72).

If a straight line is drawn between any two points on the spectrum locus, that is, between real light sources, points on this line represent two-component mixtures. Thus points lying within the curve represent real colours, capable of being produced by an additive combination of light from real sources. Points lying outside the curve represent imaginary chromaticities, that is, colours that cannot be reproduced by additive combinations of real lights. Thus the C.I.E. unit primaries (X), (Y), and (Z), as plotted in Fig. 6-12, are all imaginary sources, as previously specified.

Whilst defining the C.I.E. primaries as imaginary stimuli to avoid the addition of negative tristimulus values, the opportunity was also taken to define mathematically the transformation to C.I.E. tristimulus values in such a way that the tristimulus value $y'$ was equal to the luminance of the test specimen, usually expressed in foot-lamberts or nits. The other two components then only affect the colour. Thus a complete C.I.E. specification includes two chromaticity co-ordinates, say $x$ and $y$, and the luminance as expressed by $y'$.

There are three ways in which a colour may be specified by C.I.E. values: by the tristimulus values $x'$, $y'$, $z'$; by the two chromaticity co-ordinates and the luminance $x$, $y$, and $y'$; or by the *dominant wavelength* $\lambda_D$, *excitation purity* $P_e$, and the luminance $y'$. If we draw a line on the C.I.E. chromaticity diagram from the chromaticity point representing white through the sample colour to be specified (C), it meets the spectrum

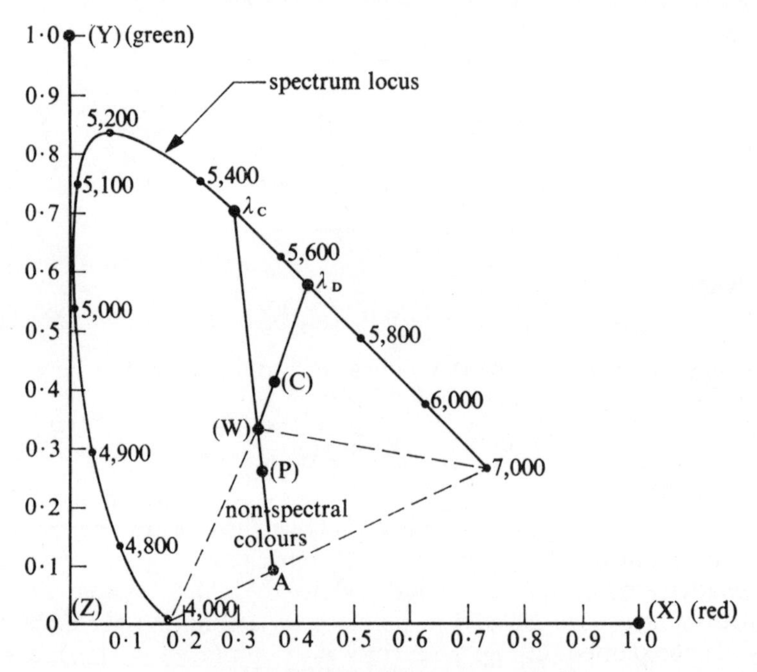


Fig. 6-12 C.I.E. chromaticity diagram

locus at a point which specifies the dominant wavelength, $\lambda_D$. This means that a mixture of white and this spectral colour, in suitable proportions, gives the colour C. The excitation purity is defined as the percentage ratio of the distance (W)(C) to the distance (W) $\lambda_D$. The white point itself is of zero purity and the purity of a spectral colour is unity.

For colours in the non-spectral, or 'purple' region, of Fig. 6-12, a slightly different convention must be adopted due to the gap in the spectrum locus. A line from the white point (W) passing through the colour to be specified, say, (P), must be produced back to cut the spectrum locus at a point $\lambda_C$ as in Fig. 6-12. The wavelength $\lambda_C$ is the *complementary wavelength* and its value is prefaced by a minus sign to show that it is different from $\lambda_D$. The excitation purity in this case is defined as the ratio of the distances (W)(P) and (W)A.

An alternative way of specifying a colour is by comparison with some type of shade card, colour atlas, or colour chart. A type of colour chart commonly used is known as the Munsell system, after its originator.

In the *Munsell system* the sample colours are arranged in the form of the colour solid (Fig. 6-13). The central vertical axis shows the neutral colours, varying from perfect black, through the greys, to perfect white, arranged on a numbered scale. The *values* range from 0 for black to 10 for

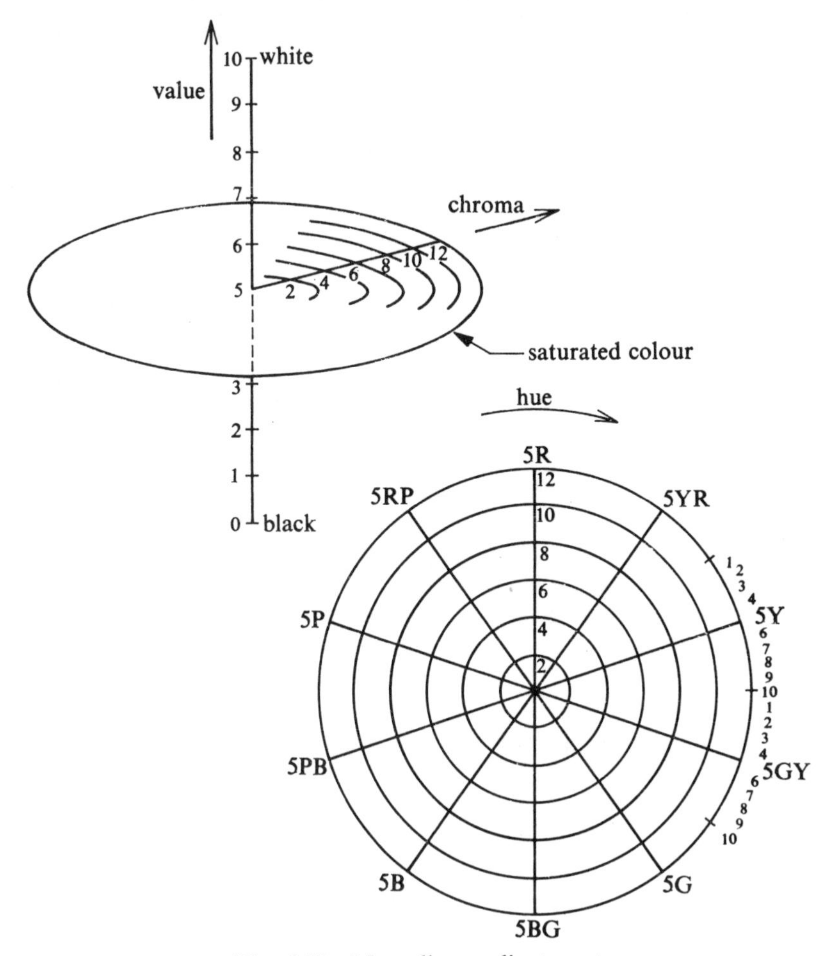


Fig. 6-13 Munsell co-ordinate system

white and so provide a measure of the brightness. The distance of a colour sample from this vertical axis is a measure of the saturation or intensity of the colour and is termed the *chroma*. This too is measured on a scale, representing equal differences in saturation, numbered from 0 at the equivalent grey on the vertical axis up to 16, depending on the particular hue and value. Horizontal circular planes centred at each of the steps on the value scale are imagined divided circumferentially into 100 scale divisions, repetitively numbered from 0 to 10. The colours red (R), yellow (Y), green (G), blue (B), and purple (P) are located in turn at scale reading 5 on each alternate set of scale readings. At scale reading 5 on the intervening sets come the intermediate colours YR, GY, BG, PB, and RP (Fig. 6-13). Other scale readings represent intermediate colours. In the Munsell system these are termed *hues*.

The colours may be arranged in convenient chart form in two different ways. Each chart can represent a different hue, that is, a vertical section through the solid. Thus a section through the 5B–5YR plane would show a central vertical column of neutral colours varying from black at the bottom to white at the top. Colours to the right of this column would represent the YR series of colours getting more intense in colour further to the right—that is, with increasing chroma—and getting brighter nearer to the top of the chart—that is, with increasing value. Similar charts could be made up for any required hue.

Alternatively, a chart could represent a horizontal section through a particular value scale point. This would have a central point that could be from near black to near white as numbered from 1 to 9. Different radial directions would represent different hues, each colour sample getting more intense with greater distance from the centre—that is, with increasing chroma.

A particular sample colour may be specified in terms of its Munsell coordinates. For example, 4YR 7/8 means a Munsell hue of 4 yellow-red, a Munsell value 7/, indicating a brightness nearer to white than to black, and a Munsell chroma of /8, indicating 8 chroma steps away from the neutral grey of the same value. The 'Munsell Book of Colour' contains over 1,200 such individual colour samples arranged in the form of charts.

However, as has already been said, the sensation of colour is purely subjective. Different observers will not agree completely on a particular colour definition or match. The best that can be achieved is a definition or match that is in agreement with that of the majority of normal observers or the mean of a large number of observations.

## PROBLEMS

**6-1** A prism has a refracting angle of 60°. If the minimum deviation it produces for sodium light is 30°, show that the refractive index of the prism material is equal to $\sqrt{2}$.

**6-2** Show that the limiting angle of a prism, such that the incident ray does not emerge when it meets the second face, is given by $\sin^{-1}[(\sin\theta)/n] + \sin^{-1}(1/n)$, where $\theta$ is the angle of incidence made by the ray with the first face and $n$ is the refractive index of the prism material.

**6-3** Explain the theory and construction of (a) an achromatic objective and (b) an achromatic eyepiece formed of two separated lenses.

**6-4** A convex crown glass lens has refractive indices 1·516 for red light and 1·524 for blue light and a mean focal length of 50 cm. What will be the mean focal length of a flint glass lens with corresponding refractive indices 1·619 and 1·637 which will correct the chromatic aberration of the convex lens? What will be the focal length of the combination? [−93·1 cm; +108 cm]

**6-5** Using the glasses of Example 6-1, calculate the external radii of a cemented achromatic doublet of focal length 15 cm if the contact face is to be plane. [3·41 cm; 7·55 cm]

**6-6** What determines the apparent colour of (a) sodium D light, (b) yellow glass, (c) yellow paint?

**6-7** Explain how a colour may be specified by the C.I.E. system.

**6-8** Explain the terms (a) spectrum locus, (b) dominant wavelength, (c) excitation purity.

# 7 Illumination and photometry

## 7-1 Introduction

The units in which light is measured have evolved over a period of time and have become possibly more complicated than they need be. For this reason formal definitions of the units have been deferred until the end of this chapter (Section 7-13). For most practical purposes it is sufficient to appreciate that the basic unit of light flux emitted by a source is the *lumen* (lm). Light falls upon a surface and produces an *illumination* measured in lumens per square metre, termed *lux* (lx), or in lumens per square foot. The $lm/ft^2$ unit was formerly known as the 'foot candle' since it was originally defined as the illumination of one square foot of surface due to a candle held one foot away. The *luminous intensity* of a source is no longer measured relative to the intensity of a candle, but in terms of the *candela* (cd). This new unit is of approximately the same size but is more rigorously defined and reproducible. The lumen and candela are small units. For example, a 100-watt electric bulb emits a total of about 1,200 lumens and has a corresponding luminous intensity of $1{,}200/4\pi = 96$ cd if the light is radiated equally in all directions. A table top in a well-lit room is illuminated by about 100 to 300 lux. Full sunlight gives an illumination of about 100,000 lux, but moonlight only about 0·1 lux. In Great Britain, the daylight illumination due to an unobstructed overcast sky is about 5,000 lux.

How bright a surface appears depends not only on the light falling on the surface, the illumination, but also on how much of the light is reflected by the surface. Thus a white surface which reflects almost all the incident light, and is thus said to have a *reflectance* of nearly 100%, appears to be much brighter than a black surface with a reflectance of about 1 or 2%. Fresh whitewash has a reflectance of up to 90% whilst white paper is generally less than 75 or 80%. Black paper has a reflectance of the order of 5% whilst black velvet may be as low as 1%. Between these extremes come the more usual surfaces. For example, a grey stone building may have a reflectance of about 20%.

The brightness, or more correctly the *luminance*, of a surface is measured in terms of candelas per square metre ($cd/m^2$), termed *nits* (nt). For example, one lux incident upon a perfectly white surface with a 100% reflectance would give a luminance for the surface of $1/\pi$ $cd/m^2$, i.e., 0·318 nit, since it may be shown that a uniform diffuser with a luminance of one candela per unit area in all directions emits $\pi$ lumens per unit area. Alternatively, if the luminance is measured in *foot-lamberts* (ft-L), one lumen per square foot incident on the same surface would give a luminance of one foot-lambert. Thus an open book on a well-lit desk, illuminated by, say, 200 lux and with a reflectance of, say, 50%, would have a luminance of 31·8 nit. In bright sunlight out-of-doors, this same surface may have a luminance of 15,000 nit giving some idea of the range of brightness to which the eye can accommodate itself and still see well.

## 7-2 Lighting

It is usually necessary for economic reasons to choose levels of illumination which are below those which would give the maximum visual performance throughout a room area. The level chosen is a policy decision based on past experience to give satisfactory conditions. For example, a watchmaker requires bright lighting for his immediate working area but a lower level suffices for his surroundings. The region with lower illumination should, however, be sufficiently well lit that people can see where they are going and so that the watchmaker on looking up from his bench does not have to undergo large changes in his eye adaption. Otherwise he will not be able to see clearly for a short period and similarly when he looks down again to his work he will again have a temporary difficulty in seeing. Small changes in adaption, however, are useful in that they counteract the strain of close work and allow periods of rest. It is also an aid to concentration if the working area, for example a desk top, is brighter than its surroundings.

Lighting should be arranged to reduce glare as much as possible. Glare is a result of the bad relative positioning of the work area and the light source, whether it be daylight or artificial lighting. It can arise from too many bright light sources, which can be seen in whichever direction a person looks, or from a source being badly positioned relative to a reflecting surface so that light is reflected into the eyes. Thus a person working at a desk with a shiny top should preferably have the light coming in from over his shoulder. Care must be taken, however, to see that in curing a source of glare, shadows are not introduced instead. Usually, glare is merely a matter of discomfort but in extreme cases it can lead to a scattering of light within the eye, and hence interfere with the ability to see clearly.

Direct and indirect lighting affect the appearance of an object in different ways. A small bright source produces sharp shadows whereas diffused lighting tends to give an absence of shadows and consequently surfaces tend to lose texture. Therefore there should be a combination of the two forms of lighting, and care should be taken to see that the ceiling is well illuminated, otherwise a room will appear gloomy.

We can now consider how these principles apply to the design of the lighting of a building. From past experience the general level of illumination that is required in different types of buildings is known. Some typical values are given in Table 7-1. A decision has to be made as to how much of this light is to be provided by natural daylight and how much by artificial lighting. In computing the amount to be supplied by daylight it is

Table 7-1 Illumination levels required in different types of building

| *Building type* | *Illumination* (lux) | |
|---|---|---|
| School classroom | 100 | minimum |
| | 300 | average |
| Factory | 60 | minimum |
| | 150 | ordinary work |
| | 450 | fine work |
| | 700 | very fine work |
| Office | 300 | clerical work |
| | 450 | drawing office |

recognized that the state of the sky is continually changing and therefore an illumination level of 5,000 lux for daylight has been adopted as a working mean. The *daylight factor* may then be calculated. This is the ratio of the illumination at a point in the room to the total amount of illumination available simultaneously out-of-doors from an unobstructed sky. For example, to obtain the minimum of 100 lux in a classroom a minimum daylight factor of $100/5{,}000 = 2\%$ is required. Knowing the daylight factor it is possible to calculate, by means beyond the scope of this book, the area of window that is required to achieve this factor, as a function of the floor area. The height of the window is limited by the required sill and ceiling heights. Thus, having chosen a convenient length of room, which will also determine the length of the windows, the total area of the window is determined since the height is known. Since the area of the window that gives the required daylight factor is a function of the floor area and the length of the room is known, the width of the room is determined. If this produces a room of unsatisfactory dimensions, then the dimensions and placing of the windows must be readjusted. For example, it might help to put windows in two walls. This may have an added advantage, since the wall surrounding the windows will be made lighter. Also it should be remembered that a tall narrow window allows light to penetrate further into a room than a long low window of the same area.

Having decided an approximate room shape and the placing and dimensions of windows, a more detailed analysis of the illumination of the room should then be carried out. This should take into account its occupation—that is, for example, the positioning in a classroom of desks and blackboard—and pay attention to such additional factors as glare and shadows. Reflecting surfaces play a large part in the final assessment. For example, a highly polished floor and light-coloured paving slabs outside the window reflect light towards the ceiling and help to raise the general level of illumination. The designer may also use models at this stage.

Making illumination by daylight a main consideration in building design often results in high narrow buildings, where each floor has a central corridor with rooms on either side illuminated by daylight, or in large-area, single-storey buildings in which interior rooms may receive daylight lighting by means of roof-lights. When it is not possible to achieve sufficient daylight due to limitations on size and position of windows or room shape, supplementary lighting is required. Supplementary lighting may also be required on extra dark days in rooms which are normally adequately illuminated by daylight. It may also be required to raise the level of illumination to counter the effects of glare caused by seeing the sky through a window. The positioning of these supplementary light sources must be integrated into the design of the room as a whole. For example, in a deep hospital ward the lighting sources for evening illumination may be different from those required for supplementary daytime lighting. During the day the supplementary lighting in the depth of the ward must balance the bright daylight seen through the windows. In the evening this outside daylight is absent and so the illumination level may be reduced. In the evening the lighting must provide a uniform illumination throughout the ward, and it must be free from glare. This usually means two complete lighting circuits with separate switches.

If the requirement for daylight is relaxed, there is far more flexibility of building design; rooms may be built in the interior of a building and lit entirely by artificial means. Here various types of light fittings may be used depending on the character of the lighting required. These may be classified as follows: *direct lighting*, in which most of the light is concentrated onto the working area; *general lighting*, in which the light is distributed in all directions as from a bare bulb or a globe fitting; and *indirect lighting*, in which light only reaches the working surface by reflection, usually from the ceiling. Direct lighting is used, for example, in window displays. Semi-direct lighting is often used in factories; with this a large fraction of the light falls on the work area and the remaining fraction goes towards illuminating the surroundings. General lighting is most commonly used for domestic purposes in conjunction with indirect lighting by reflection from walls and ceiling. Completely indirect lighting eliminates all glare and shadows, leading to a general lack of modelling and texture within a room, and is therefore generally avoided.

The chief types of light sources for interior use are incandescent-filament lamps, hot-cathode and cold-cathode fluorescent tubular lamps, and fluorescent mercury-vapour lamps. Due to poor colour rendering the last of these types is limited to industrial uses.

Incandescent-filament lamps have an efficiency of about 12 lm/W and a lifetime of about 1,000 h. Fluorescent tubular lamps have an initial efficiency of about 20 to 60 lm/W and a lifetime of about 10,000 h. The hot-cathode type loses efficiency continuously throughout its life and is normally discarded before complete failure. The cold-cathode type maintains its initial efficiency for a much longer period. Fluorescent tubes are coated internally to adjust their colour rendering properties and this also affects their efficiency. Table 7-2 lists the characteristics of the various types of fluorescent tube. The efficiency is also affected by the length and

Table 7-2 Average efficiency over a lifetime of 5,000 h for 20-watt (2 ft long) to 125-watt (8 ft long) fluorescent tubular lamps (BSI 1270), and their characteristics

| *Type* | *Efficiency* (lm/W) | *Characteristics* |
|---|---|---|
| White or Warm White | 35–60 | Good general source |
| Daylight or Cool | 34–58 | Similar to average daylight but does distort most colours |
| Natural | 27–44 | Better colour rendering than Daylight type |
| Colour Matching or North Light | 25–40 | Efficiency sacrificed to obtain colour rendering similar to North sky light |
| De Luxe Warm White | 22–36 | Efficiency sacrificed to obtain special colour rendering properties |

diameter of the tubes. Generally the greater these dimensions, the greater is the efficiency.

Having decided the type of illumination and knowing the required illumination in the different parts of the room, the size, type, and positioning of the sources may be calculated. The actual details of the calculations are beyond the scope of this book, but basically they are extensions of the method discussed in Section 7-6 for a specific simple case. They must take into account the reflectance of the various surfaces and a 'maintenance factor', which allows for loss of efficiency due to the light fittings getting dirty and depends on the type of fitting, its situation, and frequency of cleaning. For example, a maintenance factor as low as 0·5 may apply in a dirty industrial area.

Little can be said here about the actual design of the light fittings, except that aesthetic considerations may be of as great an importance as the efficiency. Generally, however, an ugly fitting cannot be justified by technical considerations.

So far we have not considered the economics of lighting. Obviously daylight itself is free, but its use must be considered in relation to the building as a whole. To obtain a satisfactory daylight factor for one room may mean a certain ceiling height in order that sufficient window area may be obtained. Although this may give economy in running cost for this particular room, it may result in all other rooms in the building having ceilings higher than necessary. This in turn would lead to higher than necessary running costs for lighting and heating, quite apart from the initial expenditure of a higher building. To make full economic use of a building site may lead to deep rooms without high ceilings. In this case window lighting is insufficient and it may be more economical and utilitarian to use permanent supplementary lighting rather than to design a building of large window area. As well as letting light in, large windows let heat out. There may thus even be a case for omitting windows altogether, as the economy achieved in heating could well outweigh the extra cost of lighting and air-conditioning plant.

## 7-3 Inverse square law

So far this chapter has discussed the general requirements of illumination. It is now necessary to examine the basic physical laws governing illumination. The first of these is the *inverse square law.*

Consider a point source at P (Fig. 7-1) emitting light of luminous intensity $I$ cd in the direction PX. Let A, B, and C be areas cut from the surfaces of concentric spheres centred at P and having radii 1 m, 2 m, and 3 m respectively. If these areas each subtend the same solid angle $\omega$ steradians at the source P, then the luminous flux, $F$, within this solid angle is

$$F = I\omega \text{ lumens}$$

(A steradian is the angle subtended at the centre of a sphere of unit radius by unit area on the surface of the sphere.)

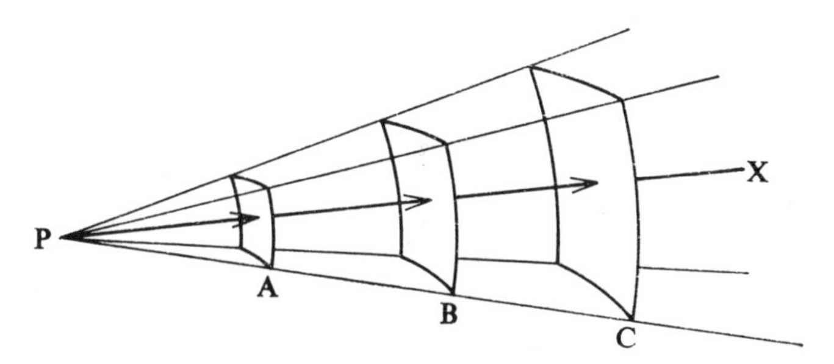


Fig. 7-1 Illustrating the inverse square law

If the illumination on each area be considered separately, then the illuminations will be in the ratio

$$E_A : E_B : E_C = \frac{F}{A} : \frac{F}{B} : \frac{F}{C}$$

$$= \frac{1}{1^2} : \frac{1}{2^2} : \frac{1}{3^2}$$

Thus

*the illumination of a surface normal to the incident light is inversely proportional to the square of the distance of the surface from the source.*

In practice, however, the surfaces involved are usually flat and so the condition of normal incidence is not fulfilled. Also, point sources not only do not exist but practical sources may be quite large. But, providing the source to screen distance is large compared with the other dimensions, little error is introduced.

It will be seen that, since the surfaces have area $\omega d^2$, where $d$ is the distance in metres between the surface and the source, and $F = I\omega$, the illumination of the surface in lux is given by $E = I/d^2$, or in lm/ft² where $d$ is measured in feet.

### 7-4 Cosine law of illumination

The inverse square law deals with the illumination of a surface by a source whose size is small compared with its distance from the surface and where the surface is normal to the incident light. In practice, however, the incident beam is generally inclined at an angle to the surface. Figure 7-2(a) shows a parallel beam of light incident on a surface of area $S$ at an

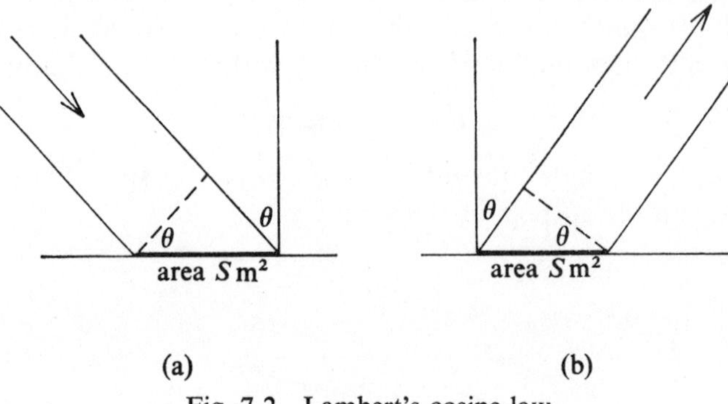


Fig. 7-2 Lambert's cosine law

angle $\theta$ to its normal. If the luminous flux in the beam is $F$ lm per unit of cross-section of the beam, then the flux passing through a normal cross-section of the beam is $FS \cos \theta$. This total flux is spread over the area $S$ of the surface. Thus, the flux falling upon unit area of the surface is $F \cos \theta$ lux. That is,

> *the illumination at a point on a surface is proportional to the cosine of the angle between the incident rays and the normal to the surface.*

This is *Lambert's cosine law*. This may be combined with the inverse square law, giving the basic law of photometry

$$E = \frac{I_\theta}{d^2} \cos \theta \qquad (7\text{-}1)$$

for the illumination at a point of the surface, distance $d$ from a small source, and where $I_\theta$ is the magnitude of the luminous intensity, $I$, in the direction considered.

Lambert's cosine law is equally applicable to light emitted from a self-luminous surface. Thus, the luminous flux emitted from a self-luminous surface of unit area in a direction $\theta$ to the normal is $F \cos \theta$. The surface could, for example, be a matt surface, one that reflects light equally in all directions and may be considered as a collection of luminous points.

The luminance of a surface, expressed in nits, is its luminous intensity in a given direction divided by the cross-sectional area of the beam.

From Fig. 7-2(b), the luminance of the area $S$ when viewed normally is $I/S$, and when viewed in a direction $\theta$ to the normal is

$$\frac{I \cos \theta}{S \cos \theta} = \frac{I}{S}$$

Thus, a surface appears equally bright when viewed from any angle. This means that a self-luminous sphere such as the sun appears as a flat disc because, although different parts of it have different inclinations to an observer, the different parts appear equally bright.

**Example 7-1** A lamp of 200 cd whose intensity is uniform in all directions is mounted 1 m below a large horizontal mirror. Determine the illumination at a point on a table 4 m below the lamp and 3 m to one side of the vertical through the lamp, if the mirror reflects 75% of the light falling on it.

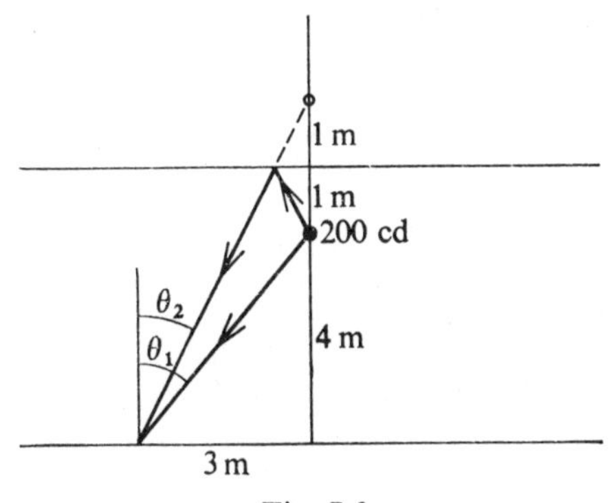


Fig. 7-3

From Fig. 7-3,

Direct illumination
$$E_D = \frac{I}{d_D^2} \cos \theta_1 = \frac{200}{5^2} \frac{4}{5} = 6{\cdot}40 \text{ lux}$$

Reflected illumination
$$E_R = 75\% \text{ of } \frac{I}{d_R^2} \cos \theta_2 = \frac{75}{100} \frac{200}{(3^2 + 6^2)} \frac{6}{(3^2 + 6^2)^{1/2}} = 2{\cdot}98 \text{ lux}$$

∴ Total illumination at the point

$$= 6{\cdot}40 + 2{\cdot}98 = 9{\cdot}38 \text{ lux}$$

## 7-5 Illumination at a point on a surface

It is convenient in dealing with calculations to write the Eqn (7-1) for the illumination at a point P on a horizontal plane in a different form (Fig. 7-4). Thus

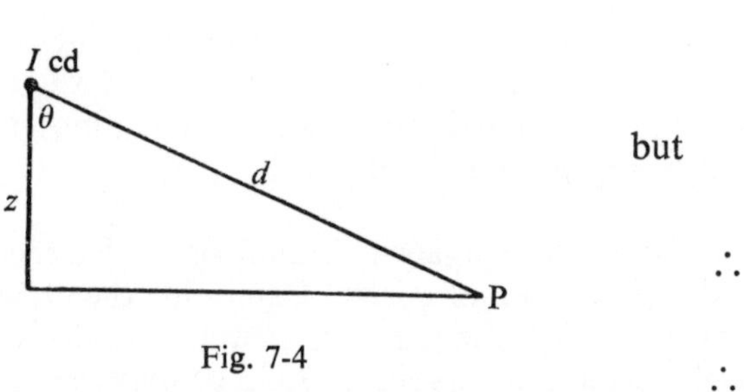


Fig. 7-4

$$E_P = \frac{I_\theta}{d^2}\cos\theta$$

but $$\frac{z}{d} = \cos\theta$$

$$\therefore \frac{1}{d^2} = \frac{\cos^2\theta}{z^2}$$

$$\therefore E_P = \frac{I_\theta}{z^2}\cos^3\theta \quad (7\text{-}2)$$

where $z$ is the height above the horizontal plane.

A common problem involves a source at ceiling height at the centre of a room and the illumination produced at floor level at a corner. Supposing the floor size is $2x \times 2y$ and the room height $z$, then the illumination at P is (Fig. 7-5)

$$E_P = \frac{I_\theta}{d^2}\cos\theta = \frac{I_\theta z}{d^3}$$

but $$d = (a^2 + z^2)^{1/2}$$
$$= (x^2 + y^2 + z^2)^{1/2}$$

i.e., $$E_P = \frac{I_\theta z}{(x^2 + y^2 + z^2)^{3/2}} \quad (7\text{-}3)$$

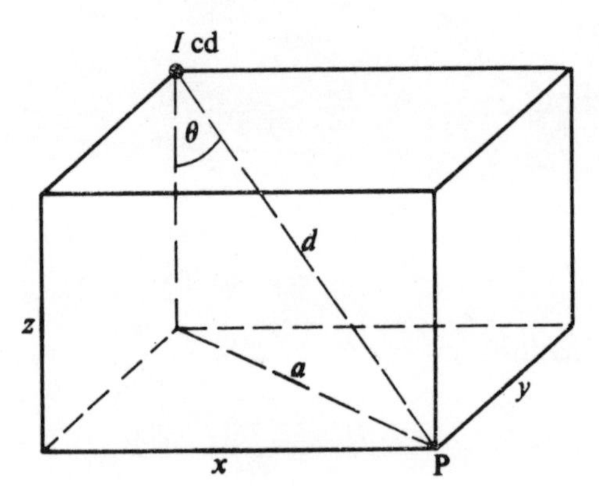


Fig. 7-5

## 7-6 Sources in a row

This type of problem occurs, for example, in street lighting and in the lighting of a long corridor. Let the sources, considered small in relation to the other distances involved, be arranged in a horizontal line and have a spacing $x$ (Fig. 7-6). Let it also be assumed that the intensity distribution below the horizontal plane of the sources is uniform. Then the illumination at some point P, directly below the line of sources and on the horizontal plane distance $z$ below the sources, is the sum of the illuminations produced at P by the individual sources. Therefore the total illumination at point P is given by

$$E_P = \frac{I}{z^2}(\cos^3 \alpha_1 + \cos^3 \alpha_2 + \cdots + \cos^3 \beta_1 + \cos^3 \beta_2 + \cdots) \quad (7\text{-}4)$$

where $I$ is the luminous intensity of a source in a direction below the horizontal.

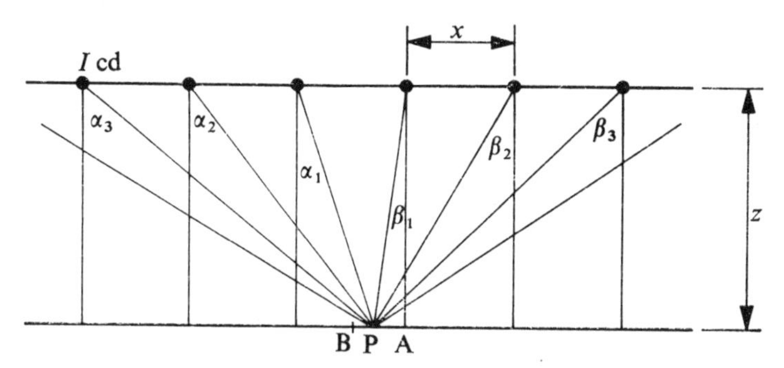


Fig. 7-6 Sources in a row

In practice, we often also want to know the maximum and the minimum illuminations that may be produced along the line of P. The maximum illumination obviously occurs when the point P is directly below one of the sources, for example, at A. Angle $\beta_1$ is then zero and $\cos^3 \beta_1 = 1$, also $\beta_2 = \alpha_1$, $\beta_3 = \alpha_2$, and so on, and hence the maximum illumination will be

$$E_A = \frac{I}{z^2}(1 + 2\cos^3 \alpha_1 + 2\cos^3 \alpha_2 + \cdots) \quad (7\text{-}5)$$

where $\cos \alpha_1 = \dfrac{z}{[x^2 + z^2]^{1/2}}, \quad \cos \alpha_2 = \dfrac{z}{[(2x)^2 + z^2]^{1/2}}$, etc.

The value of $E_A$ is best computed using a tabular form for the $\cos^3$ terms. It is found that, after only a small number of terms, their contribution becomes negligible and may be disregarded.

Similarly, since there cannot be two points of minimum illumination between two sources, the minimum illumination must occur when P is midway between two sources, for example, at B. Hence $\beta_1 = \alpha_1$, $\beta_2 = \alpha_2$, and so on, and the minimum illumination is

$$E_B = \frac{2I}{z^2}(\cos^3 \alpha_1 + \cos^3 \alpha_2 + \cdots) \quad (7\text{-}6)$$

where $\cos \alpha_1 = \dfrac{z}{[(x/2)^2 + z^2]^{1/2}}$, $\cos \alpha_2 = \dfrac{z}{[(3x/2)^2 + z^2]^{1/2}}$, etc.

As before it is found that the $\cos^3$ terms soon become sufficiently small to be negligible.

**Example 7-2** A long corridor is illuminated by 200-cd lamps spaced centrally at 4-m intervals and at a height of 3 m. If each lamp has a uniform intensity in all directions below the horizontal, calculate the minimum illumination at floor level along the centre of the corridor.

| | | Sum |
|---|---|---|
| $\cos \alpha_1 = z\Big/\left[\left(\frac{x}{2}\right)^2 + z^2\right]^{1/2}$ | | |
| $= 3/(2^2 + 3^2)^{1/2}$ | | |
| $= 0{\cdot}8320$ | $\cos^3 \alpha_1 = 0{\cdot}5759$ | 0·5759 |
| $\cos \alpha_2 = z\Big/\left[\left(\frac{3x}{2}\right)^2 + z^2\right]^{1/2}$ | | |
| $= 3/(6^2 + 3^2)^{1/2}$ | | |
| $= 0{\cdot}4472$ | $\cos^3 \alpha_2 = 0{\cdot}0894$ | 0·6653 |
| $\cos \alpha_3 = z\Big/\left[\left(\frac{5x}{2}\right)^2 + z^2\right]^{1/2}$ | | |
| $= 3/(10^2 + 3^2)^{1/2}$ | | |
| $= 0{\cdot}2874$ | $\cos^3 \alpha_3 = 0{\cdot}0237$ | 0·6890 |
| $\cos \alpha_4 = z\Big/\left[\left(\frac{7x}{2}\right)^2 + z^2\right]^{1/2}$ | | |
| $= 3/(14^2 + 3^2)^{1/2}$ | | |
| $= 0{\cdot}2095$ | $\cos^3 \alpha_4 = 0{\cdot}0092$ | 0·6982 |
| $\cos \alpha_5 = z\Big/\left[\left(\frac{9x}{2}\right)^2 + z^2\right]^{1/2}$ | | |
| $= 3/(18^2 + 3^2)^{1/2}$ | | |
| $= 0{\cdot}1644$ | $\cos^3 \alpha_5 = 0{\cdot}0044$ | 0·7026 |

There is obviously no point in going beyond $\alpha_5$ since the $\cos^3$ term is making negligible contribution. Therefore the minimum illumination is

$$E_B = \frac{2I}{z^2} \times 0{\cdot}703$$

$$= \frac{2 \times 200 \times 0{\cdot}703}{3^2}$$

$$= 31{\cdot}2 \text{ lux}$$

In practice, of course, it would be desirable to have even illumination along the line of P—that is, $E_A/E_B = 1$. To achieve this without using a large number of closely spaced sources requires a lateral distribution of intensity considerably greater than the intensity in a downward direction. This may be brought about by the use of suitable shades which reflect or refract the light in the required directions. Such shades are called luminaires.

The variation of luminous intensity with direction for a given source is conveniently shown on a polar diagram. For example, Fig. 7-7 shows the light distribution for a headlight bulb fitted with an axial filament.

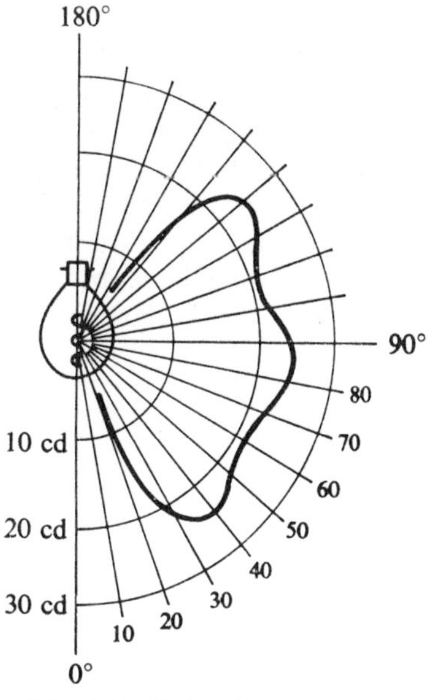


Fig. 7-7 Polar diagram of a headlight bulb

## 7-7 Photometers

These devices are used to measure the luminous intensities of sources. To measure the intensity of a source in candelas we compare it with some standard source of known intensity. In the laboratory, two types of standard are used; the *substandard* source—that is, one that has been calibrated by direct comparison with the primary platinum standard (see definition of the 'candela', Section 7-13); and the *working standard*—that is, one that has been calibrated against the substandard. The substandard source is normally a specially constructed and aged tungsten-filament lamp, and the working standard is preferably a lamp with similar performance characteristics to the lamp under test.

To make a measurement we illuminate the photometer head with light from the two sources and view adjacent illuminated areas. The distance between each source and its illuminated area is then adjusted until the areas have equal luminosity. (Of course, only light from the two sources is considered and all extraneous and reflected light is excluded.) Then the illumination at the photometer is

$$E = \frac{I_1}{d_1^2} = \frac{I_2}{d_2^2}$$

i.e.,

$$I_1 = \frac{d_1^2}{d_2^2} I_2 \tag{7-7}$$

where $I_2$ is known.

Having found the value of $I_1$, the efficiency of the source may also be found. The *efficiency of a source* of light may be defined as the ratio of the intensity, $I$, to the material or energy consumed per second. Thus, the efficiency of an electric lamp is $I$/(watts supplied to it), and of a gas flame $I$/(volume of gas consumed per minute). The efficiency is also commonly expressed, for example, in terms of lumens/watt.

## 7-8 Bunsen's grease-spot photometer

This is one of the simplest photometers and consists of an opaque white screen of stiff paper (Fig. 7-8) with a small grease spot at the centre, making a small translucent area. This spot reflects part and transmits part of the light falling on it. Thus, when the screen is illuminated from one side only, it appears bright with a dark spot when viewed from the same side as the source, but dark with a bright spot when viewed from the other side. If the screen is illuminated from both sides, the positions of the

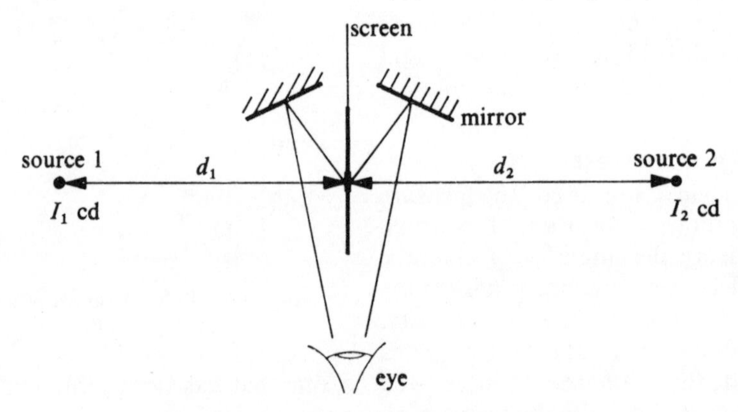


Fig. 7-8 Bunsen's grease spot photometer

sources can be adjusted so that the luminosities of the screen and the spot appear equal. To find this position accurately, it is necessary to view both sides of the screen at once. This is accomplished with the aid of the mirrors. It is also necessary that the grease spot should be on the line joining the sources and that both sources produce light of the same colour. The distances $d_1$ and $d_2$ are then measured and the intensity of Source 1 in terms of the intensity of Source 2 calculated.

## 7-9 Lummer-Brodhun photometer

In this more precise type of photometer (Fig. 7-9), light from the two sources falls on either side of a matt, white, magnesium carbonate screen S. Scattered light from the two sides of this screen is reflected by prisms onto the Lummer-Brodhun cube, consisting of two right-angled prisms, A and B. The hypotenuse face of A is rounded off, except for a circular centre portion which is placed in optical contact with the hypotenuse face of B. Light from Source 1 passes through this central disc and into the telescope. Light from Source 2 is reflected into the telescope by the outer ring of the hypotenuse face of B not in contact with A, and the remainder of the light from Source 2 passes through the central disc and is lost. The image formed in the telescope consists of a circular area illuminated by Source 1, surrounded by an annulus illuminated by Source 2. (Some photometers have contact areas with a more complicated shape. Various areas of the hypotenuse face of A are etched away with acid, leaving only the remaining areas as contact areas.) The sources are moved along a line normal to the screen until both areas have equal

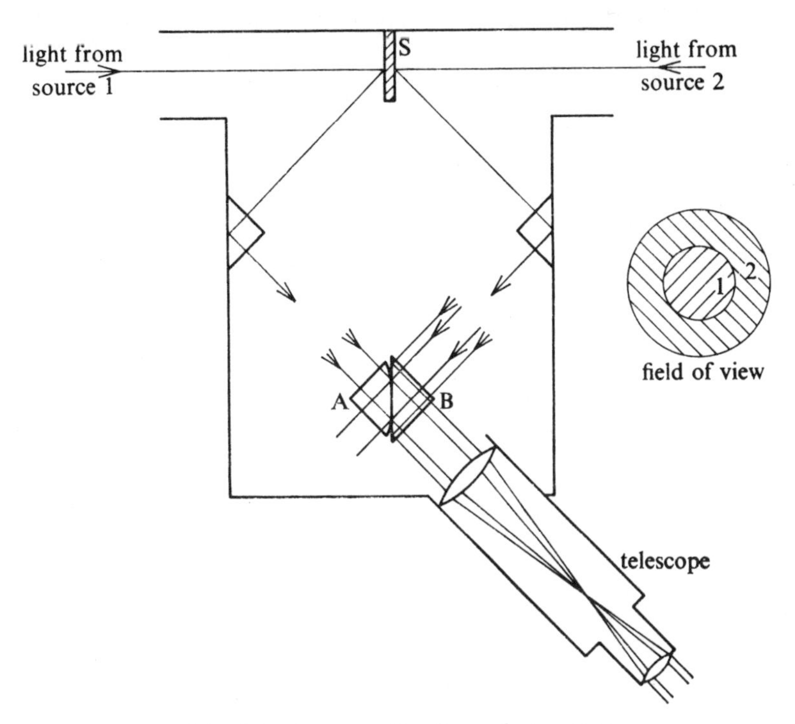


Fig. 7-9 Lummer–Brodhun photometer

luminosity. To obviate errors arising from inequalities in the reflecting power of the two sides of the screen, the readings should be repeated with the instruments rotated through 180° about a vertical axis through S. Again the intensity of one source in terms of the other is calculated, using the mean values for $d_1$ and $d_2$.

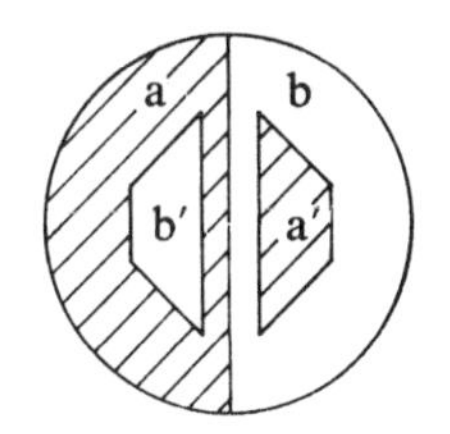


Fig. 7-10

Lummer and Brodhun also devised a contrast version of their photometer. In this, the contact faces of the prisms were modified to give a field of view as shown in Fig. 7-10. One source illuminated the areas a and a′, whilst the other source illuminated the areas b and b′. A pair of thin glass plates was inserted into the beams so that in the central trapezoidal areas the luminosities were reduced by about 8%. Balance is obtained when the left- and right-hand sides of the field appear the same.

## 7-10 Flicker photometer

The photometers already described are only accurate if both sources have the same colour characteristics, producing the same colour in both the comparison areas in the photometer head. To compare sources of different colour, such as a fluorescent tube and a tungsten filament lamp, it is better to use a flicker photometer. In this type of photometer, the field of view is illuminated by the two sources alternately in rapid succession. When the luminosities at the photometer head are unequal and differently coloured, it is found that, as the frequency of flicker is gradually raised, the flicker due to colour differences disappears before the flicker due to luminosity differences. To compare two coloured sources, therefore, the flicker frequency is adjusted until the effect due to colour difference disappears and then the distance to the sources is adjusted to obtain a luminosity balance. The setting tolerance at the luminosity balance increases with increasing flicker frequency, and the positions for luminosity balance should therefore be found at the lowest frequency that removes colour differences. However, the flicker sensitivity of the eye also increases as the field luminosity increases. These two effects lead to an optimum frequency of about 15 to 25 hertz (Hz). The use of such a photometer causes the observer some strain, with a consequent falling off in accuracy. This strain can be alleviated by surrounding the flickering field of view with a steady field of about the same luminosity.

The **Guilds' Photometer** (Fig. 7-11) involves these principles. A white screen, S, of magnesium oxide is mounted inside a black box. Light from Source 1 is directed onto the screen by a totally reflecting prism, A. A rotating sectored disc, B, coated with magnesium oxide and driven by a clockwork motor, is illuminated by Source 2. Thus, the screen and the sectored disc appear alternately as the disc rotates. They are viewed

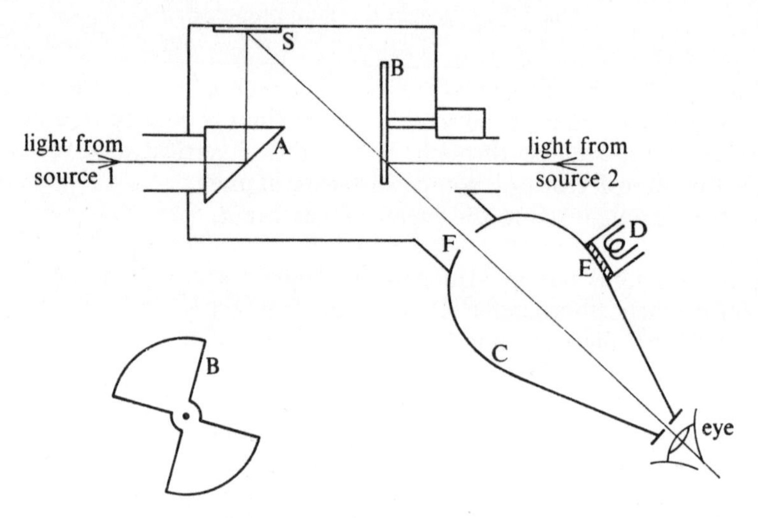


Fig. 7-11 Guilds' photometer

through a diffusing chamber, C, coated matt white on the inside, and with a small lamp, D, with a variable voltage control and a diffusing screen, E, mounted on the side. The lamp is used to illuminate the interior of the chamber so as to obtain a steady field surrounding the flicker field. The diameter of the opening, F, to the chamber is such that an angle of 2° is subtended at the viewer's eye. This means that the peripheral part of the retina, with its higher sensitivity to flicker, is not utilized.

To use the instrument, the surrounding steady field is adjusted to a convenient luminosity, similar to that of the flicker field; the two sources are then moved to give approximately equal luminosities; the flicker frequency is raised until the colour difference is eliminated; and, finally, one of the sources is moved to obtain a balance at minimum flicker.

## 7-11 Light meters

In addition to the comparison-type photometer, calibrated direct-reading light meters are used. These include photoemissive cells, photoresistive cells, and photovoltaic cells, with the latter having the advantage that they do not require an external source of e.m.f. The photoemissive cell

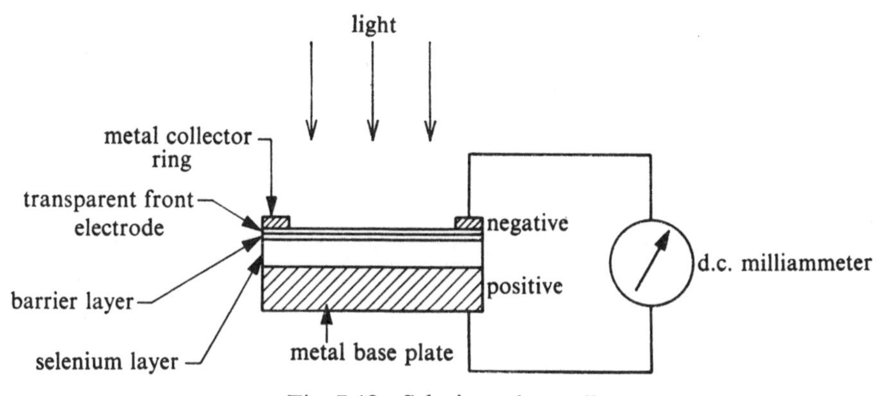


Fig. 7-12 Selenium photocell

is the most accurate, due to its stability, but it has the disadvantage that its associated electronics make it really suitable only for laboratory use. The photovoltaic cell—for example, the selenium photocell (Fig. 7-12)—is the simplest and therefore the most portable. The selenium photocell consists of a thin layer of selenium deposited on a metal base plate. On top of the selenium layer is a barrier layer of molecular thickness and, on top of this, a thin layer of precious metal. This top layer is sufficiently thin to be transparent and acts as a front negative electrode. The metal base plate forms the positive electrode. Light falling on the cell passes through to the selenium layer and causes electrons to be freed from selenium atoms. These electrons migrate through the barrier layer and cause the front electrode to become negatively charged. At the same time, positive 'holes' produced in the selenium layer migrate to the metal base plate which, therefore, becomes positively charged. The electrons passing into the front electrode are unable to leak back into the selenium due to the high inverse resistance of the barrier layer. Thus, a DC voltage is generated across the cell and a current will flow through a meter connected between the metal base plate and the front electrode. The current, and hence the meter deflection produced, varies approximately exponentially with the illumination and also depends on its exposed area and external load resistance. Such photocells deteriorate slightly with age and are affected by temperature changes.

## 7-12 Luminous intensity transmitted by a glass plate

The illumination produced at a surface, due to a source intensity $I$ at distance $d$ from the surface, is $E = I/d^2$. If two sources are compared using a photometer, then at balance

$$E = \frac{I_1}{d_1^2} = \frac{I_2}{d_2^2}$$

i.e.,

$$\frac{I_1}{I_2} = \frac{d_1^2}{d_2^2} \tag{7-8}$$

If a sheet of glass or other absorbing medium is placed between the source of intensity $I_1$ and the photometer head, the effective intensity of this source is reduced to some value $I_1'$. To obtain luminosity balance the source of $I_1'$ must be at $d_1'$ such that

$$\frac{I_1'}{I_1} = \frac{d_1'^2}{d_1^2} \tag{7-9}$$

That is, the effective intensity of $I_1$ is reduced to $d_1'^2/d_1^2$ of its former value. In other words, the sheet of glass is transmitting $d_1'^2/d_1^2 \times 100\%$ of the incident luminous flux.

## 7-13 Photometry units

We may all enter a room and decide whether or not it is sufficiently well illuminated. But this is purely a personal choice, depending on a number of factors. For example, the eye will adapt itself to various illuminations by opening or contracting the pupil, controlling the amount of light entering the eye. Whether we decide the illumination is sufficient or not will thus be partially dependent on whether we enter it from brighter or darker surroundings. In other words, our decision is based on a sensation produced by the impact of light radiation on a sense organ, the eye. Unfortunately, the eye is incapable of giving any numerical magnitude to this sensation. Therefore, unlike the measurements of length and time, for example, which can be expressed in what are called physical units, since they are independent of the peculiarities of the person measuring them, the magnitude of the illumination of our room must be determined in non-physical units by comparison with artificial standards. To do this, a number of terms* and units must be defined.

* BSI 233, *Glossary of Terms used in Illumination and Photometry*.

The amount of light emitted by a point source is its *luminous intensity*, and the amount of light from this source that falls on a surface is the *illumination* of the surface. Luminous intensity is measured in *candelas* (or in *candle power*) and the illumination of the surface is measured in *lux* (or in *lumens per square foot*). An illuminated surface appears 'bright' and can act as a source. The amount of light emitted by such a surface source is its *luminance*, which is measured in *nits* (or in *foot-lamberts*).

To define these terms more fully it is necessary to define a number of other terms. The first of these is

**Radiant flux** ($P$). *Power emitted, transferred, or received in the form of radiation.*

This refers to all types of radiation, including, for example, heat. Therefore, to be more precise, we can narrow down our definition of flux to

**Luminous flux** ($F$). *That quantity characteristic of radiant flux which expresses its capacity to produce visual sensation.*

Even so, this is still not very precise. We cannot give a number to it and talk about a certain amount of luminous flux. To do this we must next define a unit of luminous flux.

**Lumen** (lm). *The flux emitted in unit solid angle of one steradian by a point source having a uniform intensity of one candela.*

One steradian, or unit solid angle, is the angle subtended at the centre of a sphere of unit radius by unit area on the surface of the sphere. An area $A$ on the surface of a sphere of radius $r$ subtends a solid angle of $A/r^2$ steradians at the centre of the sphere.

Thus, if we know the measure of the intensity of our source of light in terms of a unit, the candela, we have a measure of the quantity of flux it is emitting. The whole solid angle about a point being $4\pi$, the total luminous flux emitted from a point source of intensity $I$ cd radiating equally in all directions is $4\pi I$ lm. Knowing this quantity of flux, our next practical interest is to be able to measure how much of this light emitted falls onto a surface. To do this, we must first define

**Illumination** ($E$), *at a point of a surface. The quotient of the luminous flux incident on an infinitesimal element of the surface containing the point under consideration, by the area of that element.*

Figure 7-13 illustrates this. In the first case the illumination of the small element $dA$, receiving a small amount $dF$ of the total flux, is $dF/dA$, whilst in the second case the illumination of the whole surface is $F/A$. We have already defined luminous flux in terms of lumens, so to obtain the illumination of a surface we merely divide the number of lumens reaching the surface by the area of the surface. This quantity can be measured in two units, depending on the units used to measure the surface area.

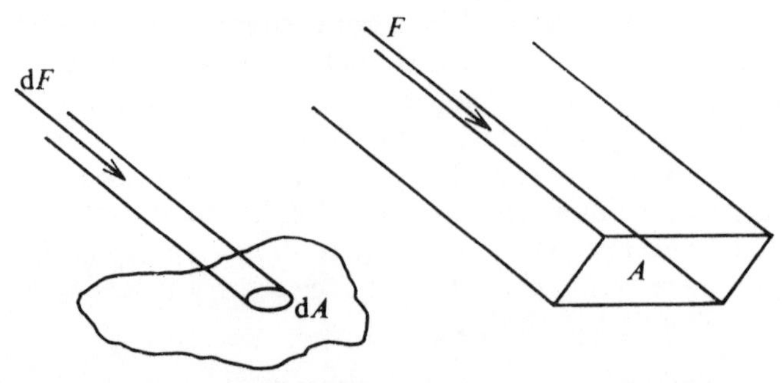


Fig. 7-13 Illumination at a point of a surface

**Lux** (lx). *An illumination of one lumen per square metre.*

**Lumens per square foot** ($lm/ft^2$) (formerly called the 'foot candle'). *An illumination of one lumen per square foot.*

These units are basically the same in form and are readily converted one to the other, since 1 $lm/ft^2 = 10{\cdot}764$ lux. Some idea of the size of these units may be obtained by reference to Table 7-1.

Illumination of a surface concerns the amount of light *incident* upon a surface; luminous emittance concerns the amount of light *emitted* from a surface. Although similar, these terms clearly differentiate between the receiver and the source.

**Luminous emittance** ($H$), *from a point of a surface. The quotient of the luminous flux emitted from an infinitesimal element of surface containing the point under consideration, by the area of that element.*

In practice the source of light can more often be treated as a point source emitting light in a given direction than as an area of surface. For this reason, we usually consider the 'luminous intensity' of a source rather than its 'luminous emittance'.

**Luminous intensity** ($I$), *in a given direction. The quotient of the luminous flux emitted by a source, or by an element of a source, in an infinitesimal cone containing the given direction, by the solid angle of that cone.*

To measure this for a particular source, we again have the problem that luminous flux cannot be defined in any absolute way. Therefore we must choose arbitrary units of luminous intensity that are practically obtainable and measurable. The original arbitrary standard of intensity was a candle manufactured and used in a prescribed manner. This, however, was not sufficiently reproducible and luminous intensity is now defined in terms of the

**Candela** (cd). *A unit of magnitude such that the luminance of a full radiator at the temperature of the solidification of platinum* (2,046°K) *is* 60 *candelas per square centimetre.*

This definition does not mention the direction in which flux is being emitted and, since sources often have directional properties, two additional units are used.

**Candlepower** (CP). *The light radiating capacity of a source in a given direction, in terms of the luminous intensity expressed in candelas.*

**Mean spherical candlepower.** *The average value of the candlepower of a luminous source in all directions.*

The amount of light emitted from a source or a diffusing surface per unit area is its brightness or 'luminance'.

**Luminance** ($L$) (formerly called 'photometric brightness'), *at a point of a surface and in a given direction. The quotient of the luminous intensity in the given direction of an infinitesimal element of the surface containing the point under consideration, by the orthogonally projected area of the element on a plane perpendicular to the given direction.*

The two units of luminance in common use depend on the units in which the area is measured. These are

**Nit** (nt). *A luminance of one candela per square metre.*

**Foot-lambert** (ft-L) (formerly called 'equivalent foot-candle'). *The luminance of a uniform diffuser emitting one lumen per square foot.*

Again these units are readily converted one to the other since 1 ft-L = 3·426 nt. Other obsolete units of luminance may be still encountered in older literature:

**Stilb** (sb). *A unit of luminance. A luminance of one candela per square centimetre.*

**Apostilb** (asb). *A unit of luminance. The luminance of a uniform diffuser emitting one lumen per square metre.*

**Lambert** (L). *A unit of luminance. The luminance of a uniform diffuser emitting one lumen per square centimetre.* Lamberts may be converted to foot-lamberts by multiplying by 929.

There is one further term that also frequently occurs. This is

**Luminosity** (formerly called 'apparent brightness'). *The attribute of visual perception in accordance with which an area appears to emit more or less light.*

This is normally used in a comparative sense in that a particular surface may appear to be brighter than another due to its colour or surface finish, and is then said to have a greater luminosity.

## PROBLEMS

**7-1** Two 10-cd sources are placed on the same side of a photometer. One source is 1 m from the photometer, the other 3 m. At what distance must a third source of 6 cd be placed to achieve a balance? [73·5 cm]

**7-2** A certain intensity of illumination is produced on a screen by a lamp situated 95 cm away. When a sheet of glass is placed between the lamp and the screen it is necessary to bring the lamp 10 cm closer to obtain the same illumination at the screen as before. Calculate the percentage of light stopped by the glass. [20%]

**7-3** Compare the illumination at a point A on a table due to a small source 2 m vertically above A, with that at a point B, 2 m horizontally from A. [2·8:1]

**7-4** A photometer whose screen reflects unequally from its two sides balances two sources of $I_1$ and $I_2$ cd at distances $d_1$ and $d_2$. On turning the photometer head through 180°, the distances for balance change to $d_1'$ and $d_2'$ respectively. Show that $d_1 d_1' I_2 = d_2 d_2' I_1$.

**7-5** Sources A and B give a photometric balance when they are 40 and 50 cm respectively from the photometer head. A plane mirror which reflects 80% of the incident light is placed 10 cm behind source A. How much nearer to the photometer head must source B be moved to restore the photometric balance? [7 cm]

**7-6** A 100-cd lamp is suspended 2 m above each corner of a 2 m × 2 m table. Calculate the illumination (a) at the centre of the table and (b) half-way along one edge. [(a) 54·4 lux; (b) 50·6 lux]

**7-7** Show that for maximum illumination at the edge of a circular table of radius $r$, a point source above the centre of the table should be at a height $0{\cdot}71r$.

**7-8** A small vertical screen is placed at the centroid of a horizontal equilateral triangle. If there are equal point sources at the vertices of the triangle, show that the illumination is always equal on both sides of the screen.

**7-9** Distinguish clearly between 'luminous intensity' and 'illumination' and their units. Describe how the intensities of two sources may be compared. How may a photometer be used to compare the reflecting powers of various surfaces?

**7-10** Determine the maximum illumination at floor level along the centre of the corridor in Example 7-2. [34·8 lux]

# 8 Interference

## 8-1 Wave properties of light

In the preceding chapters we have assumed that light travels in straight lines—that is, we have been concerned with *geometric optics*. This assumption is an adequate basis for discussing aspects such as reflection and refraction. There are many other phenomena, however, that cannot be treated in this way. To understand interference, diffraction, and polarization, for example, we must treat visible light as a form of wave motion; that is, as part of the spectrum of electromagnetic radiation that ranges from gamma rays to long-wave radio bands (Fig. 8-1). This study of light as wave motion is *physical optics*. There are yet other phenomena, such as emission and absorption, which cannot be treated in either of these ways. Since these involve interactions between atoms and molecules, we must discuss them in terms of quantum mechanics. This is *quantum optics* and it is beyond the scope of this book.

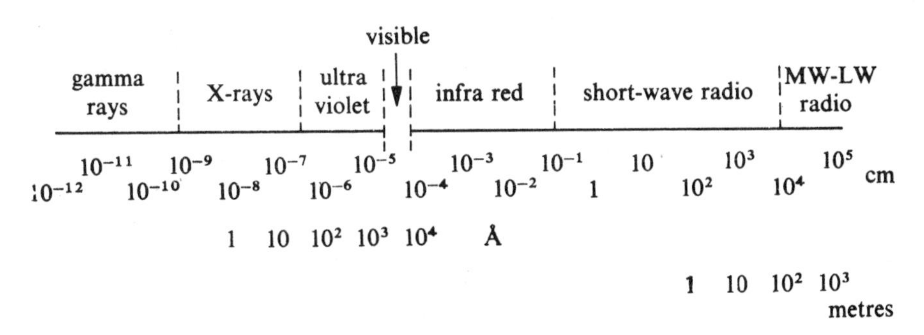


Fig. 8-1 The electromagnetic spectrum

## 8-2 Principle of superposition

When a still surface of water is disturbed at two points, circular waves travel out from the points and form complicated patterns where they cross. Beyond the region where they cross the waves travel on as before. Similarly, where two or more light waves cross they *interfere* with each other to produce a new motion in this region. The waves may be assumed to have the same velocity, form, and wavelength, and therefore can only differ in amplitude and phase. The phase difference between the waves merely decides how much they are out of step with each other. The *principle of superposition* states that

*in the region of crossing the resultant displacement at any point is the sum of the individual displacements due to the separate waves.*

The separate waves may be represented on a vector diagram (Fig. 8-2). If the two waves have amplitudes $r_1$ and $r_2$ and phase angles $\phi_1$ and $\phi_2$ respectively, then the resultant motion in the interference region may be represented by a wave of amplitude $R$ and phase $\phi$. Two important cases arise. Firstly, if the waves are in phase, that is $\phi_1 = \phi_2$ (or differ in phase by a whole number multiple of $2\pi$), then the resultant wave motion has a maximum amplitude which is the sum of $r_1$ and $r_2$, and consequently the intensity of light, which is proportional to the square of the resultant amplitude, is a maximum. Secondly, when the phase difference is $\pi$, the resultant amplitude is the difference between $r_1$ and $r_2$ and consequently

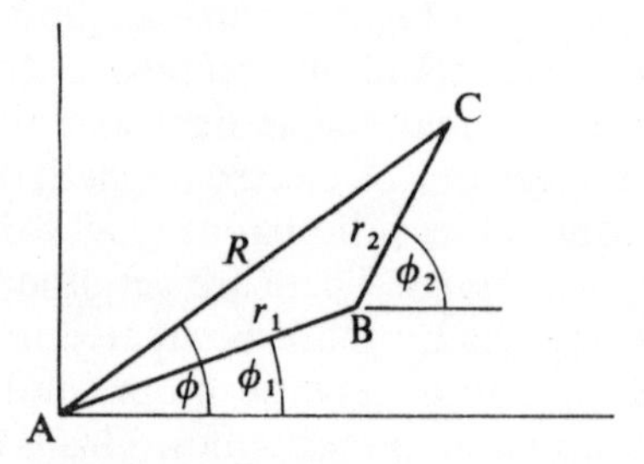


Fig. 8-2 The resultant of two wave motions represented vectorially

the resultant intensity of light is a minimum, falling to zero when $r_1 = r_2$. In other words, whenever two waves are completely in step or, what amounts to the same thing, a *whole* number of wavelengths are out of step, the resultant amplitude, and consequently the intensity, is a maximum and *constructive interference* is said to occur. Conversely, whenever two waves are completely out of step, that is, whenever they are a whole number and a *half* wavelengths out of step, the resultant amplitude, and consequently the intensity, is a minimum and *destructive interference* occurs.

In Chapter 11, the composition of such wave motions is dealt with in greater detail. The above account is sufficient at this stage to account for all optical interference phenomena.

## 8-3 Huygens' principle

Interference phenomena cannot be observed with light from two similar monochromatic light sources. This is because these sources do not send out continuous trains of waves but waves broken into short trains, of time duration of the order of $10^{-8}$ s, by abrupt changes of phase. To obtain observable interference phenomena, the phase difference between two wave trains must be constant. Normally,* this can only occur if they come

* Interference fringes have been obtained for very short periods using two independent laser sources.

from *one* source so that they are in a continuous point-to-point correspondence. Although the waves may originally be derived from one source, once the wave trains exist they may be treated as if originating from two sources, even if these are only imaginary. Two such sources are said to be *coherent*, since they have a constant phase relationship. Before obtaining the two 'sources', it is convenient first to consider Huygens' principle.

*By Huygens' principle, each point on a wave front may be considered as a new source of secondary wavelets.*

Thus, for example, if the lines in Fig. 8-3 represent corresponding points on successive waves, for example, the crests, a plane wave front is seen to approach a small slit aperture. Then, by Huygens' principle all the points on the wave front, including the part passing through the aperture, act as

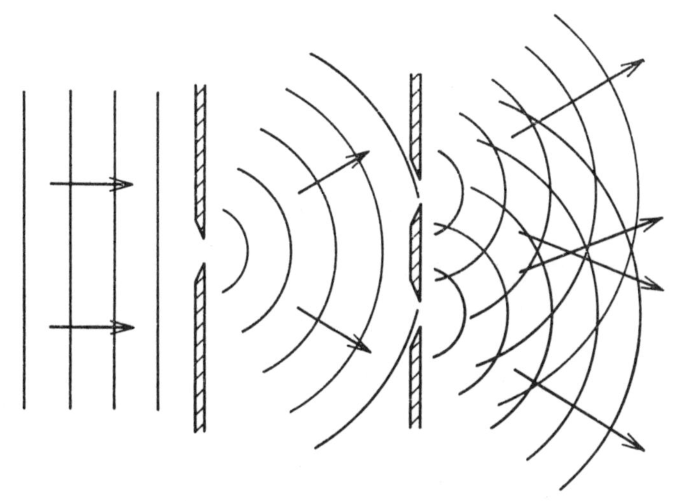

Fig. 8-3 Huygens' principle

sources of secondary wavelets. Since the aperture acts as a line 'source', a cylindrical wave front is produced. Similarly, a pinhole aperture produces a spherical wave front. Thus, the light spreads into the region not in the direct path of the incoming rays. This phenomenon is called *diffraction.*

Each point on the cylindrical wave front can act as a source of secondary wavelets. Figure 8-3 shows how the wave front reaches two more slit apertures and produces two further sets of wavelets with cylindrical wave fronts. In this way we can produce what are effectively two coherent sources, since as they are derived from a common original source there is a constant point-to-point phase relation between the two sets of waves.

Huygens' principle of secondary wavelets is generally applicable and can be used to explain other optical effects, for example, Snell's law of refraction. Figure 8-4 shows a ray of light incident at an angle $\theta$ to a boundary between two media of different refractive indices. AB represents a plane wave front making contact with the interface at A. Point A is then a source of wavelets forming a spherical wave front in the second medium. Similarly, every point on the incident plane wave front becomes, in turn, a source of secondary wavelets in the second medium. In particular, by the time $t$, at which the end B of the incident wave front reaches the interface, the spherical wave front which originated at the point A will have a radius AD, where

$$t = \frac{\text{AD}}{c_2} = \frac{\text{BC}}{c_1}$$

i.e.,

$$\text{AD} = \text{BC}\cdot\frac{c_2}{c_1}$$

where $c_1$ and $c_2$ are the velocities of light in the first and second medium respectively, and where $c_2$ is less than $c_1$. The tangent plane CD, drawn

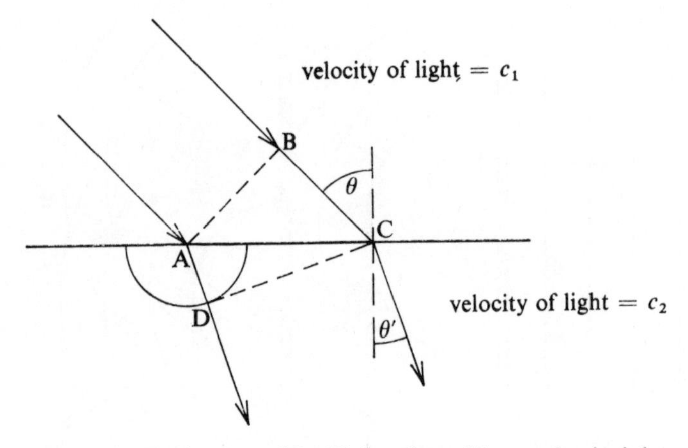


Fig. 8-4 Derivation of Snell's law from Huygens' principle

from C to this wavelet, represents the new plane wave front and forms an envelope to all the separate wavelets produced by the incident plane wave front. Then

$$\frac{\sin\theta}{\sin\theta'} = \frac{\text{BC/AC}}{\text{AD/AC}} = \frac{\text{BC}}{\text{AD}} = \frac{c_1}{c_2} = \text{const.} \tag{8-1}$$

in agreement with the general law of refraction.

Similarly, the other results of geometric optics may be derived using Huygens' principle and consequently we are not invoking anything new or particular in our production of two coherent sources and our explanation of interference phenomena.

## 8-4 Young's double-slit experiment

Interference was first successfully demonstrated by Thomas Young in 1801. In his experiment, two slits were used as the sources of Huygens' secondary wavelets as in Fig. 8-5. The thick lines represent the crests of waves and the thin lines the troughs. At the points where two crests or two troughs cross the waves are exactly in phase (or differ in phase by a multiple of $2\pi$). Since the waves are travelling outwards from the slit sources, the in-phase points are constantly moving along the paths shown dotted in Fig. 8-5. At the points where these lines meet the screen the waves are always in phase and, by the principle of superposition, the resultant amplitude and consequently the light intensity is a maximum. Similarly, at the midway points—that is, along lines in which crests always coincide with troughs—there is a minimum of intensity. Thus, evenly

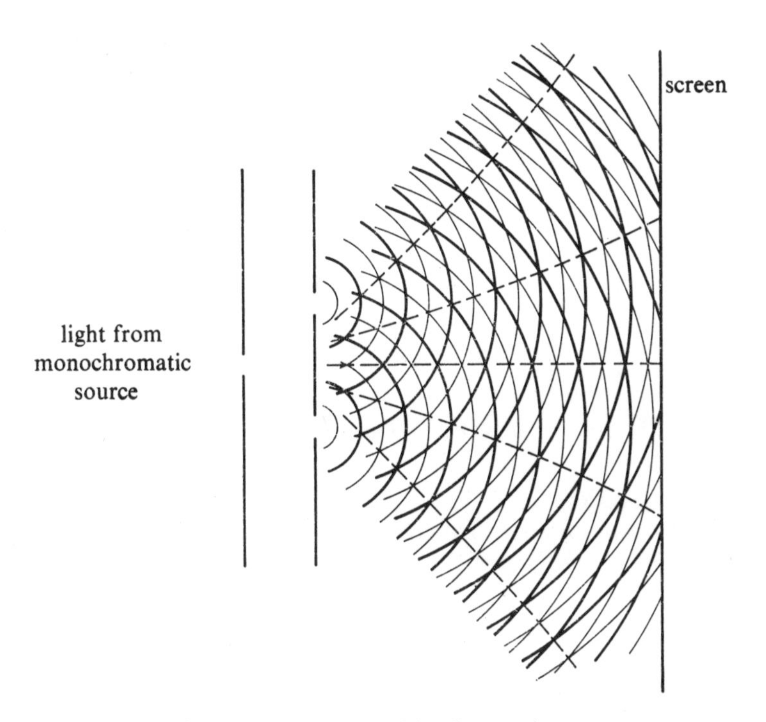


Fig. 8-5 Young's double-slit experiment

spaced light and dark bands or fringes form on the screen. It is essential to use monochromatic light, otherwise fringes of different spacing occur and tend to cause even illumination of the screen, due to the overlapping of fringes. Of course, although bright fringes are formed on the screen, no extra light is gained. The light energy falling on the screen is merely re-arranged so that reduced intensity in one place leads to increased intensity in another. The average intensity over the screen is exactly the same as it would have been without interference. The number of fringes is also limited. The central fringes appear to be of approximately equal intensity but the intensity of the outer ones rapidly falls off. This is due to diffraction fringes produced by the slit apertures occurring at the same time as the interference fringes. We shall deal with diffraction in the next chapter.

For the present diffraction may be considered as an effect due to a limited part of the wave front, whilst interference is due to the combination of two or more separate beams of light that originate from one source.

The separation of the fringes may be deduced by considering the geometry of the experiment (Fig. 8-6). Let the slit sources, $S_1$ and $S_2$, be a distance $d$ apart and $D$ from the screen. Then, if P, distance $x$ from the axis, is a point on the screen corresponding to the position of a bright

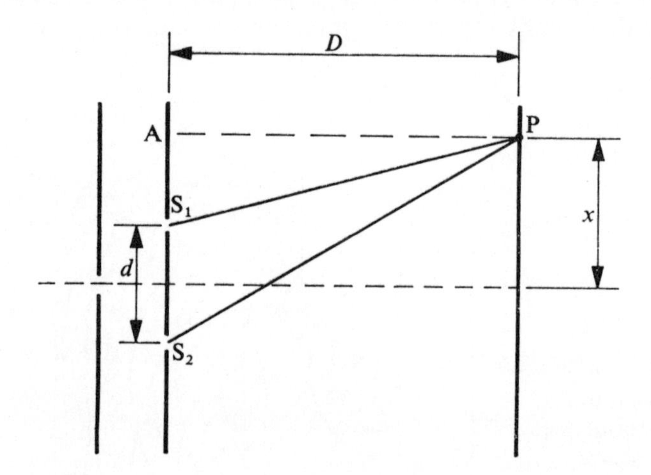


Fig. 8-6 Geometry of Young's double-slit experiment

fringe, the path difference $S_2P - S_1P$ is equal to a whole number of wavelengths ($m\lambda$)—that is, waves from $S_1$ and $S_2$ arrive at P in phase or differing in phase by a multiple of $2\pi$. In the triangle $APS_2$

$$(S_2P)^2 = D^2 + \left(x + \frac{d}{2}\right)^2 \qquad (8\text{-}2)$$

and in the triangle $APS_1$

$$(S_1P)^2 = D^2 + \left(x - \frac{d}{2}\right)^2 \qquad (8\text{-}3)$$

Subtracting

$$(S_2P)^2 - (S_1P)^2 = \left(x + \frac{d}{2}\right)^2 - \left(x - \frac{d}{2}\right)^2$$

$$= 2xd \qquad (8\text{-}4)$$

$$\therefore\ S_2P - S_1P = \frac{2xd}{(S_2P + S_1P)}$$

In practice, the slit separation $d$ is usually less than 1 mm (with a slit width of about 0·1 mm). Similarly, $x$, the distance of a bright fringe from the axis, may be up to a few millimetres, whilst the distance, $D$, to the screen is usually of the order of a metre. Therefore, replacing $(S_2P + S_1P)$ by $2D$ introduces only negligible error.

$$\therefore\ S_2P - S_1P = \frac{2xd}{2D} = \frac{xd}{D}$$

$$= m\lambda \tag{8-5}$$

Thus, the condition that a bright fringe be formed at a distance $x$ from the axis is that

$$x = \frac{m\lambda D}{d} \tag{8-6}$$

For a dark fringe

$$x = \frac{(m + \frac{1}{2})\lambda D}{d} \tag{8-7}$$

where $m$ indicates the order of the fringe and equals 0, 1, 2, . . . and $\lambda$ is the wavelength of the light used. Thus, the zero-order bright fringe always occurs on the axis—that is, when $x = 0$—and the fringes will have a constant spacing, $\lambda D/d$.

In Fig. 8-2 two waves, whose amplitudes and phases were respectively $r_1\phi_1$ and $r_2\phi_2$, are compounded vectorially (see also Chapter 11). By applying the cosine rule, we can find the resultant amplitude $R$, since

$$R^2 = r_1^2 + r_2^2 - 2r_1r_2 \cos ABC$$

but

$$\cos ABC = \cos (180° + \phi_1 - \phi_2)$$

$$= -\cos (\phi_1 - \phi_2)$$

$$\therefore\ R^2 = r_1^2 + r_2^2 + 2r_1r_2 \cos \delta \tag{8-8}$$

where $\delta$ is the phase difference, $\phi_1 - \phi_2$. In practice, since the slits $S_1$ and $S_2$ are of equal width and close together, the amplitudes $r_1$ and $r_2$ are equal, therefore

$$R^2 = 2r^2 + 2r^2 \cos \delta$$

$$= 4r^2 \cos^2 \frac{\delta}{2} \tag{8-9}$$

Since the vibrational energy of the wave motion is proportional to the square of the amplitude, the intensity of the light at any point on the screen is proportional to the square of the resultant amplitude at that point. Thus, there is a cosine-squared distribution of intensity across the screen, i.e.,

$$I \propto \cos^2 \frac{\delta}{2} \tag{8-10}$$

In other words, whenever the phase difference $\delta = \phi_1 - \phi_2$ is equal to $0, 2\pi, 4\pi, \ldots$ corresponding to a path difference $0, \lambda, 2\lambda, \ldots$ (that is, where $x = 0, \lambda D/d, 2\lambda D/d, \ldots$), the intensity is a maximum. Similarly, at the mid-positions, where $\delta = \pi, 3\pi, 5\pi, \ldots$ corresponding to a path difference $\lambda/2$, $3\lambda/2$, $5\lambda/2, \ldots$ (that is, where $x = \lambda D/2d$, $3\lambda D/2d$, $5\lambda D/2d$, $\ldots$), the intensity is zero. At intermediate points, the intensity varies between the alternating maxima and minima as the square of the cosine of half the phase difference.

Corresponding phase differences and path differences are connected by the equation

$$\text{phase difference} = \frac{2\pi}{\lambda}\cdot(\text{path difference}) \qquad (8\text{-}11)$$

Objections were raised to Young's double-slit experiment. It was argued that the bright fringes might be due to some complicated form of diffraction rather than to interference. However, to settle the problem other demonstration experiments were made, for example, by Fresnel, using a biprism, and by Lloyd, using a mirror.

## 8-5 Fresnel's biprism

The glass biprism (Fig. 8-7) has a vertex angle of nearly 180°. Light from the slit S is refracted by the two halves of the prism and appears to originate from two closely adjacent slit sources $S_1$ and $S_2$, which act effectively as two coherent sources. Two types of fringes are formed on the screen. Narrow, equally spaced interference fringes due to the compounding of waves from the two coherent sources occur in the region of overlap, area bc on the screen. In addition, broad, unequally spaced diffraction fringes are produced over the entire area ad; because they are less intense than the interference fringes the diffraction fringes modulate the intensity slightly over the region bc. The diffraction fringes are caused by the common vertex edge of the two prisms acting as a diffracting straight edge. Thus Fresnel demonstrated interference without using diffraction to produce his two sources, and so was able to produce both types of fringes at the same time.

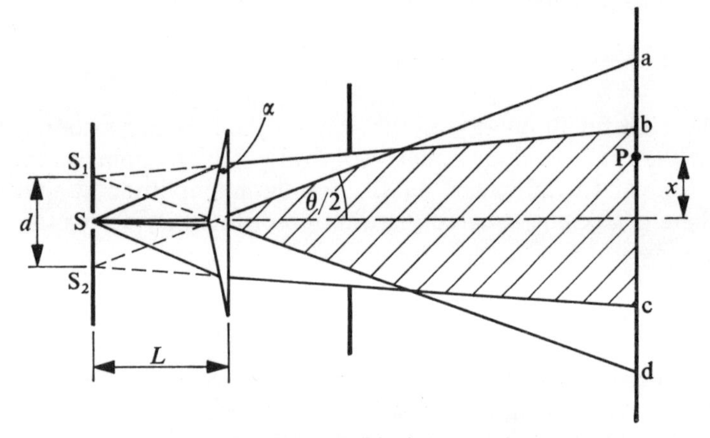


Fig. 8-7 Fresnel's biprism experiment

The condition for a bright or a dark fringe at a point P, distance $x$ from the axis, is exactly the same as in Young's double-slit experiment. Equations (8-6) and (8-7) state these conditions, where $D$ is the source to screen distance and in this case is measured from the slit source S to the screen. With Young's double slit the separation $d$ of the slit sources may be measured directly with a travelling microscope. This is not possible with the biprism as only images of the slit source are formed. Instead, a lens may be used (Fig. 8-8) to form a real image of the two slit images. $d$ is then seen as a separation $l_1$, which may be measured with a low-power travelling microscope, and

$$a/b = l_1/d \tag{8-12}$$

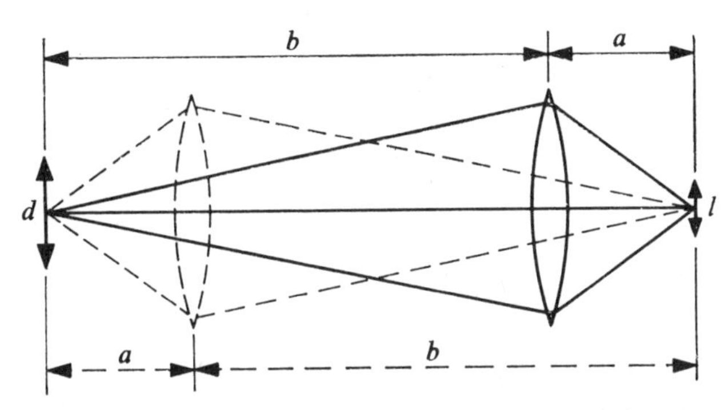


Fig. 8-8 Measurement of the source separation

If the lens is moved to its second conjugate position (shown dotted in the figure) an image is formed in the same place. This time $d$ is seen as a distance $l_2$, and may be measured as before, and

$$b/a = l_2/d \tag{8-13}$$

Multiplying Eqn (8-12) by (8-13) gives

$$d = \sqrt{(l_1 l_2)} \tag{8-14}$$

This method could, of course, be used equally well with Young's double slit.

Alternatively, since the biprism consists of two thin prisms with small refracting angles, $\alpha$, the deviation due to one prism is given by [Eqn (6-31)]:

$$\theta/2 = (n - 1)\alpha$$

The refractive index $n$, and refracting angle $\alpha$, may be measured on a spectrometer and $d$ calculated, since $d = L\theta$ (Fig. 8-7).

## 8-6 Lloyd's mirror

Lloyd demonstrated diffraction and interference fringes having obtained two coherent sources by reflection, instead of refraction (Fig. 8-9). A slit source, S, illuminates the screen directly over the region ad. Light reflected from the mirror appears to come from a source $S_1$. Since $S_1$ is an image of

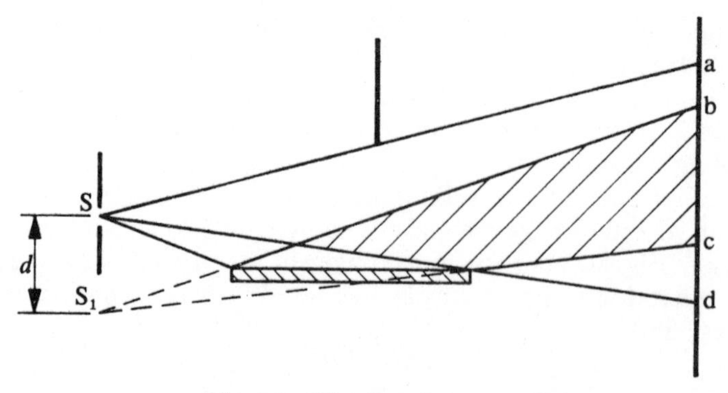


Fig. 8-9 Lloyd's mirror experiment

S, the two sources must be coherent and wherever the light from the two sources overlaps, interference will occur. Thus, as with Fresnel's biprism, interference fringes occur over the overlap region bc, and diffraction fringes, due to diffraction from the edges of the slit, occur over the whole area ad and are thus superimposed on the interference fringes in the region bc. Again, the equations that relate the spacing of the interference fringes to the geometry of the experiment are identical to those derived for Young's double slit.

If the screen is brought closer to the source, until it is in contact with the end of the mirror, the edge of the reflecting surface coincides with the centre of a *dark* fringe. Since this fringe is the one corresponding to zero path difference a bright fringe would be expected. It must therefore be assumed that one of the rays has undergone a phase change of $\pi$ to make the rays half a wavelength out of step. Rays come directly from source S and therefore there is no reason why these should have changed phase. The light from $S_1$, which has been reflected from the mirror surface, must therefore have undergone a phase change of $\pi$.

## 8-7 Interference by thin films

So far we have considered interference arising from the interaction of light waves from two coherent sources; this is called *double beam* interference or interference by *division of wave front*, since the two sources were produced by a wave front being split into two parts (Fig. 8-3). A further type of interference due to thin films is called *multiple beam* interference, or interference due to *division of amplitude*. This is the cause of the colour effects in soap bubbles or thin films of oil on water.

When light strikes a thin parallel-sided film at an angle, some of the light is refracted into the film and some is reflected (Fig. 8-10). The refracted ray is again split into a reflected and a refracted part at the lower surface. The part reflected in turn divides at the upper surface. This repeated division at the upper and lower surfaces continues until the remaining intensity falls to zero. If now the parallel rays from either the upper or the lower surface are brought together by means of a lens, interference can take place. Whether the interference is constructive or destructive depends

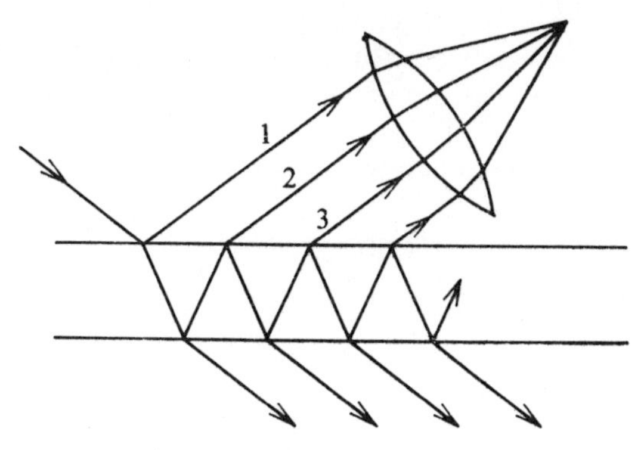


Fig. 8-10 Multiple reflections occurring at a thin film

on the phase differences arising from the different path distances travelled by the rays.

Figure 8-11 shows light from a monochromatic source incident at A at an angle $\theta$ onto a parallel film of thickness $t$ and refractive index $n'$. Let the refractive index of the surrounding medium be $n$. Then from the dividing point A, ray 1 takes the path AB in the medium of refractive index $n$, while ray 2 takes the path ACD in the film. From the wave front BD on, no further path differences, and hence phase differences, are introduced between the two rays. Hence the path difference between the two rays is equal to

$$n'(\mathrm{ACD}) - n(\mathrm{AB}) \tag{8-15}$$

i.e.,

$$\text{path difference} = n'(\mathrm{ECD}) - n(\mathrm{AB})$$
$$= n'(\mathrm{EF} + \mathrm{FD}) - n(\mathrm{AB})$$

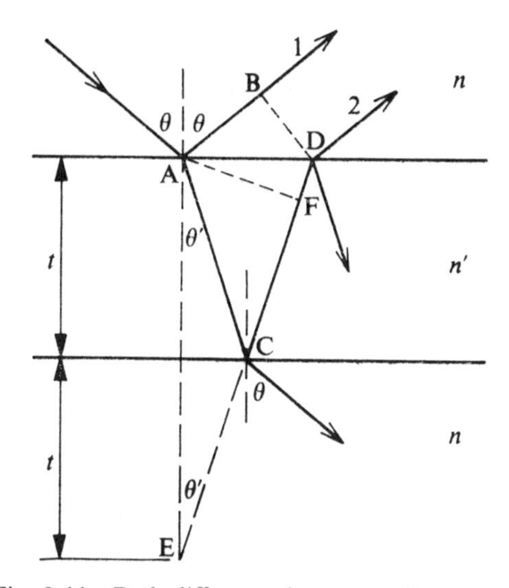


Fig. 8-11 Path difference between adjacent rays

But optical paths must be the same by any ray drawn between two wave fronts. This is the theorem of Malus and means that, in Fig. 8-12, all of the *optical paths* that may be drawn between the wave fronts (dotted) are equal. An 'optical path' is the sum of the products of the geometric distances and the corresponding refractive indices. Therefore, in Fig. 8-11,

$$n'(\mathrm{FD}) = n(\mathrm{AB}) \tag{8-16}$$

Therefore, for rays 1 and 2,

$$\begin{aligned}\text{path difference} &= n'(\mathrm{EF}) \\ &= 2n't\cos\theta'\end{aligned} \tag{8-17}$$

Whenever this path difference is equivalent to a whole number of wavelengths, one would expect the interference to be constructive and maxima to occur. But a phase change of $\pi$ occurs on reflection (Section 8-6).

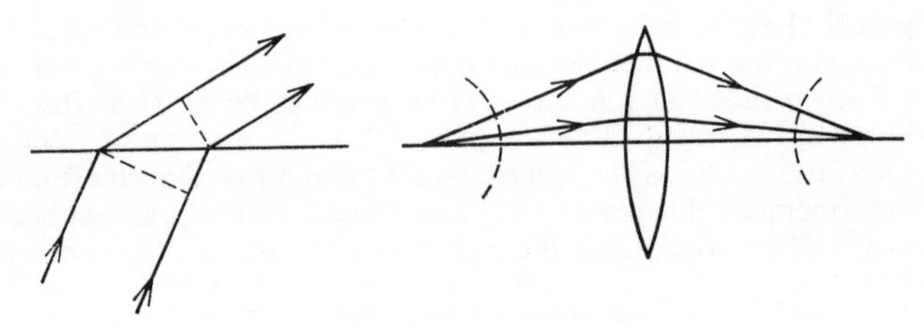

Fig. 8-12 Illustrating the theorem of Malus

If the film has a higher refractive index than its surroundings, ray 1 undergoes a phase change on reflection, but not ray 2 at its internal reflection. *A phase change of $\pi$ only occurs when light strikes a boundary from the side of lower refractive index*. Thus, when

$$2n't\cos\theta' = m\lambda \tag{8-18}$$

the phase change in ray 1 causes destructive interference between rays 1 and 2 and minima occur. For constructive interference for these two rays, producing maxima,

$$2n't\cos\theta' = (m + \tfrac{1}{2})\lambda \tag{8-19}$$

However, rays 2, 3, 4, . . . all approach the reflecting boundary from the side of high refractive index and consequently have not undergone a phase change. Equations (8-18) and (8-19) for destructive and constructive interference are therefore reversed for these rays. But, since ray 1 is much brighter than ray 2, ray 2 is much brighter than ray 3, and so on, the interference between rays 1 and 2 predominates and Eqns (8-18) and (8-19) apply.

If the film is of lower refractive index than its surrounding medium, ray 1 does not undergo the phase change, but rays 2, 3, 4, . . . do. Nevertheless, the same phase difference of $\pi$ exists between rays 1 and 2 and, since these are the two brightest rays, Eqns (8-18) and (8-19) still apply.

The transmitted rays also interfere but there is now no phase difference introduced between the rays due to the reflections and, therefore, the conditions are reversed. Equation (8-18) now becomes the condition for maxima and Eqn (8-19) for minima.

If the film is thick, the interfering rays appear well separated unless viewed nearly normal to the film. To observe interference we use a lens to bring the rays together at the eye. A lens is not necessary if the film is thin or can be viewed along the normal. For nearly normal viewing $\theta = \theta' = 0$ and hence $\cos \theta' = 1$.

If monochromatic light from an extended source falls onto a thin transparent film, the resultant intensity produced by interference varies with the viewing angle $\theta$. The extended source ensures that all values of the angle $\theta$ are available. The interference minima fall almost to zero for the reflected rays but not for the transmitted rays, since the minima are viewed against the bright source behind. If an extended white light source is used, the condition for maxima are satisfied at different angles by different wavelengths of light, producing different-coloured maxima as the viewing angle is changed.

## 8-8 Interference by wedge films

The condition for minima [Eqn (8-18)] caused by thin film interference is that

$$2n't \cos \theta' = m\lambda$$

Therefore, the thickness of the film

$$t = \frac{m\lambda}{2n' \cos \theta'}$$

Consider a wedge film of small angle $\alpha$ (Fig. 8-13). Every time the thickness of the wedge changes by an amount $\Delta t$ such that the order, $m$, changes by one, the condition for a minimum occurs, giving another dark fringe. Thus, a wedge film formed between two flat glass plates and illuminated by a monochromatic source appears as a series of dark fringes, lying parallel to the line of intersection of the plane surfaces and with a fringe being formed every time the wedge changes thickness by an amount

$$\Delta t = \frac{\lambda}{2n' \cos \theta'} \tag{8-20}$$

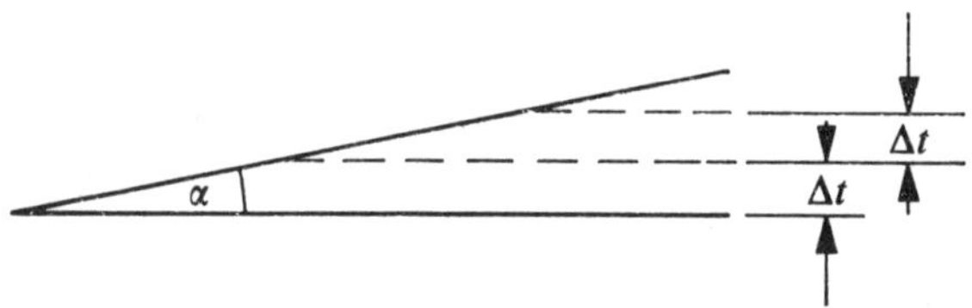


Fig. 8-13 Interference due to a wedge film

This lends itself to the accurate measurement of small angles. For example, suppose the spacing of the fringes formed by a wedge film in mercury green light of wavelength $\lambda = 5{,}461$ Å $= 0{\cdot}000021$ inches is measured with a travelling microscope focused on the wedge film and looking along the normal to the wedge, so that $\theta' = 0$. A dark fringe is formed each time the thickness of the wedge has increased by

$$\Delta t = \frac{5{,}461 \times 10^{-7}}{2 \times 1 \times 1} = 0{\cdot}000273 \text{ mm}$$

$$= 0{\cdot}0000107 \text{ in}$$

since the refractive index of the air film is taken as unity. Thus, if there are 10 fringes in a distance of 25 mm, the thickness of the wedge has increased by $10\Delta t$, and the angle of the wedge

$$\alpha = \frac{10 \times 0{\cdot}000273}{25} = 0{\cdot}000109 \text{ radians}$$

$$= 22 \text{ seconds of arc}$$

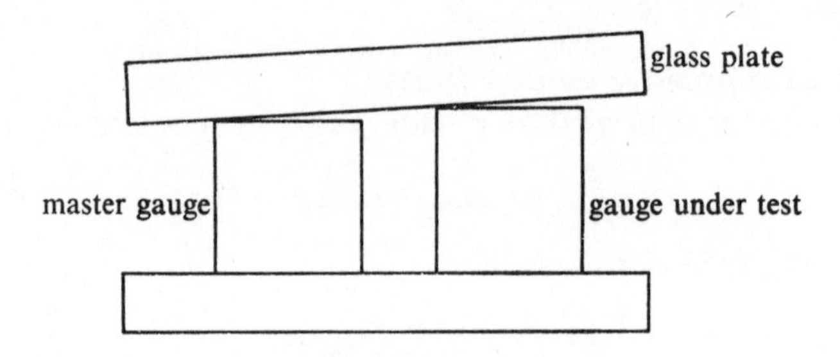


Fig. 8-14 Testing of slip gauges

This method is used to test engineers' slip gauges. The gauge to be tested and master gauge are 'wrung' down side-by-side into good contact with a flat plate (Fig. 8-14). A flat glass plate is then laid across them and illuminated with monochromatic light. If the gauges are of unequal height, a wedge-shaped air film is formed and, from the spacing of the resulting fringes, the difference in height of the gauges may be calculated.

## 8-9 Newton's rings

Newton's rings are a special case of interference by thin films. They are produced by a wedge film formed between a spherical and a plane surface (Fig. 8-15). Light from a monochromatic source, S, is reflected by a glass plate, G, onto a plano-convex lens resting on a flat glass plate. The fringes in the wedge between the convex surface and the bottom plate are in the form of concentric circles joining points of constant thickness.

To examine these fringes we use a microscope focused on the wedge. Let a particular fringe, radius $r$, be formed at a point where the thickness

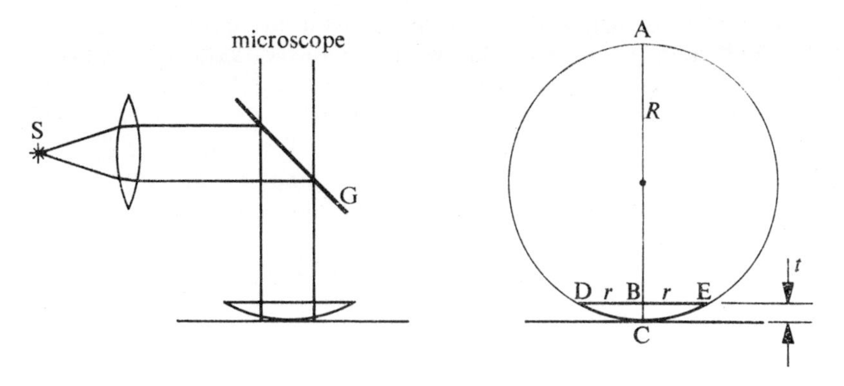


Fig. 8-15 Newton's rings

of the film is $t$ (Fig. 8-15). If the convex surface has a radius of curvature $R$, then by the theorem of the rectangular property of a circle

$$AB \times BC = DB \times BE$$

i.e.,

$$(2R - t)t = r^2$$

$$\therefore \ t = \frac{r^2}{2R - t} \tag{8-21}$$

But $t \ll R$,

$$\therefore \ t = \frac{r^2}{2R} \tag{8-22}$$

Since there is a phase change of $\pi$ on reflection, the condition for a dark fringe to occur at a point where the thickness of the wedge is $t$ is that [Eqn (8-18)]

$$2n't \cos \theta' = m\lambda \qquad \text{minima}$$

But, since the radius of curvature $R$ of the convex surface is large (e.g., 1 m) and the fringes are close to the pole of contact, $\theta'$ is virtually zero when the wedge is viewed normally. Therefore

$$2n't = m\lambda \qquad \text{minima}$$

Substituting for $t$ from Eqn (8-22) gives

$$r^2 = \frac{Rm\lambda}{n'} \tag{8-23}$$

as the condition for a dark fringe, and also

$$r^2 = \frac{R(m + \frac{1}{2})\lambda}{n'} \tag{8-24}$$

as the condition for a bright fringe. In practice, $R$ may be measured with a spherometer and $r$ measured with a travelling microscope. With an air film, $n' = 1$, and hence the wavelength, $\lambda$, of the monochromatic light may be determined, or, if we know the wavelength, the refractive index $n'$ may be determined, for example, for a liquid introduced into the wedge. Difficulty may arise in knowing the correct value of the order $m$. If the surfaces are dirty, the path difference at the pole of contact may not be zero. There may also be some distortion due to pressure of the contact. For these reasons it is better to modify the form of Eqns (8-23) and (8-24) in terms of diameters of rings. For example, if $d_m$ is the diameter of the dark ring of order $m$ and $d_{m+s}$ is the diameter of the dark ring of order $(m + s)$, then by Eqn (8-23)

$$d_{m+s}^2 - d_m^2 = \frac{4R(m + s)\lambda}{n'} - \frac{4Rm\lambda}{n'}$$

$$= \frac{4Rs\lambda}{n'} \qquad (8\text{-}25)$$

We do not therefore need to find the centre of the ring system or to know the correct order $m$; the diameter of any ring and another ring, say, 20 fringes further out ($s = 20$) is all that is required.

If the surfaces are clean and are pressed into contact, there is no path difference at the pole of contact and consequently one might expect a bright fringe at this point. Actually, a dark fringe is produced, showing that a phase change of $\pi$ must occur on reflection (Sections 8-6 and 8-7). This does not indicate at which surface the phase change occurs but, in fact, it occurs at the lower (air to glass) surface (Section 8-7). In white light, the central spot is still dark, but around this the interference conditions are satisfied by different ring radii for different wavelengths, resulting in a small number of coloured rings due to superposition.

## 8-10 Examination of surfaces by interference

Interference provides the most important method for the testing and absolute measurement of flatness. A test plate is used as a standard. This is usually made from fused silica glass which is hard and therefore resistant to wear and scratching. It has a low coefficient of thermal expansion and therefore is less susceptible to local deformation due to temperature gradients. These plates are usually made in sets of three. They are all worked together until each plate when matched against the other two produces a wedge film with straight fringes or, what is a more accurate comparison, it produces a parallel film which in white light shows a single fringe of one colour over the whole field. This condition can be satisfied for two plates if they both have matching curved surfaces, but it can only be satisfied for three plates if they are all flat. The test plate is laid on the surface under test which should be a reflecting surface free of any dust. The test plate is pressed down on one side to produce a wedge film and viewed with monochromatic light; any deviation from a straight-line fringe shows that the test piece is not flat. It is easy to measure a deviation of a tenth of a fringe spacing, which is equivalent to a height variation in the surface of $27 \times 10^{-6}$ mm or $10^{-6}$ in, in mercury green light. Whether the height variation is an elevation or a depression can be decided by observing whether the fringe deviation is towards or away from the direction of higher order fringes. If the test surface is spherical, the fringes produced are circular, forming Newton's rings and hence the radius of curvature can be measured.

The test plate need not be flat. Concave and convex plates are often used—for example, in the manufacture of ball-bearings and lenses. If the lens being tested and the test plate have the same radius of curvature, the parallel film of negligible thickness which forms when they are brought together shows as a uniform interference colour in white light. A difference in radii produces a wedge film and Newton's rings. The difference in radii may be deduced from the number of rings. If the lens is not spherical, the rings are distorted.

The method described for testing flatness can also be used to test surface finish. Surface finish is, however, often tested by a stylus method. This is similar in principle to the gramophone pick-up. Variations in the topography of the surface under test cause vertical movements of the stylus, which may be amplified electronically and shown on a chart recorder. Such methods have the disadvantages that the stylus itself may damage the surface, that the examination is restricted to a line, and that a false or non-representative reading may be obtained, depending, for example, on whether the path traced by the stylus is parallel or normal to the finishing direction of the worked surface. Optical interference methods have none of these disadvantages: the test plate need not be in contact with the surface and an area is examined. However, the area is small and therefore localized, and consequently the two methods may be regarded as complementary.

For the interference method we use a special microscope objective (Fig. 8-16) with a beam-splitting cube cemented to the front lens. Two prisms with their hypotenuse face partially silvered are cemented together to form the cube. Rigidly mounted to the side of the cube is a small reference surface which must be flat and free from any surface blemishes. It is therefore usually an optically worked glass surface with a thin layer of aluminium evaporated onto it. The position of the reference flat is such that it is always in focus. Thus, when the surface to be tested is brought

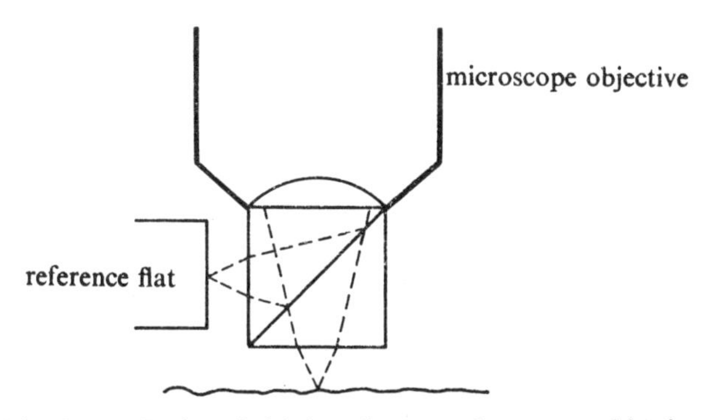


Fig. 8-16 Surface finish interference microscope objective

into focus, its image is superimposed on the image of the reference surface. With monochromatic light, thin-film interference occurs. The specimen table can be tilted to produce a wedge film and straight-line fringes or a parallel film and a uniform fringe. Any scratches or surface blemishes show up as distortion or displacement of fringes; these can be measured and the depth of a scratch calculated. The magnification of such objectives is governed by the working distance available to mount the beam-splitting device but, using various arrangements, objectives are available with magnifications up to $\times 40$; with a $\times 10$ eyepiece this provides a total magnification of $\times 400$. This method is ideally suited for examining very fine finishes on small areas—for example, for the examination of razor blade edges and watch pivots—but it can also give much useful information by the localized examination of larger areas.

## 8-11 Michelson interferometer

An interferometer is a device which uses interference for measurement. With the Michelson interferometer length can be determined with great accuracy in terms of the wavelength of light. For many years standard units of measure were defined in terms of objects, such as metal bars. These are not good standards, since, apart from the fact that they may be lost or damaged, physical changes in the bar material may occur with time causing a slow change in length. For this reason, standards of length are now defined in wavelengths of light. For example, the standard metre has been defined since 1960 as 1,650,763·7 wavelengths in vacuo of the orange-red line of the krypton isotope with mass 86 ($Kr^{86}$). This has a wavelength $\lambda_{vac} = 6{,}057{\cdot}80211$ Å when obtained from a specified lamp operated in a prescribed manner. This is a standard that is constant and can be reproduced anywhere in the world. A Michelson interferometer is a convenient instrument for measuring in the first place the standard bar metre in terms of wavelength, and now, calibrating secondary standard bars in terms of the standard metre *defined* in terms of wavelength.

The Michelson interferometer is shown diagrammatically in Fig. 8-17. A and B are two plane-parallel plates of glass of exactly equal thickness. Light from the extended source is split into reflected and transmitted rays at the back surface of plate A, which is usually partially silvered to produce equal intensities. The transmitted ray then passes through the second plate, B, is reflected by the front silvered mirror, $M_1$, and then reflected into the telescope by the back surface of A. The reflected ray is reflected again by the front silvered mirror, $M_2$, passes back through plate A, and then into the telescope. The second plate, B, acts as a compensator plate. Thus, the reflected and transmitted rays have each travelled through two thicknesses of glass from their point of division. Identical paths in glass are essential if white light is used, but not for monochromatic light fringes. When the two rays are viewed by a telescope set to receive parallel rays, the mirror $M_1$ is imaged at $M_1'$, thus forming the equivalent of a parallel air film. Mirror $M_2$ can be moved parallel to itself by a fine thread screw, to adjust the thickness of the parallel air film. When $M_2$ and the image of

$M_1$ at $M_1'$ are parallel, the fringes seen satisfy the condition of Eqn (8-18), i.e.,

$$2t \cos \theta' = m\lambda \tag{8-26}$$

For a given $t$, $m$, and $\lambda$, the angle $\theta'$ is a constant and circular fringes are formed concentric with the axis of the telescope. It can be shown that the radii of the circular fringes are proportional to the square roots of the integers and hence these fringes resemble Newton's rings. As $M_2$ approaches $M_1'$, each fringe, characterized by a particular $m$, must get smaller, since $\theta'$ must get smaller for the product $2t \cos \theta'$ [Eqn (8-26)] to

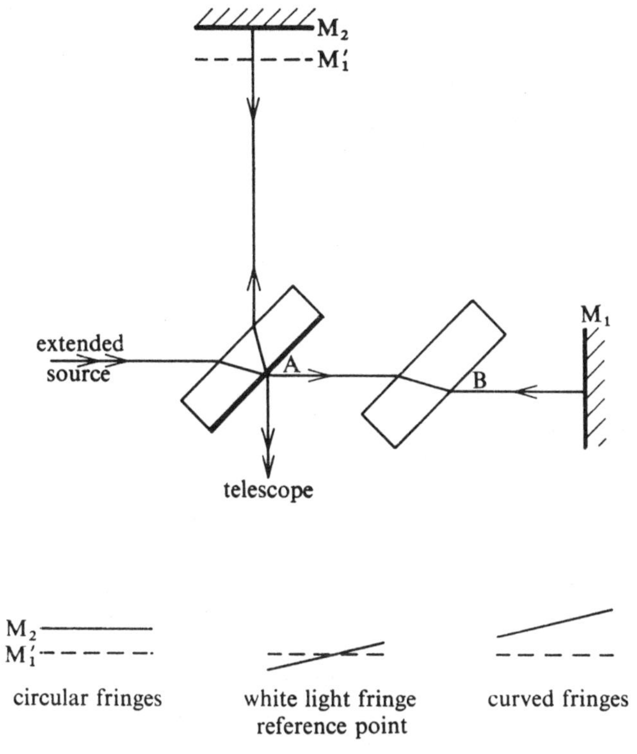


Fig. 8-17 The Michelson interferometer

remain constant. The fringes, therefore, shrink and disappear in turn at the centre. They also become more widely spaced until, when $t = 0$, one fringe fills the field.

To see fringes we must use an extended source so that a full range of values of $\theta'$ is available. With conventional monochromatic sources, fringes can be obtained with values of $t$ up to about 25 cm, but with white light $t$ must be almost zero.

White light fringes are best observed when one mirror is slightly tilted to produce a wedge film. The fringes are then localized and the telescope must be refocused onto a point near to the mirror $M_2$. By adjusting the separation, the mirror $M_2$ and the imaged mirror $M'_1$ may be made to intersect along their centre line, thus forming two wedge films. The line of zero path difference produces a straight-line fringe which may be black or white and acts as a reference point of zero path difference. If the back surface of the glass plate is not partially silvered, the usual phase change of $\pi$ will result in a black fringe for zero path difference but, if it is partially silvered, a bright fringe may result. The central fringe is not coloured in white light because here the interference condition is satisfied for all wavelengths. On either side of the zero-order fringe, the fringes are coloured as the condition is satisfied for the different wavelengths in turn but, beyond the first few fringes, there is sufficient overlap to result in white. With monochromatic light, fringes may be obtained with $t$ up to several millimetres. With tilted mirrors the zero-order fringe is straight but fringes on either side become progressively more curved, since there is some variation of path difference with angle due to the method of viewing. They are always convex towards the thin end of the wedge.

To measure length with an interferometer a microscope is fitted to the moving carriage $M_2$. The microscope is focused on one end of the specimen and the carriage then traversed until the other end of the specimen is in focus. At the same time the number of fringes crossing a graticule line in the telescope eyepiece is noted, and hence the distance traversed calculated from Eqn (8-26), since $\cos \theta' = 1$ at the centre. Each fringe shift is thus equivalent to a change in $t$ of half a wavelength. Since one-tenth of a fringe can easily be estimated and, with care, one-fiftieth of a fringe, it is possible to measure a change in $t$ of a hundredth of a wavelength ($5 \times 10^{-6}$ mm or $2 \times 10^{-7}$ in with mercury green light).

The length of the standard metre bar was first measured in terms of light wavelengths by Michelson and Benoit. They used lines of the cadmium spectrum; nowadays the metre is defined in terms of a krypton line. Since it was practically impossible to obtain and count the fringes due to moving the mirror $M_2$ over the full length of the metre, they made use of etalons and a modified arrangement of the interferometer. The arrangement is shown in Fig. 8-18. An etalon is a block fitted with two front silvered mirrors which are adjusted to be accurately parallel. Nine such etalons were used, each being approximately twice as long as the previous one, up to a maximum length of about 10 cm. The two shortest etalons were first arranged in the interferometer so that the four surfaces $M_1$, $M'_1$, $M_2$, $M'_2$, were visible in the field at the same time. Then, using white-light fringes, the front mirrors $M_1$ and $M_2$ and the movable mirror M were arranged to be optically coincident. Changing to monochromatic light, the movable mirror M was traversed to position B and the fringes passing the graticule line in the telescope counted. At B, the mirror M was optically coincident with the second mirror $M'_1$ of the shortest etalon. Exact coincidence was achieved by changing back to white light, as only with white light can a zero-order fringe be recognized. It is a little easier if the mirrors are very slightly inclined so that a straight-line fringe is obtained for the zero order, all other orders having curved fringes. The length of the shortest etalon, $M_1M_1'$, was therefore determined as a number of wavelengths. Next, without disturbing any other setting, the shortest etalon was moved back so that $M_1$ now occupied the same position as previously occupied by $M_1'$, that is, $M_1$ was now optically coincident with M at its new position B, as judged using white-light fringes. The movable mirror M was then moved again, to position C, without counting fringes, so as to be coincident with the new position of $M_1'$, as again judged using white-light fringes. The small number of fringes was then counted, using monochromatic light, and white light for the exact coincidence, to make M and $M_2'$ coincident. Thus, the length of the second etalon was two times the length of the first, plus these few extra half wavelengths.

The length of the third etalon was then measured in terms of the length of the second etalon plus the few half wavelengths for exact coincidence

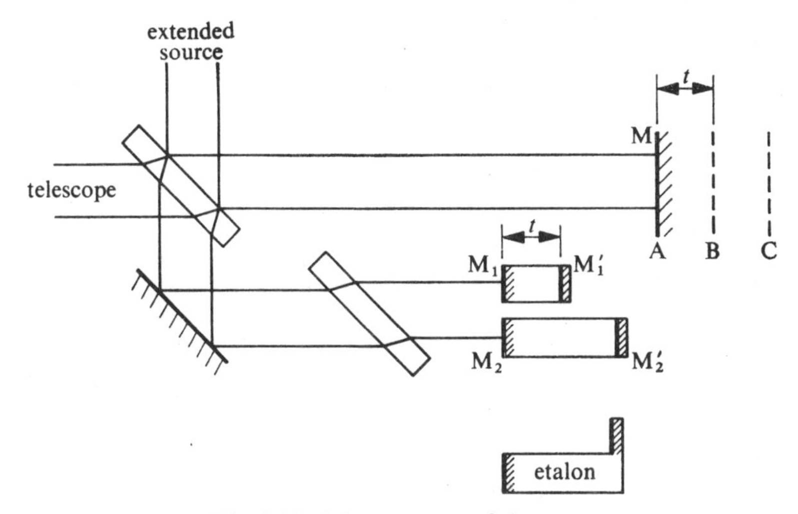


Fig. 8-18 Measurement of the metre

again, by stepping-off in a similar manner. The process was then repeated until the length of the ninth etalon was obtained. Thus, a length of approximately 10 cm was determined by counting only the number of half wavelengths in the shortest etalon, approximately 0·39 mm long, repeatedly stepping-off, and counting the small number of fringes difference for an exact coincidence at each stepping-off.

To measure the metre, a microscope fixed to the moving carriage M was focused onto the mark at one end of the metre bar. Using white-light fringes, coincidence was obtained between M and the first mirror of the 10-cm etalon. M was then traversed for coincidence with the second mirror of the etalon, which in turn was moved back to obtain coincidence between the new position of M and the first mirror of the etalon again. This stepping-off process was repeated ten times. The extra small number of fringes was then found, using monochromatic light, to line up the microscope cross-line with the mark at the other end of the standard metre bar, thus determining the number of half wavelengths and hence the number of wavelengths of the monochromatic light in the standard metre.

The cumulative error due to the stepping-off process involving the longest etalon is much smaller than that involved in setting the microscope on the end marks of the metre bar. The error involved in stepping-off the shorter etalons is not cumulative as the fractions of fringes are only needed to arrive at the correct whole number of fringes at the next doubling.

Other supplementary standards, such as the yard, are now determined in relation to the fundamental unit, the metre. For example, since 1959 the International Yard has been defined to be 0·9144 metre.

## 8-12 Non-reflecting films

Suppose light of a particular wavelength in a medium of refractive index $n$ is incident normally onto a surface of a medium of refractive index $n'$ (Fig. 8-19). On passing through the second medium, the light is then incident, again normally, on a medium of refractive index $n''$. Reflection and transmission take place at the two surfaces and the reflected rays interfere. If $n < n' < n''$, phase changes of $\pi$ occur at *both* reflections. Therefore, $2n't = (m + \frac{1}{2})\lambda$ is the condition for destructive interference.

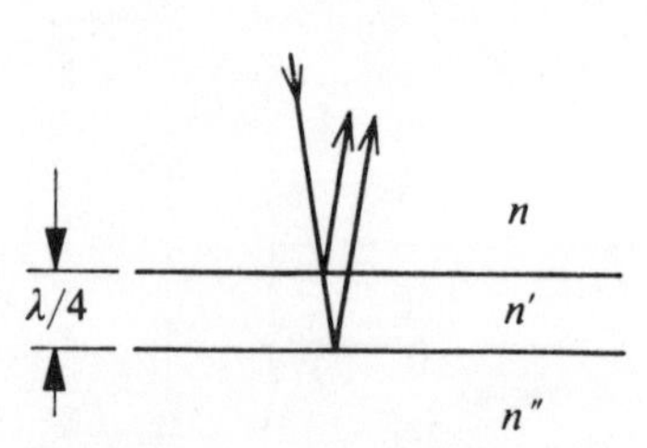


Fig. 8-19 Non-reflecting film

The smallest thickness of film, $t$ (when $m = 0$), which satisfies this condition is

$$t = \frac{\lambda}{4n'} \tag{8-27}$$

This produces a minimum for the reflected light. The value is not zero unless the interfering rays have equal intensity. The reflected intensities are equal if the reflection coefficients of the two interfaces are the same value. Therefore, equating the two reflection coefficients

$$\left(\frac{n' - n}{n' + n}\right)^2 = \left(\frac{n'' - n'}{n'' + n'}\right)^2 \tag{8-28}$$

From which

$$n' = \sqrt{nn''} \tag{8-29}$$

is the condition for equality.

For white light, it is usual to satisfy the condition of Eqn (8-27) for green light ($\lambda = 5{,}500$ Å). Very little green light is reflected but some blue and red is; these extremes of the spectrum together produce a purple colour. Putting down a non-reflecting film of a quarter wavelength optical thickness [Eqn (8-27)] is therefore termed 'blooming', since the resultant colour in white light is similar to the purple bloom on some types of ripe plum. As has already been said, the blooming has its greatest effect if Eqn (8-29) is also satisfied. In practice, the non-reflecting film is normally applied to a glass surface and then used in contact with air—that is, $n = 1$ and $n''$ is usually about 1·5. Therefore, by Eqn (8-29), $n'$, the refractive index of the film, should be about 1·22. The film material should also be easy to deposit onto glass by evaporation and produce a film that is both stable and durable in addition to being transparent and having the very low refractive index of 1·22. These conditions are difficult to satisfy. The best practical approximation is magnesium fluoride with an index of 1·38, giving a film thickness for green light of

$$t = \frac{5{,}500 \times 10^{-8}}{4 \times 1{\cdot}38}$$

$$= 10^{-5} \text{ cm}$$

or cryolite (sodium aluminium fluoride) with an index of 1·36. Thus, blooming is the most effective with glasses of high refractive index, since the condition of Eqn (8-29) is more nearly satisfied. The condition is exactly satisfied for cryolite with a glass of 1·85 refractive index.

A crown glass of refractive index 1·5 has a reflection coefficient of 0·040—that is, 4% of the incident light is lost by back reflection. This 4% loss occurs at every air-glass surface. Thus, for example, in a four-component optical system having eight air-glass surfaces, a total of 27·8% of the incident light is lost. This loss not only leads to a dimmer image but also to a loss in contrast due to light scatter. With a dense flint glass of refractive index, say, 1·7, the reflection coefficient is up to 0·067 and there is a light loss of 6·7% at each air-glass surface. With blooming these reflection losses at each surface can be reduced to a fraction of 1%.

The thickness condition of Eqn (8-27) applies to normally incident light. With a curved surface the light is incident at other angles but this does not introduce much error as the cosine changes only slowly in the neighbourhood of 0°.

By laying down a film of low refractive index and then a film of high refractive index, high reflection occurs. If the film thicknesses are such that the three reflected components are in phase, almost 50% of the light is reflected and virtually all the remainder is transmitted. This forms a high-efficiency beam splitter which has an obvious advantage over the conventional partially reflecting layer of aluminium in which 30% of the light may be lost by absorption.

## 8-13 Interference filters

We have shown (Section 8-7) that the condition for an interference maximum for light transmitted normally through a thin transparent film is

$$\lambda = \frac{2n't}{m} \tag{8-30}$$

Thus, if white light is incident on the film, there is a transmitted maximum for any wavelength that satisfies the condition for an integral value of $m$. Other wavelengths are reflected. If, however, the film is very thin so that only one or two maxima can occur for wavelengths in the visible range, the film acts as a filter transmitting only one or two narrow wavelength bands. Nearly monochromatic light may thus be obtained with bandwidths of about 100 Å.

The filters are made by evaporating a semi-transparent metal film on glass; followed by a transparent dielectric layer, such as silicon dioxide; and then a further semi-transparent metal layer. The dielectric layer has a thickness, given by Eqn (8-30), for the required wavelength and for a particular $m$. If $m = 2$ for the required wavelength, $\lambda$, the film has an optical thickness of one wavelength. Another maximum also occurs at $2\lambda$, for $m = 1$, but this may be absorbed with a conventional coloured glass or gelatin filter.

## PROBLEMS

**8-1** Two pieces of optically flat glass 10 cm long are separated at one end by a piece of paper. If the distance between fringes produced by the wedge of air between the plates, in light of wavelength 6,000 Å, is 0·06 cm, find the thickness of the paper. [0·05 mm]

**8-2** Two flat rectangular glass plates, in contact along one edge and with the opposite edges separated by a thin piece of foil, are illuminated vertically by sodium light of wavelength $5{\cdot}89 \times 10^{-5}$ cm. Calculate the thickness of the foil if thirty bright fringes are seen up to the edge of the foil. [0·0088 mm]

**8-3** Describe and account for what is seen when light from a closely spaced pair of coherent slit sources is incident on a screen.

Two slits 5 mm apart are placed 50 cm from a screen, and illuminated by light of wavelength $5{\cdot}89 \times 10^{-5}$ cm. Find the distance between the respective centres of the fifth dark and the tenth bright interference bands. [0·324 mm]

**8-4** In a Young's double-slit experiment, light from a monochromatic source of wavelength $5{\cdot}9 \times 10^{-5}$ cm passes first through a narrow slit and then through two narrow slits $S_1$ and $S_2$. If $S_1$ and $S_2$ are 2 mm apart, describe and account for the appearance observed on a screen placed 100 cm from the slits. [fringe separation = 0·295 mm]

**8-5** A Fresnel biprism is used to form interference fringes, using light of wavelength $5{\cdot}89 \times 10^{-5}$ cm. In a separate experiment using a supplementary lens, two positions of the lens were found in which double images of the slit were formed at the screen. If the separations of the slit images were 3·6 mm and 0·4 mm when the distance between the slit and the screen was 80 cm, what was the fringe separation? [0·39 mm]

**8-6** The upper glass surface in a Newton's rings experiment has a radius of 100 cm. What are the diameters of the fifth and tenth bright rings for light of wavelength 5,460 Å? [3·13 mm; 4·56 mm]

**8-7** In a Newton's rings experiment, two bright rings of orders $m$ and $m + 10$ had diameters of 5·72 mm and 8·10 mm respectively when green light of wavelength $5{\cdot}46 \times 10^{-5}$ cm was used. A drop of liquid was then placed between the lens surface and the glass plate, and the corresponding rings were then found to have diameters 4·95 mm and 7·02 mm. Calculate the radius of curvature of the lens surface and the refractive index of the liquid. [150·6 cm, 1·327]

**8-8** What is the wavelength of the visible light that is transmitted with maximum intensity by a semi-transparent metal film on glass, which is covered in turn by a transparent dielectric layer 0·00030 mm thick and a further semi-transparent metal film. The refractive index of the dielectric layer is 1·55 and is assumed to be independent of wavelength. [4,650 Å]

**8-9** A transparent plate of refractive index 1·5 and thickness 0·5 micron is illuminated normally by a beam of white light. What wavelengths in the visible region will be most strongly reflected? [4,285 Å, 6,000 Å]

**8-10** A biprism is 90 cm from a screen and 10 cm from a source emitting light of wavelength 5,016 Å. Calculate the separation of the fringes produced on the screen if the biprism has a refractive index 1·65 and $\frac{1}{2}°$ refracting angles. [0·442 mm]

# 9 Diffraction

## 9-1 Introduction

We have shown how Huygens' principle, that each point on a wave front may be considered as a new source of secondary wavelets, may be used to account for the formation of interfering rays. It was also shown that this principle was quite general and not just applicable to interference, since it could account for the paths of reflected and refracted rays. However there are certain inconsistencies in the principle as first formulated, which were subsequently resolved by Fresnel.

Figure 9-1 shows an infinite plane wave front ABC travelling in the direction of its normal OP from a point source at infinity. All points on this plane wave front have the same phase, amplitude, and frequency.

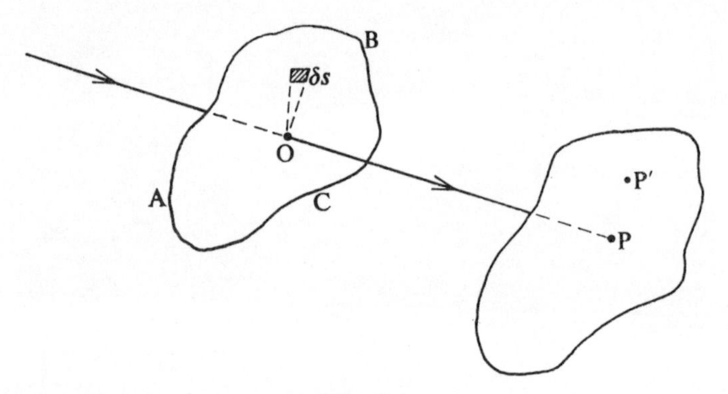


Fig. 9-1

By Huygens' principle, a small area $\delta s$ acts as a source of secondary wavelets and produces a disturbance at some point P. By the inverse square law and since the energy of the wave motion is proportional to the square of its amplitude, the amplitude of the disturbance produced at P is inversely proportional to the distance of P from the area $\delta s$, and directly proportional to the area $\delta s$. To find the total amplitude of the disturbance at P due to all such elemental areas $\delta s$, it is necessary to integrate over the total area of the plane wave front, taking into account that the areas are at different distances from P and therefore the wavelets arrive at P with different phases.

In the same way we can determine the amplitude and phase of the disturbance produced at some other point P′ lying on the plane through P parallel to the wave front. Since the plane wave front ABC is considered to be infinite in extent, the total disturbance produced at P′ or at any other point in this plane is identical to that produced at P. Hence a new plane wave front has been produced through P, parallel to the original plane wave front ABC. We have thus shown that light travels in straight lines by applying Huygens' principle to an infinite wave front. Similarly, we can show that a spherical wave is propagated as a spherical wave. A fuller treatment shows also that Huygens' secondary wavelets do not lead to the propagation of a back-wave in a direction opposite to that of the wave-front propagation and that the amplitude of the secondary wavelet is not constant in all forward directions. In a direction making an angle $\theta$ with the incident ray direction, the amplitude of the secondary wavelet varies as $(1 + \cos\theta)$. This is termed the *obliquity factor*. Thus in Fig. 9-2,

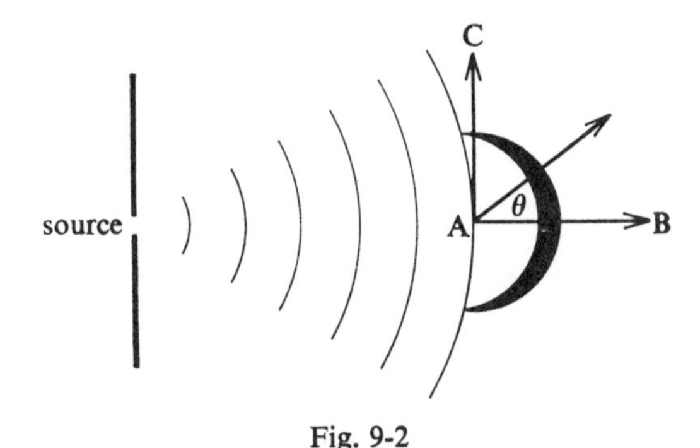


Fig. 9-2

since the intensity varies as the square of the amplitude, the intensity in the forward direction AB is proportional to 4, dropping to 1 in the direction AC, and zero in all backward directions.

Suppose our plane wave front is incident normally on a diaphragm. Suppose the diaphragm is infinite in extent, and has an aperture. If the aperture is very large compared to the wavelength of light, so that it may also be considered infinite, it allows a plane wave front to pass through to form a parallel-sided beam of light. If the aperture is very small, a fraction of the wavelength of light, the part of the plane wave incident at the aperture is sufficiently small to behave as a single point source for Huygens' secondary wavelets. The wavelets then spread to form a *spherical* wave. For apertures between these two extremes, most of the light forms a plane wave front moving in a straight line—that is, as a parallel-sided beam of light—and the rest is propagated to the sides, into the geometrical shadow region. This apparent deviation of the light as it passes an obstacle, in this case the edge of the aperture, is termed *diffraction*. It is referred to as 'apparent' deviation because it is in fact not deviated but is the normal case of propagation from a Huygenian point source, as opposed to 'recti-linear' propagation, which is a special case occurring when the wave front may be considered as infinite in extent. The actual effect of an obstacle is not to deviate the light but to limit the area of the wave front, preventing the special case for rectilinear propagation.

An aperture, then, causes a redistribution of the light falling on a screen to produce a bright area approximately equal to that of the aperture, surrounded, in the case of a circular aperture, by a number of concentric rings, rapidly decreasing in intensity with increasing radius. The larger the aperture with respect to the wavelength of light, the nearer are the form and area of the bright area on the screen to those of the aperture, and the less is the energy removed from the bright area and distributed into the diffraction pattern.

Diffraction phenomena are the result of interference due to wavelets arriving from different parts of a short wave front; interference phenomena are due to interference between two or more separate wave fronts. Diffraction phenomena may be divided into two classes: in *Fraunhofer diffraction* the incident light is effectively parallel and the diffraction pattern is observed on a screen which is effectively at an infinite distance; in *Fresnel diffraction* the wave fronts are spherical or cylindrical and so the source or screen or both are at finite distances. In this book the discussion of diffraction phenomena is limited to Fraunhofer diffraction because it is of the greater practical importance and easier to deal with mathematically.

## 9-2 Fraunhofer diffraction due to a single slit

Since it is Fraunhofer diffraction that is being considered, it is necessary to consider plane wave fronts and parallel rays, which implies the use of lenses. In Fig. 9-3 a monochromatic source is at the first principal focus of the first lens and the screen is in the second principal plane of the second lens. The slit is long so that diffraction due to points along lines parallel to the slit edges may be neglected. We have to determine the distribution of intensity at the screen, along a line $P_0P'$ perpendicular to the direction of the slit.

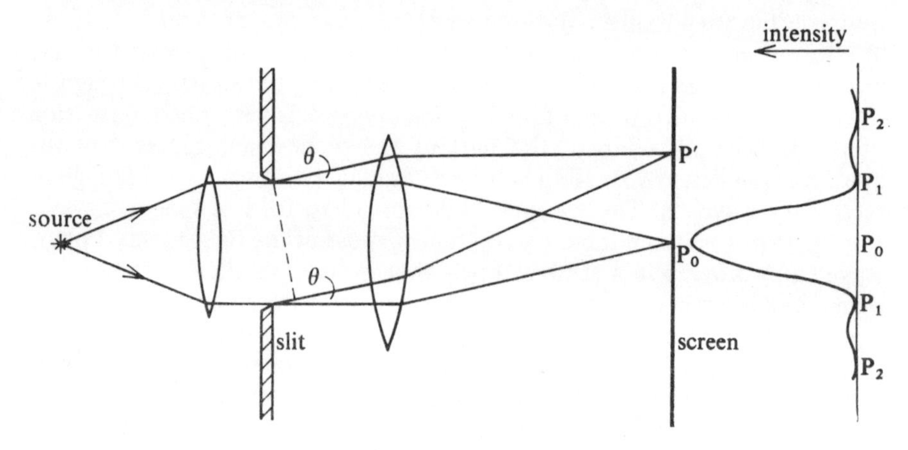


Fig. 9-3 Diffraction due to a single slit

Every point on the plane wave front incident onto the slit may be considered to be a source of Huygens' secondary wavelets at the instant it occupies the plane of the slit. Suppose the slit is divided into a large number of elemental strips parallel to the slit edge, then each strip may be regarded as a source of the wavelets. If all these elemental strips have the same width, and since the plane of the incident plane wave front is parallel to the plane of the slit, all of the secondary wavelets start off in phase and with the same amplitude, which is proportional to the width of the elemental strip. After refraction by the second lens (Fig. 9-3) all the rays from the elemental strip sources arrive at the axial focal point $P_0$ having travelled exactly the same optical path distances [theorem of Malus (Section 8-7)]. Thus all the rays arriving at $P_0$ are in phase, as they were when they started at the plane of the slit. The sum of their amplitudes gives a maximum of intensity, the *principal maximum*, at this point, or since we are dealing with a slit, along a line through $P_0$ parallel to the slit.

At some other point P′ on the screen, the rays from the elemental strip sources in the plane of the slit are inclined at some angle $\theta$ to the direction of the perpendicular to the plane of the slit. The rays arriving at P′ no longer all have the same phase, since they have travelled different optical path distances, and hence their separate contributions to the resultant amplitude are not equal. Their contributions, however, may be added as follows:

Imagine the slit is divided into two equal parts [Fig. 9-4(a)], where $a$ is the slit width and E is the foot of the perpendicular drawn from the slit edge at A to the ray from C, by the opposite slit edge, as shown. Then a ray from C travels a distance CE greater than a ray from A, by the top edge, where

$$CE = a \sin \theta_1 \tag{9-1}$$

If now this path difference is such that CE = $\lambda$, then a ray from the midpoint of the slit, B, has a path difference of $\lambda/2$ with the ray from the top edge of the slit, A. Thus the two rays, one from A and one from B, are

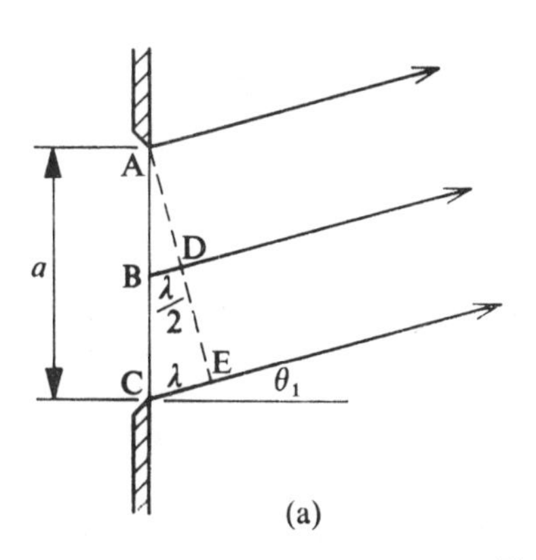


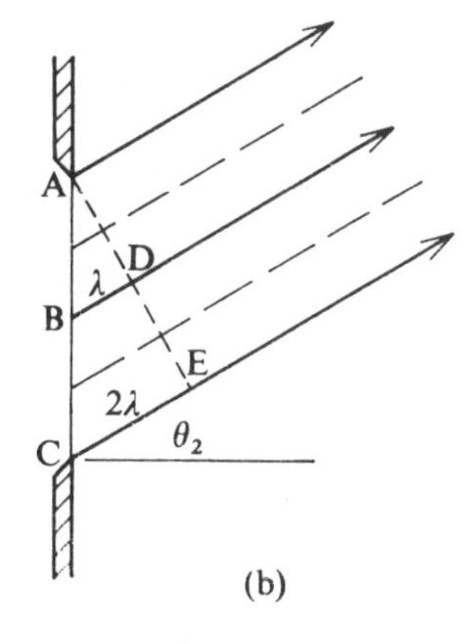


Fig. 9-4

opposite in phase and cancel each other out, producing zero amplitude and hence no intensity at the screen. Consider the elemental strips parallel to the slit edge; the ray from the element of slit immediately below A cancels out the ray from the element immediately below B, since they both have a path difference of $\lambda/2$. Similarly, the rays from the next elements below these have the same path difference and likewise cancel out, and similarly for all the elemental strips right across the slit.

A point $P_1$ on the screen, defined by an angle $\theta_1$ such that

$$\sin\theta_1 = \frac{\lambda}{a} \tag{9-2}$$

[Eqn (9-1)], is a point of zero intensity.

Consider another angle, $\theta_2$, such that the extreme path difference CE $= 2\lambda$ and thus BD $= \lambda$ [Fig. 9-4(b)]. Then, by considering AB divided into two parts, rays from elemental strips across AB may be imagined as cancelling in pairs, as they did over AC in Fig. 9-4(a). Similarly, rays from C have a path difference greater by $\lambda$ than those from B. Therefore by considering BC divided into two equal parts, the rays from this region also cancel in pairs. Thus there is another point $P_2$ on the screen which is a point of zero intensity, where $P_2$ is defined by the angle $\theta_2$ such that

$$\sin\theta_2 = \frac{2\lambda}{a} \tag{9-3}$$

Similarly, if CE is $3\lambda$, then by dividing up the slit width into three sections, the rays from elemental strips within each of these sections may be imagined as cancelling in pairs as before so that the intensity at the screen due to each section is separately zero. Thus another point of zero intensity, $P_3$, is formed such that

$$\sin\theta_3 = \frac{3\lambda}{a} \tag{9-4}$$

Generally then whenever the diffracted rays make an angle $\theta$ with the normal to the plane of the slit such that

$$\sin\theta = \frac{m\lambda}{a} \tag{9-5}$$

where $m = 1, 2, 3, \ldots$, the rays cancel out to give a point of zero intensity on the screen. When $m = 0$, $\theta = 0$ and the central intensity maximum at $P_0$ occurs. Between the points of zero intensity $P_1$, $P_2$, $P_3$, etc., *secondary maxima* occur, as in the intensity distribution curve of Fig. 9-3. These secondary maxima, *approximately* midway between the minima, are much weaker than the principal maximum at $P_0$. For example, consider a point approximately midway between $P_1$ and $P_2$ defined by an angle such that CE $= 3\lambda/2$ (Fig. 9-4). Then by dividing the slit AC into three parts, rays from two of the parts may cancel each other in pairs, since they have path differences of $\lambda/2$, leaving the third part. Thus only a third of the slit is effective in producing an intensity at the screen, and since the various rays from this third do not arrive in phase, the resultant intensity is even further reduced.

Since $\theta$ is small, the positions of the minima are given by

$$\theta = \frac{m\lambda}{a} \tag{9-6}$$

If the imaging lens is close to the slit, the distance of the screen to the slit is approximately equal to $f$, the focal length of the lens, so that $\theta = x/f$, where $x$ is the linear distance measured on the screen from the axial point $P_0$. Thus

$$x = \pm\frac{mf\lambda}{a} \tag{9-7}$$

where $m = 1, 2, 3, \ldots$, gives the distances of the minima from the centre at $P_0$, measured in either direction. The width of the principal maximum—that is, the distance between the first minima from $-f\lambda/a$ to $+f\lambda/a$—is twice the width of the secondary maxima—that is, the distance $f\lambda/a$ between neighbouring minima. The widths of the principal and secondary maxima are inversely proportional to the slit width.

However, to know only the positions of the minima is not sufficient. To know also the intensity distribution across the screen we find the sum of the equal amplitude contributions from each of the elemental strips, taking into account the different phases due to the different optical path distances. This is most easily done using a graphical method. Each of the finite number of elemental strip sources contributes a wavelet such that each of the wavelets has the same small amplitude. There is a constant phase difference $\alpha$ between wavelets from neighbouring elemental strips, since the distance of each strip from some particular point on the screen changes by constant steps. We can use vectors (Fig. 9-5) to find the resultant amplitude, $A_\theta$, and corresponding phase angle. When the phase difference $\alpha$ between the successive wavelets is zero, the resultant amplitude is simply the sum of the elemental amplitudes, $A_0$ [Fig. 9-5(a)], corresponding to the principal maximum at $P_0$ on the screen. When the phase difference $\alpha$ is not zero, the resultant is as given in Fig. 9-5(b), the vibration polygon. If the number of elemental strips is increased so that the amplitudes due to each of the strips becomes infinitesimal, the vibration polygon becomes the vibration curve of Fig. 9-5(c), in which the length of the arc $A_0$ equals the sum of the elemental amplitudes. If the radius of the arc is $r$ and the arc subtends an angle $2\beta$ at the centre of the circle, then there is a difference of $2\beta$ in initial phase between the wavelets coming from the first and the last of the elemental strips—that is, between those wavelets from the top and bottom edges of the slit. Thus $\beta$ is a convenient variable to describe the shape of the intensity distribution curve [Fig. 9-5(d)], where the length of the arc, $A_0$, remains constant, and the intensity, $I$, is proportional to the square of the resultant amplitude, $A_\theta$.

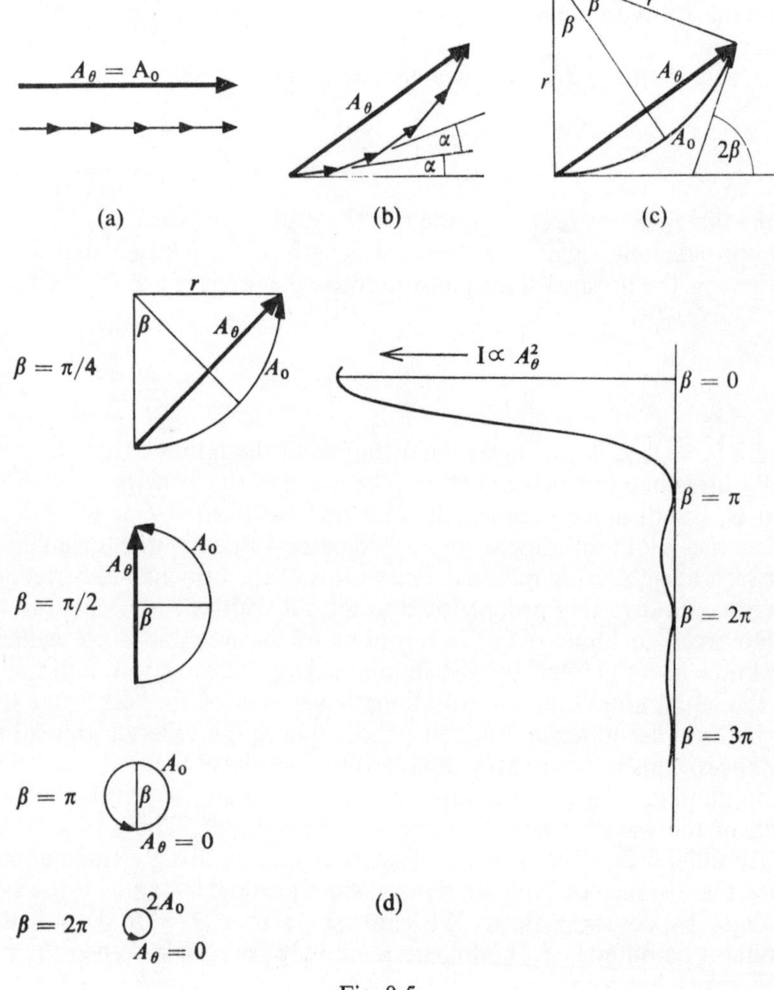


Fig. 9-5

Also, it will be seen from Fig. 9-5(c) that

$$\sin \beta = \frac{A_\theta/2}{r}$$

i.e.,

$$A_\theta = 2r \sin \beta \tag{9-8}$$

Also

$$\frac{\text{chord}}{\text{arc}} = \frac{A_\theta}{A_0} = \frac{2r \sin \beta}{2\beta r}$$

$$= \frac{\sin \beta}{\beta} \tag{9-9}$$

so that the intensity $I$ is proportional to

$$A_0^2 \frac{\sin^2 \beta}{\beta^2} \tag{9-10}$$

From Eqn (9-1), the path difference for those wavelets travelling from the top and bottom edges of the slit to a point on the screen is $a \sin \theta$. Therefore, the phase difference between these wavelets on arrival at the screen is [Eqn (8-11)]

$$\frac{2\pi}{\lambda} \cdot (\text{path difference}) = \frac{2\pi}{\lambda} a \sin \theta = 2\beta$$

i.e.,

$$\beta = \frac{\pi a \sin \theta}{\lambda} \tag{9-11}$$

Thus Eqn (9-10), with (9-11), gives the variation of intensity across the screen with direction $\theta$, and Eqn (9-7) gives the linear positions of the minima on the screen, as measured from the axis.

## 9-3 Diffraction by a circular aperture

Fraunhofer diffraction due to a circular aperture is of greater importance that that due to a slit, because most optical instruments involve circular apertures and it is diffraction that limits their resolving power. Unfortunately, it is a far more difficult problem mathematically and consequently cannot be dealt with fully here. The diffraction pattern formed has rotational symmetry with a bright disc-like area centred on the axis, surrounded by light and dark rings. The central bright region or principal maximum is known as *Airy's disc* and contains approximately 84% of the light energy transmitted by the aperture. The first bright ring, that is, the second maximum, contains only 7%, with successive maxima having 3%, 1·5%, respectively.

Table 9-1

| | *Circular aperture* | | *Slit aperture* | |
|---|---|---|---|---|
| | $m'$ | *Relative intensity* | $m$ | *Relative intensity* |
| Principal maximum | 0 | 1 | 0 | 1 |
| 1st minimum | 1·22 | 0 | 1 | 0 |
| 2nd maximum | 1·64 | 0·0175 | 1·43 | 0·0472 |
| 2nd minimum | 2·23 | 0 | 2 | 0 |
| 3rd maximum | 2·67 | 0·0042 | 2·46 | 0·0168 |
| 3rd minimum | 3·24 | 0 | 3 | 0 |
| 4th maximum | 3·69 | 0·0016 | 3·47 | 0·0083 |
| 4th minimum | 4·24 | 0 | 4 | 0 |
| 5th maximum | 4·72 | 0·0008 | 4·48 | 0·0050 |
| 5th minimum | 5·24 | 0 | 5 | 0 |

To sum the respective phases and amplitudes over the area of the screen involves Bessel functions. This summation was carried out by Airy in 1835 and showed that the successive maxima and minima are formed in a direction $\theta$ with the axis, where

$$\sin \theta = \frac{m'\lambda}{a} \tag{9-12}$$

where $a$ is now the diameter of the circular aperture. $m'$ no longer has integral values for the minima, which are also no longer equally spaced, but has the values given in Table 9-1 for the first few rings. Table 9-1 also gives comparative values of $m$ for diffraction by a single slit.

## 9-4 Rayleigh criterion for resolution

At the limit of resolution of an instrument a small object can at the best be seen only as a diffraction image of a central disc surrounded by circular fringes. If two neighbouring small objects are viewed through a circular aperture by, say, a telescope, they each produce their own diffraction patterns. If the two objects come closer together, so do their diffraction patterns, until these eventually merge. When the two Airy discs begin to merge, they first form a dumb-bell shape. If the objects come any closer together, the Airy discs appear as a slightly elongated region and an observer is unable to decide whether there were originally two point objects or one elongated object.

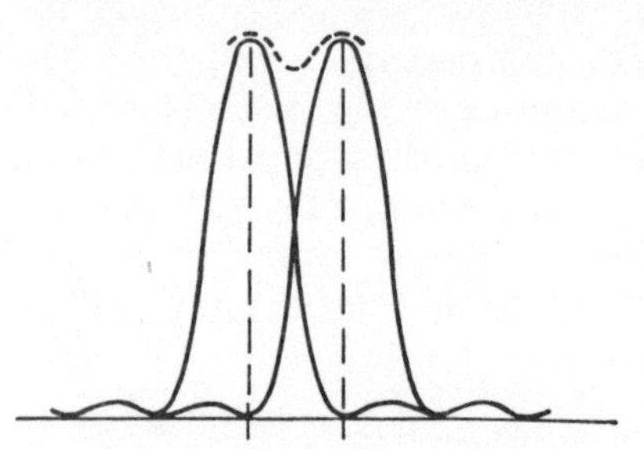

Fig. 9-6 Rayleigh criterion for resolution

Rayleigh suggested the criterion that two point objects, for example, two neighbouring stars, would just be resolvable as two objects if the two diffraction patterns overlapped such that the central maximum of one pattern coincided with the first minimum of the second pattern (Fig. 9-6). The dotted line shows the resultant intensity due to both patterns together. In practice this criterion agrees very well with observed limits of resolution, but only for two point objects that are equally bright. To resolve detail in an extended object contrast is important and the Rayleigh criterion may not apply.

The Rayleigh criterion is thus a rough, but nevertheless very useful, measure of the limit of resolution that may be applied to all types of optical instruments. If the resolution limit is set by diffraction, no more information can be obtained by magnifying the image further.

## 9-5 Resolving power of telescopes and microscopes

Table 9-1 shows that the angular radius $\theta$ of the first dark ring, or minimum, in the diffraction pattern due to a circular aperture is given by

$$\sin \theta = \frac{1{\cdot}22\lambda}{a} \tag{9-13}$$

where $a$ is the diameter of the aperture, in this case the objective of the telescope. Generally $\theta$ is small and can therefore replace the sine. For light of wavelength, say 5,000 Å, the angular radius of the first minimum is

$$\theta = \frac{1{\cdot}22 \times 5 \times 10^{-5}}{a}$$

$$= \frac{6{\cdot}10 \times 10^{-5}}{a} \text{ rad}$$

where $a$ is measured in centimetres. By the Rayleigh criterion this is also the angular separation of two just-resolvable point objects, since it is the angle between the two principal maxima, as the principal maximum of one pattern is to coincide with the first minimum of the other.

Thus a telescope with an objective 6 cm in diameter is capable, if sufficiently free from aberrations, of resolving two stars of angular separation

$$\frac{6{\cdot}10 \times 10^{-5}}{6} = 1{\cdot}01 \times 10^{-5} \text{ rad}$$

$$= 2{\cdot}1 \text{ seconds of arc}$$

A star may be treated as a point object, since the angular diameters of stars are generally less than about 0·05″—much less than the angular diameter of the Airy disc. If this telescope objective has a focal length of 50 cm, the Airy disc has a linear radius of $1{\cdot}01 \times 10^{-5} \times 50 = 5{\cdot}05 \times 10^{-4}$ cm, that is, a diameter of 0·010 mm. Thus a single point object such as a star would be imaged as a disc of diameter 0·01 mm, whilst two stars could just be resolved when the distance between the centres of their images is 0·005 mm. The eyepiece only magnifies these distances, without any additional information being gained.

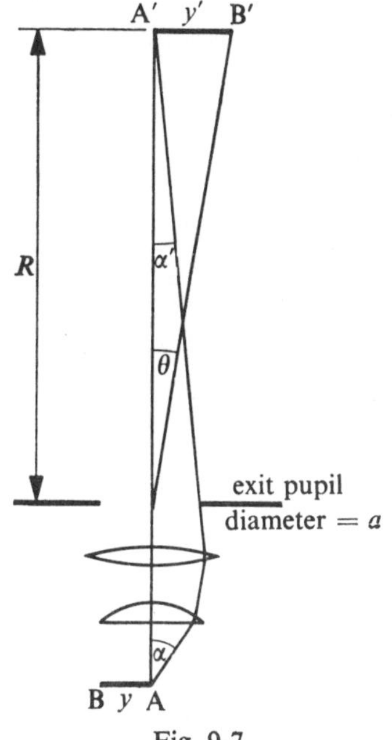

Fig. 9-7

The resolving power of the microscope may be determined by a similar application of the Rayleigh criterion. With a microscope we are interested in the smallest linear separation between two small objects very close to the front of the objective. The objective therefore subtends a large angle at the object plane. In Fig. 9-7, $a$ is the diameter of the exit pupil of the objective and is a distance $R$ from the objective image plane. Since the distance $R$ is large compared with the distance between the objective exit pupil and the objective or the object plane, this exit pupil may be regarded as the aperture which produces the diffraction. Two small self-luminous point objects A and B situated a distance $y$ apart are imaged as two circular diffraction patterns centred at A′ and B′, a distance $y'$ apart. When A and B are only just resolvable as two point objects, the central maximum of one diffraction pattern, say, at A′, must coincide with the position of the first minimum of the other pattern centred at B′. That is, the distance A′B′ = $y'$ must be equal to the radius of Airy's disc for A and B to be just resolvable. The angular radius of the Airy disc is given by

$$\sin\theta = \theta = \frac{1{\cdot}22\lambda}{a}$$

from Eqn (9-13), since $\theta$ is small. Also, from Fig. 9-7

$$\theta = \frac{y'}{R} \tag{9-14}$$

and therefore

$$y' = \frac{1{\cdot}22\lambda R}{a} \tag{9-15}$$

If the extreme rays through the objective make angles $\alpha$ and $\alpha'$ with the axis (Fig. 9-7), then $\alpha' = (a/2)/R$ and therefore

$$y'\alpha' = 0{\cdot}61\lambda \tag{9-16}$$

Also, by the sine condition of Section 4-3

$$ny\sin\alpha = y'\sin\alpha' \tag{9-17}$$

where $n$ is the refractive index of the object space (about 1·5 for an oil immersion objective) and where the image space is in air. The distance between the objects A and B for them to be just resolvable is then

$$y = \frac{y'\sin\alpha'}{n\sin\alpha}$$

i.e.,

$$= \frac{y'\alpha'}{n\sin\alpha} \tag{9-18}$$

since in this case the angle $\alpha'$ is small.

$$\therefore\; y = \frac{0{\cdot}61\lambda}{n\sin\alpha}$$

$$= \frac{0{\cdot}61\lambda}{NA} \tag{9-19}$$

where $NA$ is the numerical aperture. This equation for resolving power has already been used in Section 5-11.

In deriving Eqn (9-19) we assumed that the point objects A and B were self-luminous. In practice objects in microscopy are not self-luminous but are illuminated by light from a condenser. Abbe investigated this different condition and concluded that the factor of 0·61 in Eqn (9-19) should be 0·5.

## 9-6 Comparison of single- and double-slit diffraction

When the Fraunhofer diffraction pattern due to two parallel slits of equal width is observed by means of a lens and screen, it is seen that the intensities in the pattern of parallel fringes appear to be modulated. Alone each slit would produce the same single-slit pattern as the other. When both slits are used the pattern does not remain the same but with double the intensity, but breaks up into narrower fringes caused by interference between the two diffracted beams. This pattern of narrow interference

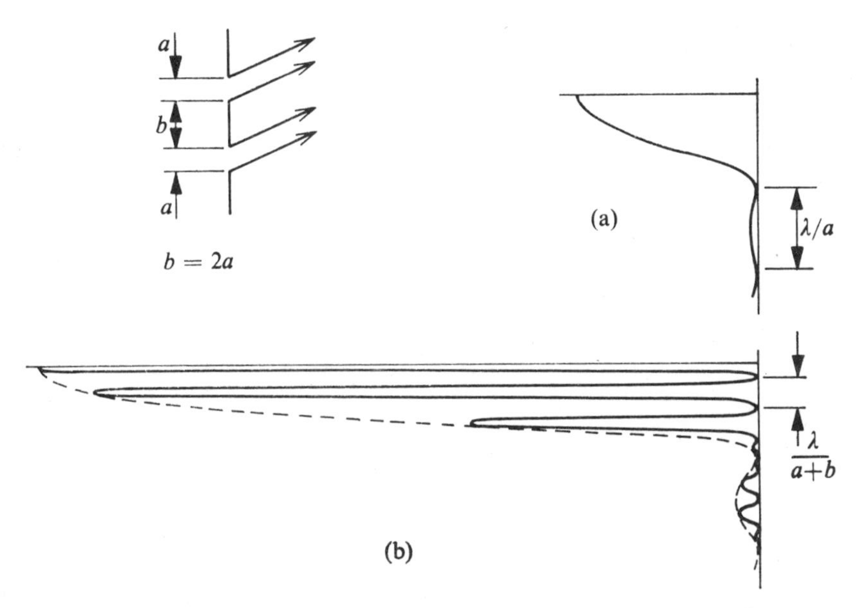


Fig. 9-8 Intensity distributions in (a) single-slit and (b) double-slit diffraction patterns

fringes has the general envelope shape of the single-slit pattern, with the principal minima in the same positions. The intensity of each of the maxima, however, is four times that of the single-slit pattern at the corresponding point. Figure 9-8 shows intensity distribution curves for single- and double-slit diffraction, where $a$ is the slit width and $b$ is the distance between the slits. Only half of each pattern is shown.

Young used two slit apertures to produce interference. To demonstrate interference rather than diffraction the slit widths should be very small, comparable to the wavelength of light, so that the central diffraction maximum from each slit subtends a large angle which may contain a large number of interference fringes. As the slit widths increase the diffraction pattern becomes narrower and modulates the interference fringes.

## 9-7 Diffraction grating

If the number of slit apertures is increased so that a grating is formed, the Fraunhofer diffraction pattern changes. The fringes become much sharper so that a grating with a dozen or so slits produces fringes like lines. There is little difference in intensity between the fringes in the central region so that the central maximum differs only very slightly from its neighbours. These fringes are all called the principal maxima. A number of secondary maxima are formed between each of the principal maxima, depending on the number of slits. All the fringes still lie within a general envelope shape

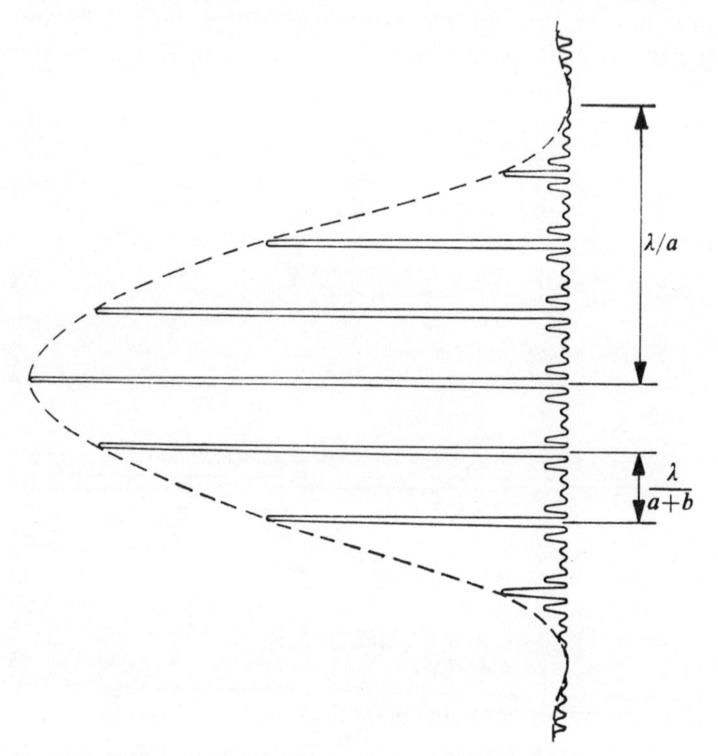


Fig. 9-9 Intensity distribution across a six-slit diffraction pattern

having the same width as the intensity distribution due to one slit (Fig. 9-9). The actual intensity distribution may be determined by the method used for the single-slit pattern.

Figure 9-10 shows an arrangement of slits for producing Fraunhofer diffraction; the rays are parallel and therefore a lens must be used to focus the diffraction pattern onto a screen or alternatively a telescope may be used to receive the rays. Rays that are diffracted in a direction $\theta = 0$—that is, rays travelling in a direction normal to the slit plane—all arrive at the screen or at the focal plane of the viewing telescope in phase. Along a line on the screen parallel to the slits the contributions from each of the slits are directly additive. Thus, if there are $N$ evenly spaced slits of constant width each contributing a wave of resultant amplitude $A_0$ in a direction normal to the plane of the grating, the resultant amplitude at the centre of the diffraction pattern on the screen is $A_0N$. At points on the screen on either side of the axis, contributions from the different slits have travelled different distances and hence have different phases, which must be taken into account in determining the resultant amplitude and hence intensity. From Fig. 9-10, the path difference between corresponding rays from successive slits in a direction $\theta$ to the axis is $(a + b) \sin \theta$. This gives a phase difference $\delta$ between these rays. From Eqn (8-11)

$$\delta = \frac{2\pi}{\lambda} \cdot (\text{path difference})$$

$$= \frac{2\pi}{\lambda} (a + b) \sin \theta \tag{9-20}$$

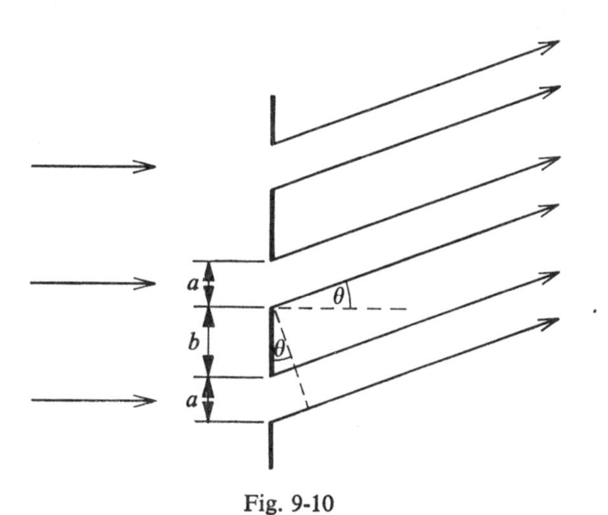


Fig. 9-10

Taking this phase difference, $\delta$, into account, the amplitude contributions in a direction $\theta$ from each slit may be summed vectorially as for a single slit.

The amplitude contributions, $A_\theta$, and phase differences, $\delta$, are shown in Fig. 9-11, where QT represents the total amplitude contribution from all the slits. Then since

$$\angle\text{RQO} = \angle\text{QRO} = \angle\text{SRO} = \psi$$

$$\delta = 180° - 2\psi \tag{9-21}$$

and

$$\angle\text{QOR} = 180° - 2\psi \tag{9-22}$$

hence

$$\angle\text{QOR} = \delta \tag{9-23}$$

and

$$\angle\text{QOP} = \frac{N}{2} \delta \tag{9-24}$$

where $N$ is the total number of slits, and hence 'chords' in the vector diagram. Thus the resultant amplitude $A$ is equal to

$$\mathrm{QT} = 2r \sin\left(\frac{N\delta}{2}\right) \tag{9-25}$$

Also, applying the sine rule to triangle QOR

$$\frac{\sin \angle \mathrm{QOR}}{\mathrm{QR}} = \frac{\sin \angle \mathrm{QRO}}{\mathrm{QO}}$$

i.e.,

$$\frac{\sin \delta}{A_\theta} = \frac{\sin \psi}{r}$$

$$= \frac{\cos (90° - \psi)}{r}$$

i.e., from Eqn (9-21)

$$\frac{\sin \delta}{A_\theta} = \frac{\cos (\delta/2)}{r} \tag{9-26}$$

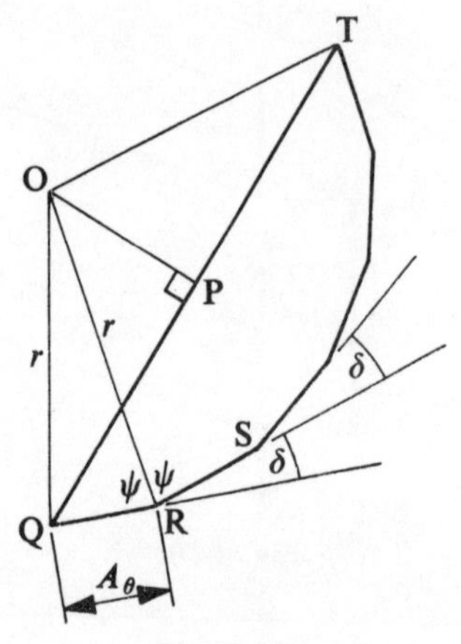


Fig. 9-11

The resultant amplitude $A$ can thus be written as

$$A = 2A_\theta \frac{\cos (\delta/2)}{\sin \delta} \sin \frac{N\delta}{2} \tag{9-27}$$

by substituting for $r$ from Eqn (9-26) in Eqn (9-25). Also,

$$\sin \delta = 2 \sin \frac{\delta}{2} \cos \frac{\delta}{2}$$

and therefore

$$A = A_\theta \frac{\sin (N\delta/2)}{\sin (\delta/2)} \tag{9-28}$$

where $\delta$ is the phase difference between corresponding successive rays, that is [Eqn (9-20)]

$$\delta = \frac{2\pi}{\lambda}(a + b)\sin\theta$$

$A_\theta$, however, is the resultant amplitude contribution in the direction $\theta$ from a single slit, as expressed in Eqn (9-9).

$$\therefore\ A = A_0 \frac{\sin\beta}{\beta}\frac{\sin(N\delta/2)}{\sin(\delta/2)} \tag{9-29}$$

where $2\beta$ is the phase difference between wavelets originating from the two edges of a single slit and travelling in a direction $\theta$, so that, from Eqn (9-11):

$$\beta = \frac{\pi a \sin\theta}{\lambda}$$

$A_0$ is the resultant amplitude, due to a single slit of width $a$, of the wavelets travelling in a direction normal to the plane of the slit. The intensity on the screen in a direction $\theta$ is then proportional to the square of the resultant amplitude, that is, to

$$A_0^2 \frac{\sin^2\beta}{\beta^2}\frac{\sin^2(N\delta/2)}{\sin^2(\delta/2)} \tag{9-30}$$

In particular it will be seen from Eqn (9-30) that for a single slit ($N = 1$) the intensity varies as

$$A_0^2 \frac{\sin^2\beta}{\beta^2}$$

as already shown [Eqn (9-10)]. Similarly, for a double slit ($N = 2$)

$$\text{intensity} \propto A_0^2 \frac{\sin^2\beta}{\beta^2}\frac{\sin^2\delta}{\sin^2(\delta/2)}$$

i.e.,

$$\text{intensity} \propto 4A_0^2 \frac{\sin^2\beta}{\beta^2}\cos^2\frac{\delta}{2} \tag{9-31}$$

since

$$\sin\delta = 2\sin\frac{\delta}{2}\cos\frac{\delta}{2}$$

Equation (9-30) shows that the intensity distribution depends on two factors: the first, $(\sin^2\beta)/\beta^2$, governs the intensity distribution pattern due to diffraction from a single slit; and the second, $\sin^2(N\delta/2)/\sin^2(\delta/2)$, governs the interference pattern due to the $N$ slits. In particular, the second term is mathematically the more quickly varying term and it is this term that leads to the sharpening of the fringes. That is, it is the effect of the number, width, and spacing of the slits that governs the sharpness.

As $\delta/2$ approaches the values $0, \pi, 2\pi, \ldots$

$$\frac{\sin^2(N\delta/2)}{\sin^2(\delta/2)} \to \frac{N^2\delta^2/4}{\delta^2/4} = N^2 \tag{9-32}$$

That is, the *interference* term in the intensity distribution [Eqn (9-30)] has its maximum values equal to $N^2$. The condition for this interference maximum from Eqn (9-20)

i.e., $$\frac{\delta}{2} = \frac{\pi}{\lambda}(a + b)\sin\theta = 0, \pi, 2\pi, \ldots \tag{9-33}$$

thus becomes

$$(a + b)\sin\theta = m\lambda \tag{9-34}$$

which is then the condition for the formation of a *principal maximum* in a direction $\theta$ with the normal to the plane of the grating, where $m = 0, 1, 2, \ldots$.

Also, $\dfrac{\sin^2(N\delta/2)}{\sin^2(\delta/2)}$ becomes zero when $N\delta/2 = 0, \pi, 2\pi, \ldots$. That is, from Eqn (9-20),

$$N\frac{\delta}{2} = N\frac{\pi}{\lambda}(a + b)\sin\theta = 0, \pi, 2\pi, \ldots \tag{9-35}$$

Thus for a minimum of the interference term

$$(a + b)\sin\theta = 0, \frac{\lambda}{N}, \frac{2\lambda}{N}, \ldots, \frac{(N-1)\lambda}{N}, \frac{N\lambda}{N}, \frac{(N+1)\lambda}{N}, \ldots \tag{9-36}$$

The values $0, N\lambda/N, 2N\lambda/N, \ldots$ also satisfy the conditions for principal maxima to occur, so that the overall effect is to reduce the principal maxima at these positions. The condition for a *minimum* to occur, due to the interference term, is thus, from Eqn (9-36)

$$(a + b)\sin\theta = \frac{p\lambda}{N} \tag{9-37}$$

where $p = 1, 2, 3, \ldots$ excluding the values $p = N, 2N, 3N, \ldots$.

By Eqn (9-34), principal maxima occur whenever $m = N/N, 2N/N, 3N/N, \ldots$ therefore, by Eqn (9-37), there are $(N - 1)$ points of zero intensity between successive principal maxima, which implies that between these minima the intensity must rise again, giving *secondary maxima*.

As well as the quickly varying interference term, there is the slowly moving *diffraction* term, $(\sin^2\beta)/\beta^2$. This governs the general envelope shape and gives a minimum of intensity whenever $\beta$ has a value $\pi, 2\pi, 3\pi, \ldots$. $\beta = 0$ is, of course, the condition for the central maximum. Thus the condition for an intensity minimum due to the diffraction term is, from Eqn (9-11)

$$\beta = \frac{\pi a \sin\theta}{\lambda} = \pi, 2\pi, 3\pi, \ldots$$

i.e., $$a\sin\theta = p\lambda \tag{9-38}$$

where $p = 1, 2, 3, \ldots$. Since the diffraction term gives the envelope shape it means that when an interference maximum happens to coincide with a minimum due to the diffraction term, the diffraction term is over-riding and a *missing order* occurs.

Thus, to summarize, the intensity variation with direction $\theta$ to the normal to the grating is [Eqn (9-30)]

$$I \propto A_0^2 \frac{\sin^2 \beta}{\beta^2} \frac{\sin^2 (N\delta/2)}{\sin^2 (\delta/2)}$$

where $\beta$ is defined by Eqn (9-11) and $\delta$ by Eqn (9-20). In particular, a principal maximum occurs in a direction $\theta$ whenever [Eqn (9-34)]

$$(a + b) \sin \theta = m\lambda$$

where $m = 0, 1, 2, \ldots$ with an intensity minimum occurring in a direction $\theta$ such that [Eqn (9-37)]

$$(a + b) \sin \theta = \frac{p\lambda}{N}$$

where $p = 1, 2, 3, \ldots$. There are thus $N - 1$ points of zero intensity between successive principal maxima, leading to secondary maxima between the zero intensity points. In addition minima occur when [Eqn (9-38)]

$$a \sin \theta = p\lambda$$

where $p = 1, 2, 3, \ldots$, due to the diffraction term. This is an over-riding effect and may cause missing orders.

## 9-8 Formation of spectra by a grating

Diffraction gratings are principally used as an alternative means of forming spectra. They have the advantage over prisms that absolute measurements of wavelength are possible without the need for calibration. When a slit aperture and a collimator are placed in front of a diffraction grating with the grating rulings parallel to the slit, principal maxima appear as sharp lines when viewed by a telescope. This is the same as replacing a prism on a spectrometer with a grating and using the spectrometer in the usual way. Equation (9-34) gives the directions $\theta$ of the principal maxima for light incident normally onto the grating. For the general case of light incident at an angle $i$ with the normal to the grating, Eqn (9-34) may be written in the form

$$(a + b)(\sin i + \sin \theta) = m\lambda \qquad (9\text{-}39)$$

where $m = 0, 1, 2, \ldots$. This is the *grating equation*. With this, knowing the angle $i$, which may conveniently be made zero, and the grating space $(a + b)$, the wavelength $\lambda$ of light can be determined by measuring the angle $\theta$ with a spectrometer. The order, $m$, of the spectrum must be determined by inspection, starting with $\theta$ as a small angle. If light of several wavelengths is present, the zero-order fringes, when $m = 0$, coincide on the axis to form the central fringe. For a white-light source, this central fringe is white, with coloured bands on either side as each wavelength satisfies the grating equation for non-zero values of $m$.

## 9-9 Dispersion by a grating

The angular dispersion of a grating is the rate of change of angle, $\theta$, with change in wavelength, $\lambda$. Obviously, the greater this dispersion, the greater is the separation between two spectral lines due to slightly different wavelengths—that is, the greater the resolution. More resolution provides more accuracy in the measurement of wavelength. To find the dispersion it is necessary to differentiate the grating equation (9-39), with respect to $\lambda$, where angle $i$ is a constant and independent of $\lambda$. Therefore

$$(a + b) \cos \theta \, \mathrm{d}\theta = \mathrm{m} \, \mathrm{d}\lambda$$

i.e.,

$$\frac{\mathrm{d}\theta}{\mathrm{d}\lambda} = \frac{m}{(a + b) \cos \theta} = \text{angular dispersion} \qquad (9\text{-}40)$$

This means that the angular separation $\mathrm{d}\theta$ between two spectral lines with a difference in wavelength of $\mathrm{d}\lambda$, is directly proportional to the order $m$. Thus, although we may not be able to resolve two lines in one order, by swinging the spectrometer telescope round it may be possible to find a higher order in which they can be resolved.

Since several orders are possible, the set of lines due to one order may overlap those due to another. Although this improves resolution because higher orders may be used, it may lead to confusion of orders and great care must be taken in deducing the correct order for a particular line. The colour of the line helps in direct observation, but not if, as usual, the spectrum is recorded photographically in monochrome. For example, a third-order, red spectral line of wavelength 7,500 Å would be in a direction $\theta$ such that

$$(a + b)(\sin i + \sin \theta) = 3 \times 7{\cdot}5 \times 10^{-5}$$

But this direction $\theta$ may also give the position of another line in the fourth order such that

$$3 \times 7{\cdot}5 \times 10^{-5} = 4 \times 5{\cdot}625 \times 10^{-5}$$

that is, a fourth-order, yellow line of wavelength 5,625 Å. Similarly, at the same position a fifth-order, blue line of 4,500 Å could occur. With visible light there is, of course, no overlapping in the first and second orders since the red end of the first-order spectrum with wavelength 8,000 Å just meets the violet end of the second-order spectrum with a wavelength of 4,000 Å.

From Eqn (9-40) the angular separation $\mathrm{d}\theta$ is inversely proportional to the grating space $(a + b)$ and hence the number of slits for a grating should be as large as practical. Also, providing $\theta$ is small, $\cos \theta$ does not vary significantly from unity. Thus, if measurements are made at small angles to the axis, the change in angles is directly proportional to the changes in wavelength within a particular order $m$. This is a *normal spectrum* and has an obvious advantage over that obtained with a prism spectrometer which has a very non-linear scale. Whether the advantages of the linear scale, absolute measurement of wavelength, and variable resolution depending on order outweigh the disadvantage of overlapping of orders, with the possible confusion of lines, depends on the complexity of the spectrum under investigation.

So far we have discussed transmission gratings. *Reflection gratings* work on exactly the same principle of introducing constant phase differences between successive rays. If the gratings are in a step form, termed 'echelette gratings' (Fig. 9-12), most of the energy can be concentrated into one particular order. This avoids any danger of confusing orders.

Gratings are now manufactured by a replica technique proposed by T. Merton in 1948 and developed at the National Physical Laboratory, Teddington. A fine helix, which has been turned on a polished cylinder by a lathe, is used as the lead-screw of the lathe and combined with a long 'nut' made from pith clamped between metal shells. The soft 'nut'

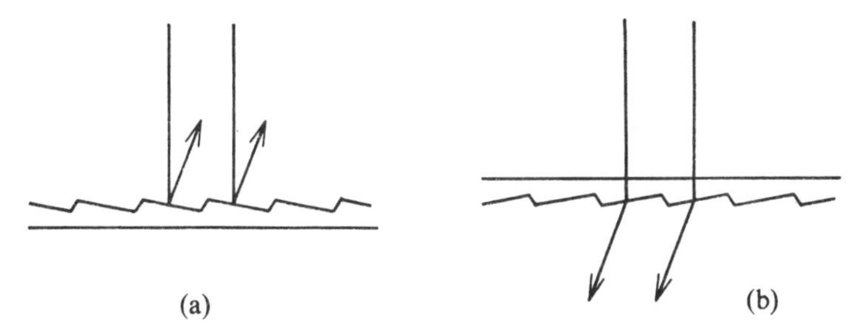


Fig. 9-12 (a) Reflection and (b) transmission gratings blazed to concentrate light into a single order

evens out periodic errors due to the original turning operation. With suitable gearing this lead-screw is used to cut a new helix of much finer pitch onto another polished cylinder, which in turn becomes the lead-screw for cutting an even finer pitched helix. This process may be repeated until a sufficiently fine, even-pitched helix is obtained. The required step form of the helix is determined by the form of the diamond cutting tool, which in this case does not remove a shaving as in conventional turning, but displaces the metal into the correct form. Having obtained the cylinder, a thin plastic replica is made by coating the cylinder, allowing the plastic to cure and cutting longitudinally to remove. This in turn is replicated in gelatine which is then hardened and mounted on a glass plate. This transmission grating may be converted to a reflection grating, by coating it with aluminium by vacuum evaporation.

## 9-10 Resolving power of gratings and prisms

The resolving power of a grating or prism is its ability to distinguish two closely adjacent spectral lines as separate lines. If two lines have wavelengths $\lambda$ and $\lambda + \delta\lambda$ and an angular separation $\delta\theta$, then by the Rayleigh criterion they are just resolvable when the diffraction maximum due to one wavelength coincides with the first diffraction minimum due to the other wavelength; that is, if the spectral line due to the wavelength $\lambda$ and in a direction $\theta$ coincides with a minimum occurring immediately adjacent to a spectral line of wavelength $\lambda + \delta\lambda$ and of the same order $m$ [Fig. 9-13(a)].

The angular separation $\delta\theta$ between the just-resolvable spectral lines is [Fig. 9-13(a)] the angular distance between a spectral line and its adjacent minimum. If $\theta$ is the direction of this spectral line and $\theta + \delta\theta$ is the direction of its adjacent minimum [Fig. 9-13(b)], the difference between the extreme path differences is $\lambda$. That a change in the extreme path difference of $\lambda$ causes a change from a maximum of intensity to a minimum has already been shown (Section 9-2). The path difference between rays from

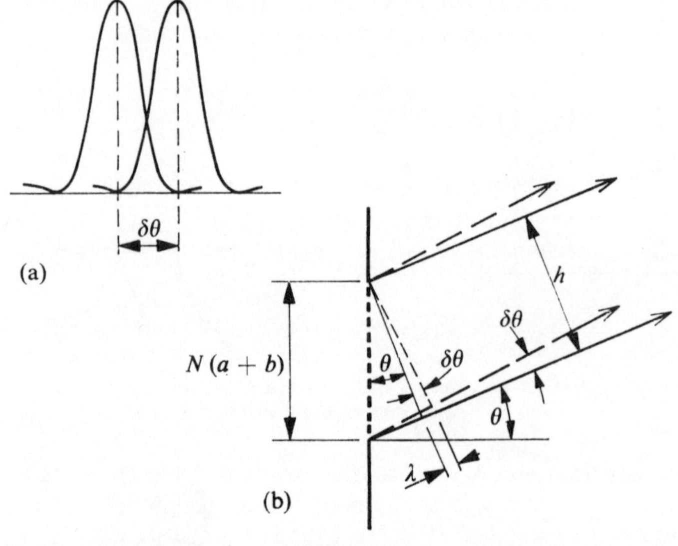


Fig. 9-13

the top edge of the grating and from the middle is $\lambda/2$ and so their effects cancel, similarly for the pair of corresponding rays just below these, and so on for all the pairs right across the grating. Thus, from Fig. 9-13(b)

$$\delta\theta = \frac{\lambda}{h} \tag{9-41}$$

for the spectral lines to be just resolvable, and where $h$ is the effective aperture of the grating. The difference in the wavelengths associated with this angle $\delta\theta$ is

$$\delta\lambda = \delta\theta \frac{d\lambda}{d\theta} \tag{9-42}$$

$$= \frac{\lambda}{h} \cdot \frac{(a + b)\cos\theta}{m} \tag{9-43}$$

by substituting from Eqns (9-41) and (9-40).

Also

$$h = N(a + b)\cos\theta \tag{9-44}$$

where $(a + b)$ is the grating space and $N$ is the number of rulings. Therefore, substituting in Eqn (9-43)

$$\delta\lambda = \frac{\lambda}{Nm}$$

i.e.,

$$\frac{\lambda}{\delta\lambda} = Nm \tag{9-45}$$

where $m$ is the order of the spectrum and $\lambda/\delta\lambda$ is the resolving power.

Thus, for example, the sodium D-lines with wavelengths 5,890 and 5,896 Å require a resolving power, $\lambda/\delta\lambda$, of approximately 1,000. To resolve these lines requires a grating with a total number of 1,000 lines in the first order, but only 500 in the second order. Very high resolving powers may be obtained using diffraction gratings. For example, a 5 cm grating with 10,000 lines to the centimetre has a resolving power of 150,000 in the third order.

We can compare the resolving power of a diffraction grating with that of a prism as follows. In Fig. 9-14 a plane wave front of height $h$ is incident

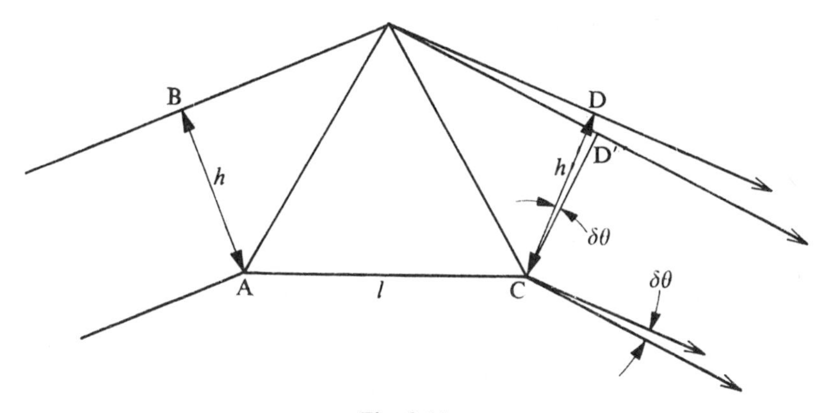


Fig. 9-14

onto a prism of base length $l$ such that the beam passes through at minimum deviation. If the incident wave front is composed of wavelengths $\lambda$ and $\lambda + \delta\lambda$, then refraction by the prism produces two wave fronts, inclined to each other with an angle $\delta\theta$. If the prism material has refractive indices $n$ and $n + \delta n$ for light of wavelengths $\lambda$ and $\lambda + \delta\lambda$ respectively, then the maximum optical path difference for rays through the prism is

$$l(n + \delta n) - ln = l\,\delta n$$

But from Fig. 9-14, this optical path difference is also equal to DD′—that is, $h\,\delta\theta$. Therefore,

$$\delta\theta = \frac{l}{h}\,\delta n \tag{9-46}$$

The prism may also be regarded as a single diffraction slit of width $h$. Then, if $\delta\theta$ is the angular separation of two spectral lines of wavelength $\lambda$ and $\lambda + \delta\lambda$, which are just resolvable—that is, $\delta\theta$ is the angle between the central maximum and the first minimum of the diffraction pattern—then by Eqn (9-41)

$$\delta\theta = \frac{\lambda}{h}$$

and hence by Eqn (9-46)

$$\lambda = l\,\delta n \tag{9-47}$$

Hence the resolving power is

$$\frac{\lambda}{\delta\lambda} = l\frac{\mathrm{d}n}{\mathrm{d}\lambda} \tag{9-48}$$

$\mathrm{d}n/\mathrm{d}\lambda$ is normally a negative quantity but must be treated here as positive since the resolving power is regarded as a positive number.

For ordinary flint glass $\mathrm{d}n/\mathrm{d}\lambda$ is of the order of 1,000 if $\lambda$ is measured in centimetres; that is, the resolving power is approximately a thousand times the length of the prism base in centimetres. Thus a prism with a base length of 1 cm can resolve the two sodium D-lines, providing the spectrometer telescope has sufficient resolution and the other optical components are sufficiently free from aberrations.

## PROBLEMS

**9-1** What is meant by (a) interference, (b) diffraction, of light? What part does each play in the production of spectra by a diffraction grating? Compare the spectrum produced by a grating with that due to a prism.

**9-2** A parallel beam of light of wavelength 5,461 Å is incident onto a slit of width 0·2 mm. Immediately in front of the slit is a converging lens of focal length 1 m. Explain the distribution of light intensity in the focal plane of the lens and calculate the width of the central bright fringe. [5·4 mm]

**9-3** Discuss the Rayleigh criterion for resolution. Explain the term *resolving power* as applied to (a) microscopes, (b) diffraction gratings, and derive expressions for its value.

**9-4** A beam of mercury light is incident normally onto a diffraction grating. Calculate the number of lines per centimetre of the grating, if the angle between the two first-order 5,461-Å lines on either side of the normal is 24° 12′.

If the grating is 10 cm wide, what is the smallest wavelength interval to an adjoining spectral line that could be resolved in this order? [3,838; 0·142 Å]

**9-5** Sketch the intensity distribution of the Fraunhofer diffraction pattern obtained with seven equally spaced slits, if the opaque intervals between the slits are equal to three times the width of each slit. What determines (a) the positions, (b) the widths, of the principal maxima?

**9-6** Light of wavelength $\lambda$ is incident normally onto a diffraction grating. If the first-order spectrum lines occur at $\pm 12°$, what will be the angular positions if the incident light is at an angle of (a) 5°, (b) 25° to the normal to the grating plane. What effect will the obliquity factor have?

[(a) 6° 56′, −17° 10′; (b) −12° 24′, −39° 5′]

**9-7** For a particular grating the fifth-order blue line of wavelength 4,600 Å coincides with a fourth-order, yellow line. What is the wavelength of the yellow line? Show that this is not possible with a 5,000-line/cm grating with the incident ray normal to the grating. [5,750 Å]

**9-8** Derive from first principles Eqn (9-31) for the intensity distribution on a screen due to Fraunhofer diffraction by two narrow slits.

**9-9** What is meant by the resolving power of a telescope? Calculate the minimum angle of resolution of a telescope with an objective whose clear aperture has a diameter of 4 cm. Assume $\lambda = 5{,}500$ Å.

If the objective has a focal length of 50 cm and the limit of resolution of the eye is taken as 100 seconds of arc, find the focal length of the eyepiece necessary for clear resolution of the image. What is then the magnification of the telescope? [3·5″, 17 mm, ×29·4]

**9-10** A glass prism has refractive indices 1·638 and 1·621 for the C and F wavelengths, 6,563 and 4,861 Å respectively. Determine the base length of the prism necessary to resolve the sodium D-lines of wavelengths 5,890 and 5,896 Å. [9·83 mm]

# 10 Polarization

## 10-1 Introduction

Interference and diffraction result from the interaction of light waves. In our discussion of these phenomena we consided only a few simple characteristics of light waves, such as amplitude and wavelength. It was not necessary to discuss the actual form of the wave. For this reason interference and diffraction may readily be observed in sound waves for example.

Polarization requires more than just a wave for its explanation; it requires that the wave be transverse. A *transverse wave* is one in which the vibrations forming the wave are perpendicular to the direction of propagation. Sound waves are *longitudinal*—that is, the vibrations are parallel to the direction of propagation—and therefore cannot be polarized. Light waves, however, are transverse. Light is an electromagnetic radiation obeying Maxwell's equations, which specifically demand transverse waves.

Electromagnetic waves are due to the vibrations of charged particles within the molecules making up the source. Vibrations of the correct frequencies produce electromagnetic waves in the form of visible light. The light wave sent out by one molecule may be considered to be *plane polarized*, since vibrations occur in only one plane containing the direction of propagation. For a large number of molecules, each of the vibrations is normally in a random direction. If a beam of light is made up of vibrations in all directions perpendicular to its direction of propagation, so that there is no preponderance of vibrations in any particular transverse direction, the light is said to be *unpolarized*. But if there is any dissymmetry, the beam is said to be *polarized*. Figure 10-1(a), for example, represents a *plane polarized* wave. The vibration in Fig. 10-1(b) is *circularly polarized*; this may be right-handed, as in the figure, or left-handed. Here the direction of the transverse vibration rotates about the direction of propagation and each vibration has a constant phase difference with the adjoining vibration. The intersection points of the helix with a perpendicular plane as the helix moves forward form a circle. If the helix is flattened, the intersection points with the plane form an ellipse and the light is then *elliptically polarized*. Again it may be right- or left-handed. This however, is only a *representation* of circular and elliptical polarization. The combination of wave motions into a circular or elliptical resultant will be discussed in detail in Sections 10-5 and 11-10.

A device that eliminates certain transverse vibrations to produce polarized light is a *polarizer*. Similarly, a device that detects polarized light is an *analyser*. Polarized light is produced by reflection; by means of dichroic crystals; and by double refraction.

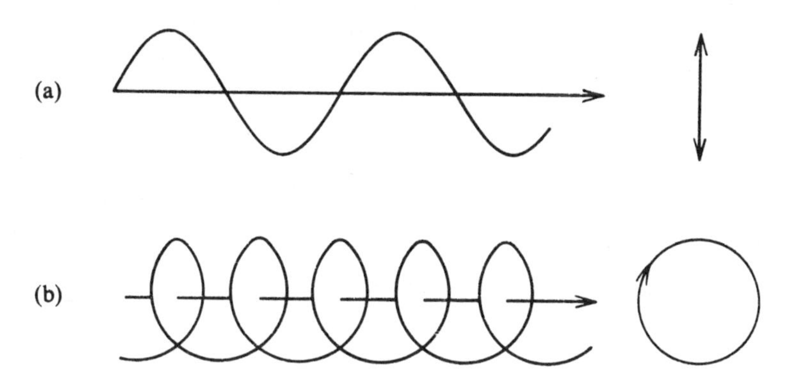


Fig. 10-1 Representation of (a) plane polarized and (b) right-hand circularly polarized waves

## 10-2 Polarization by reflection

Every electromagnetic wave has associated magnetic-field, electric-field, and electric-displacement vectors (**H**, **E**, and **D**). The magnetic- and electric-field vectors and the direction of energy propagation, which in the case of light waves is the ray direction, are mutually perpendicular. In an isotropic material, such as unstrained glass, the electric-field and displacement vectors coincide. Thus light, being an electromagnetic wave, has associated with it these three vectors of which the electric-displacement vector, that is the electric vibration direction, is the most important.

Figure 10-2(a) shows a plane-polarized beam of light incident onto a glass plate at a particular angle $\theta$. Suppose that the direction of the electric-displacement vector is parallel to the plane of the paper—that is, it is in the plane of incidence, containing both the incident ray and the normal to the surface. In the figure the electric displacements or vibrations, which are also normal to the direction of the beam, are shown as short arrowed lines. At the surface of the glass, both reflection and refraction are possible and with this special value of $\theta$ the reflected and refracted rays are at right angles to each other. This means that the electric vibrations in the refracted ray are parallel to the direction of the reflected ray. Thus, as the incident ray enters the surface, there is no component of electric vibration at right angles to the required direction of the reflected ray. But according to electromagnetic theory, light must be a transverse vibration and therefore the reflected ray cannot exist.

Therefore, for this particular angle of incidence $\theta$, the *polarizing angle*, incident plane-polarized light having its electrical vibration in the plane of incidence cannot be reflected, only refracted. If the electric-displacement

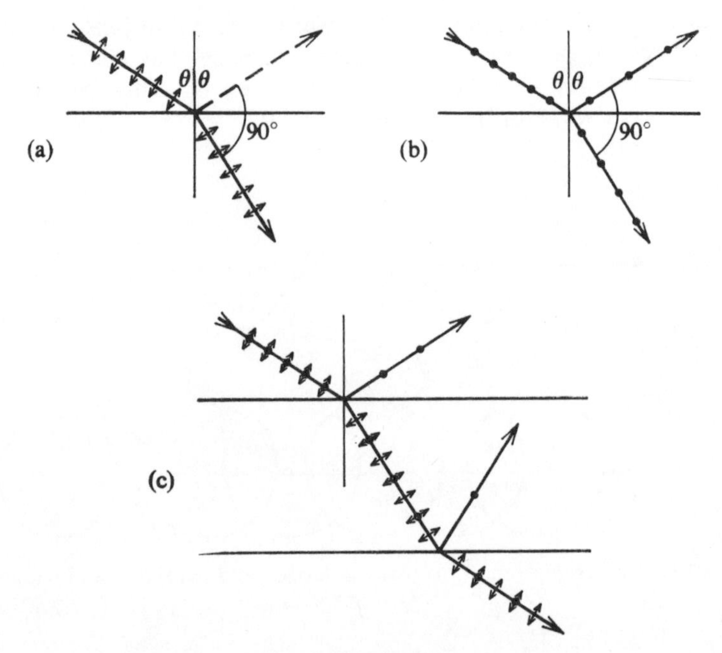


Fig. 10-2

vector is normal to the plane of incidence, both reflection and refraction are possible, since the vibrations are normal to the directions of both the reflected and refracted rays. This is shown in Fig. 10-2(b) in which the electric vibrations are indicated by dots, since their direction is perpendicular to the plane of incidence. If the incident beam is unpolarized, it may be imagined resolved into two equal components, one with its electric vibration in the incident plane and the other with its electric vibration normal to this plane. The first of these components is completely refracted, with no reflection, whilst the second is partially refracted and partially reflected. The reflected light is completely plane polarized, since it comprises only vibrations normal to the plane of incidence. The refracted light, however, is partially polarized, since it contains only some of the component with electric vibrations normal to the plane of incidence and

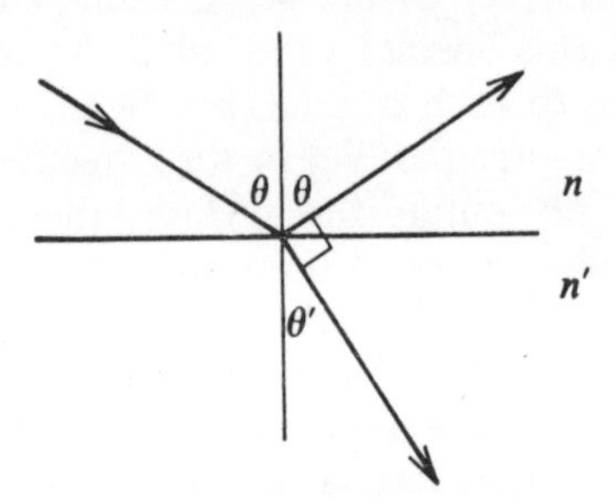


Fig. 10-3 Brewster's angle

all of the component with vibrations in the plane of incidence [Fig. 10-2(c)]. The plane-polarized reflected light for glass contains about 8% of the total incident energy—that is, about 15–16% of the component vibrating normally to the plane of incidence. The transmitted partially polarized beam, therefore, contains about 92% of the total incident energy and 100% of the component vibrating parallel to the plane of incidence.

The polarizing angle, $\theta$, which gives reflected and refracted rays at right angles to each other and hence no reflection for electric vibrations in the plane of incidence, is also called the *Brewster angle*. In Fig. 10-3, by Snell's law

$$\frac{n'}{n} = \frac{\sin\theta}{\sin\theta'}$$

$$= \frac{\sin\theta}{\cos\theta} = \tan\theta \qquad (10\text{-}1)$$

Thus, the polarizing angle depends on refractive index, which in turn is dependent on wavelength; the reflected beam can be plane polarized only when monochromatic light is used. For glass of refractive index 1·5 for a particular wavelength, the polarizing angle is approximately 56·3°, rising to 61° for a refractive index of 1·8.

For unpolarized light incident at angles other than the polarizing angle, the refracted beam has neither of the two components into which the incident beam was considered resolved exactly parallel to the reflected ray direction. Therefore, both components in turn produce transverse components of vibration in the reflected ray, so that it is no longer completely plane polarized.

If a pile of reflecting plates is used, with unpolarized light incident at the polarizing angle, then at each reflection, a further 15% or so of the remainder of the component vibrating normally to the plane of incidence is added to the reflected beam. After a number of reflections the intensity of the reflected plane-polarized beam has much increased, and the transmitted ray is very nearly completely plane polarized, since a fraction of the perpendicular component has been removed at every reflection. The process is limited by the loss of intensity that occurs at each reflection. The intensity of the plane-polarized reflected beam must in any case be less than 50% of that of the incident beam.

The production of a plane-polarized reflected beam by reflection was first demonstrated by Malus in 1808 using an arrangement like the one shown in Fig. 10-4. A monochromatic beam AB is incident at the polarizing angle, $\theta$, onto the lower glass plate. The reflected ray BC is then plane polarized. The upper glass plate reflects this plane-polarized beam in the direction CD. If the upper plate is rotated through 90° about BC as an axis, there can be no reflection, since the plane of incidence is now parallel to the plane containing the electric-vibration direction of the beam BC. A further 90° rotation brings the incident plane perpendicular to the electric-vibration direction again, and the reflected intensity rises again

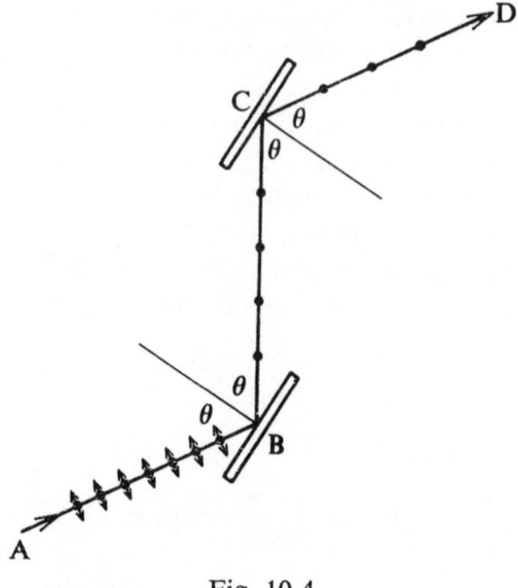


Fig. 10-4

to its maximum. Another 90° rotation and the reflected intensity again becomes zero, and another 90° brings the plate back to its starting point. In this way the lower plate acts as a polarizer and the upper plate as an analyser. When the polarized beam is transmitted by the analyser (as in the figure) the polarizer and analyser are said to be 'parallel', and in the cut-off conditions, they are 'crossed'.

## 10-3 Polarization by dichroism

Most methods for producing polarized light involve producing unpolarized light, dividing it into two components having their electric vibrations at right angles, and then eliminating one of these components. Polarization by dichroism is no exception, since it involves selective absorption depending on the direction of the electric vibration.

Imagine a beam of unpolarized light incident on a fine wire grid (Fig. 10-5). The beam is imagined divided into two components one of which has electric vibrations parallel to the wires of the grid, and the other has electric vibrations at right angles to this. The parallel component has vibrations in the direction of the wires and thus produces an electric

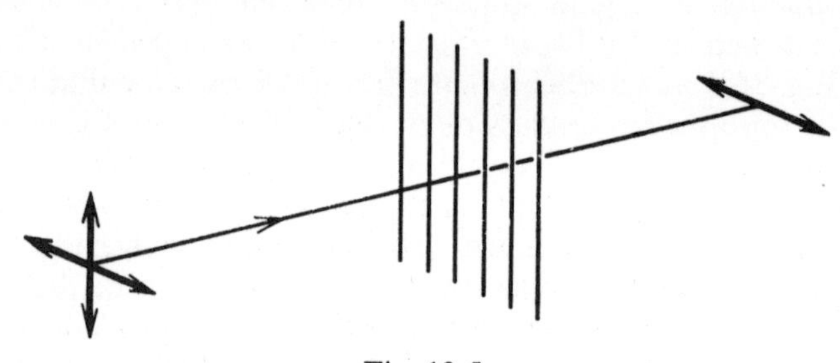

Fig. 10-5

current in them, which in turn produces heat; in this way, the energy of this vibration is used up. The other component, with vibrations at right angles to the direction of the wires, cannot produce any current in the wires and so does not lose energy; these vibrations pass through the grid to form a plane-polarized beam. (Incidentally, the often quoted analogy of a vibrating rope passing through a picket fence is thus 90° wrong!)

Dichroic polarizers work on this principle, but instead of wires, long-chain molecules are used. These are aligned parallel, absorbing the component with its electric vibrations parallel to the molecule chains and passing the other component. This principle is used in Polaroid film, invented by E. H. Land. A transparent sheet of plastic, such as polyvinyl-alcohol, is warmed and then rapidly stretched in one direction. This aligns the long-chain molecules parallel with the stretch direction. To prevent shrinkage the stretched sheet is usually cemented to a rigid plastic sheet and then it is dipped into an iodine solution. The iodine diffuses into the polyvinyl-alcohol layer where the iodine atoms tend to arrange themselves along the long chains. After washing and drying, the sheet behaves as a polarizer with iodine chains acting as the 'wires' and absorbing parallel electric vibrations. The component with its electric vibrations perpendicular to the stretch direction is transmitted. The intensity transmitted must, of course, be less than 50% of the incident light, and with losses by reflection at the surfaces and absorption within the film, is usually less than 40%.

Some naturally occurring crystals are dichroic—for example, crystals of tourmaline and of herapathite—but these tend to be small and expensive and not particularly good polarizers.

## 10-4 Polarization by double refraction

A ray of unpolarized or polarized light incident onto an isotropic medium such as glass obeys the laws of refraction, including Snell's law (Chapter 1). If, however, a ray is incident onto an anisotropic medium—for example, a rhombohedral crystal of calcite—the refraction laws are not simply obeyed. The atoms that make up the crystal lattice of an anisotropic material are arranged differently in different directions through the crystal. Thus a ray of light inside the crystal 'sees' a different configuration of atoms depending on the direction in which it is travelling. Moreover, the electric vibrations that make up the ray of light interact differently with the atoms of the crystal lattice, depending on the relative orientation of the electric-vibration direction with the crystal lattice.

A crystal of anisotropic material splits an unpolarized beam into two components that have their directions of electric vibration at right angles to each other. The different interactions of these vibrations with the crystal lattice generally cause the two components to take different ray paths. The two rays are both plane polarized. One ray, termed the *Ordinary* or *O-ray*, obeys the ordinary laws of refraction. This component always has its electric-vibration direction perpendicular to a particular direction in the crystal, called the *optic axis*. In crystals such as calcite there is only one direction of the optic axis and such crystals are therefore termed *uniaxial*. Another group of crystals having two optical axes, *biaxial* crystals, produce more complicated optical effects which are beyond the scope of this book. It should, however, be emphasized that the optic axis is a direction through the crystal, not a line, as its name might imply.

The direction of electric vibration of the other component of the original ray, the *Extraordinary* or *E-ray*, is inclined at an angle to the optic axis which depends on the crystal face and angle of incidence used. The E-ray does not obey the simple laws of refraction: it may not be refracted in the plane of incidence and does not obey Snell's law. It has different velocities depending on its direction of travel, with its extreme velocity occurring when its electric-vibration direction is parallel to the direction of the optic axis. In calcite, the O-ray has a velocity of $1{\cdot}808 \times 10^8$ m/s for yellow light of wavelength 5,893 Å, and the E-ray has velocities varying from $1{\cdot}808 \times 10^8$ to $2{\cdot}017 \times 10^8$ m/s depending on direction. The refractive index—that is, the ratio of the velocity of light in vacuum to the velocity in the medium—is thus 1·658 for the O-ray and between 1·658 and 1·486 for the E-ray. Thus in one direction, in fact, the direction of the optic axis, both the O- and E-rays travel with the same velocity and along the same path.

The refractive index for the O-ray is called the *major principal refractive index* and designated as $n_O$. The extreme value of the refractive index for the E-ray is termed the *minor principal refractive index*, designated $n_E$. The difference $(n_E - n_O)$ is $(1{\cdot}486 - 1{\cdot}658) = -0{\cdot}172$ (for yellow light $\lambda = 5{,}893$ Å). Calcite is therefore said to be a negative crystal, with a *birefringence* of 0·172. Quartz is a positive crystal, since $n_E > n_O$.

Since the E-ray travels with different velocities in different directions, the direction of energy flow, which is the ray direction, is generally different from the direction of the wave normal (Fig. 10-6). For an electromagnetic wave, the magnetic- and electric-field vectors, **H** and **E**, and the direction of energy propagation, which is the ray direction, are always mutually perpendicular. Also, the magnetic-field and electric-displacement vectors lie in the wave front. For an isotropic medium the electric-field

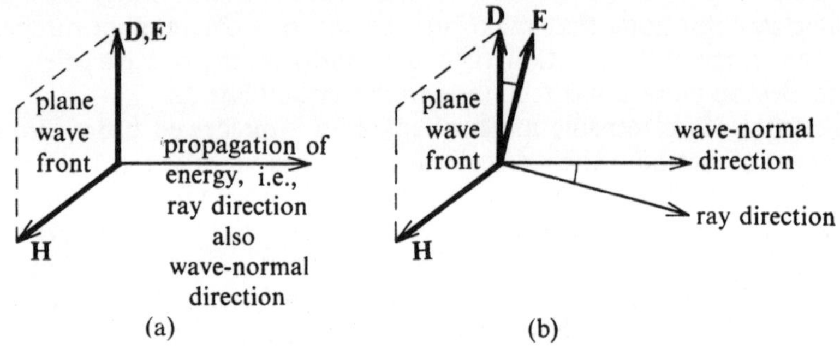


Fig. 10-6 Electromagnetic wave in (a) isotropic and (b) anisotropic media

and electric-displacement vectors, **E** and **D**, coincide [Fig. 10-6(a)]; that is, the directions of the normal to the wave front and the ray are the same. For an anisotropic medium the vectors **E** and **D** do not coincide. **H** and **D** still lie in the wave front, thus defining the wave normal direction. Also, from electromagnetic theory, it is known that the electric-field vector **E** lies in the plane of the electric-displacement vector **D** and the wave normal direction. Since the vectors **H** and **E** and the direction of energy propagation, or ray direction, are mutually perpendicular, the ray direction is defined [Fig. 10-6(b)]; that is, the ray direction is different from the direction of the wave normal.

Figure 10-7 shows the shape of the wave fronts in uniaxial crystals. The waves may be imagined radiating from a point source inside the crystal, with the position of the O- and E-wave fronts as shown after a short time of travel has elapsed. The O-wave front is a sphere, since the velocity is the same in any direction, but the E-wave front is an ellipsoid of revolution about the optic axis. With a negative crystal, such as calcite, the ellipsoid encloses the sphere, whereas the sphere encloses the ellipsoid in a positive crystal, such as quartz.

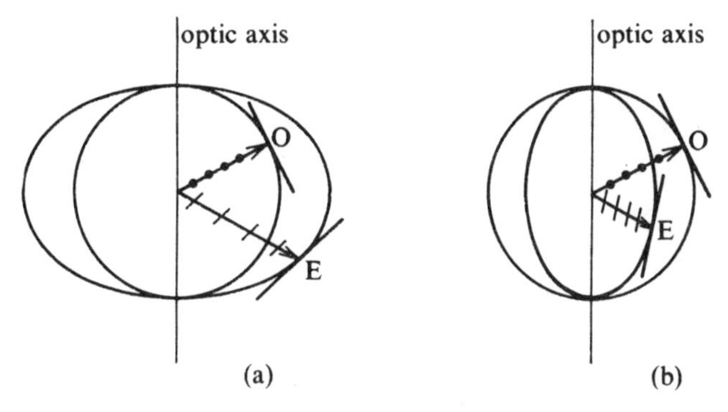


Fig. 10-7 Wave surfaces in (a) negative and (b) positive uniaxial crystals

To summarize: the electric vibrations of the O-ray are perpendicular to the principal plane of the O-ray—that is, the plane containing the O-ray and the optic axis—and parallel to the tangent to the O-wave surface (Fig. 10-7). The electric vibrations of the E-ray are always at right angles to those of the O-ray—that is, the rays are always plane polarized with their planes at right angles. The E-ray vibrations are in the E-ray principal plane—that is, the plane containing the E-ray and the optic axis—and parallel to the tangent to the E-wave surface. They are thus perpendicular to the wave normal direction.

In a polarizing prism one of the rays is removed by total internal reflection, leaving a single plane-polarized beam. The earliest design was the Nicol prism [Fig. 10-8(a)]. A rhombohedral crystal of calcite is cut in two along the diagonal AB, and with the direction of the optic axis as shown. The cut faces are polished flat and cemented together again with

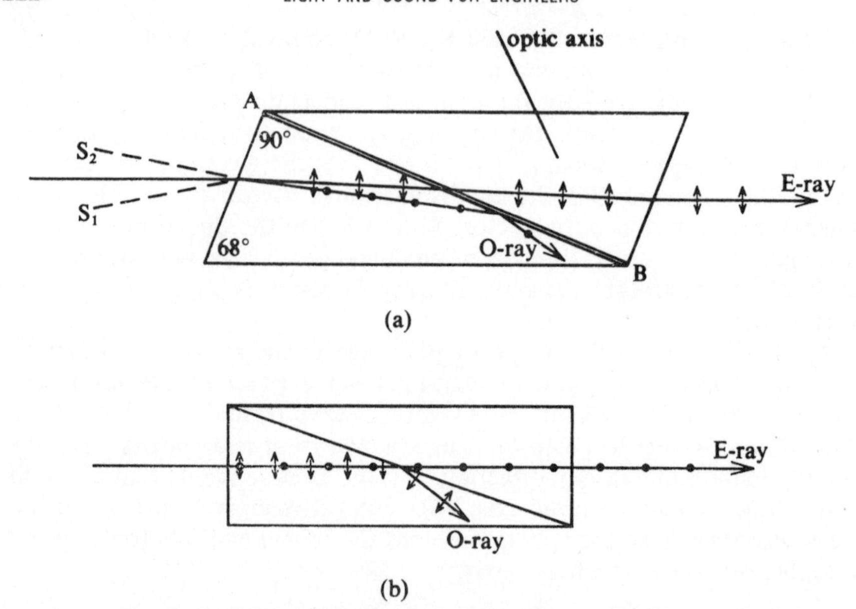


Fig. 10-8 (a) Rays through a Nicol prism and (b) arrangement of a Glan Thompson prism

Canada balsam. The relevant refractive indices (for yellow light, $\lambda = 5{,}893$ Å) are

| | | |
|---|---|---|
| Refractive index, | O-ray | 1·6584 |
| | Canada balsam | 1·55 |
| | E-ray | 1·4864 |

Since the refractive index of the balsam is higher than that of calcite for the E-ray, but lower than that for the O-ray, for large angles of incidence the O-ray is totally reflected and only the E-ray is transmitted. The prism may not function for rays entering the prism at angles different from that shown in the figure. Rays from the direction $S_1$ may be such that the critical angle of incidence for the O-ray is not reached and the O-ray is transmitted. Rays from the direction $S_2$ may be such that since the refractive index of the E-ray varies with direction its critical angle may be reached and it will also be totally reflected by the balsam. The end face of the calcite crystal is cut back when making the prism so that these two extreme rays, from $S_1$ and $S_2$, are equally inclined to the geometric axis of the prism. With a prism having a side to length ratio of about $1:2\frac{1}{2}$, the angle of useful aperture is about 22°. However, the rhombic cross-section and the slanting end faces are a disadvantage; the slanting end faces, particularly, because they produce slight depolarization. This prism is therefore only of limited use.

There are other designs of prism in which different angles are made with the optic axis. Of the different designs, the Glan Thompson is the best. This has a rectangular cross-section with the end faces perpendicular to the lengthwise direction [Fig. 10-8(b)]. It is cut from the calcite crystal so that the direction of the optic axis is perpendicular to the section shown in the figure. It gives complete polarization of the transmitted light and with a side to length ratio of about 1:3, has a field of vision of about 30°.

## 10-5 Wave plates

Wave plates or 'retarders' convert one form of polarized light to another. For example, a beam of plane-polarized light may be converted into circularly or elliptically polarized light, or vice versa, or the plane of polarization may be rotated. Unfortunately, wave plates suffer from chromatism, that is, the magnitude of the change depends on wavelength. Our discussion of wave plates will therefore be restricted to monochromatic light.

Wave plates divide incident plane-polarized light into two components, slow down one component relative to the other—that is, they introduce a phase difference between the components—and then allow them to recombine. For example, Fig. 10-9 shows a double-refracting plate cut with its optic axis as shown, and with a plane-polarized beam incident from the left. This beam splits into two components on entering the crystal plate. One component, the O-ray, has an electric-vibration direction perpendicular to the optic axis—that is, in a vertical direction. The other component, the E-ray, has its vibration perpendicular to this—that is, in a horizontal direction, parallel to the optic axis. The two rays then travel at different speeds. If the plate is made of calcite, the O-ray travels slower and the direction of its vibration is therefore termed the *slow axis* as in the figure. The other axis is correspondingly termed the *fast axis.* Let the thickness of the plate be such that the slow ray is retarded exactly half a wavelength relative to the fast ray. A plane-polarized beam incident normally onto the retardation plate, with its plane of polarization at 45°

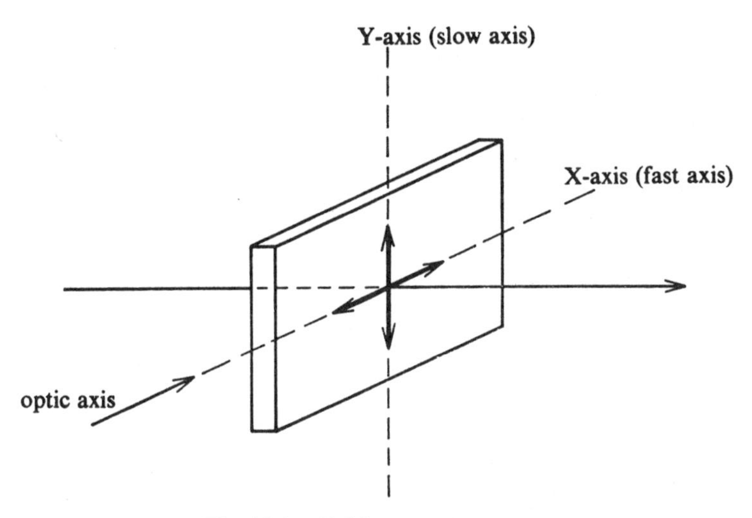


Fig. 10-9 Calcite retardation plate

to the plate axes, may be considered as made up of two equal components at right angles but of the same phase [Fig. 10-10(a)]. After passing through the retardation plate the slow ray has been retarded half a wavelength, and is lagging 180° in phase; it may therefore be represented vectorially as in Fig. 10-10(b), where the **Y**-component is shown rotated through 180°. Thus the resultant transmitted beam has had its vibration direction—that is, its plane of polarization—rotated through 90°. Similarly, any other angle $\theta$ of the incident vibration direction with one of the plate axes is rotated to a symmetrical position at $-\theta$ to the axis. The plates also convert right-handed circular or elliptical polarized light into the left-handed circular or elliptical form, and vice versa. Plates that introduce 180° phase changes, or half-wave retardations, are termed *half-wave plates*. In practice, calcite is so highly birefringent that the plates are very thin and fragile and therefore quartz or mica is used.

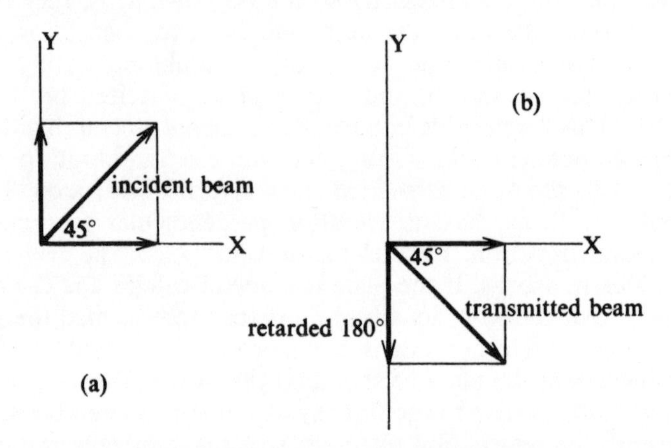


Fig. 10-10 Effect of half-wave retardation plate

*Quarter-wave plates* introduce a 90° phase difference. The transmitted beam is the resultant of two simple harmonic motions at right angles to each other which are 90° out of phase. The resultant vibration is an ellipse and becomes a circle only if the two components have equal amplitude as when the plate axes make angles of 45° with the vibration direction of the incident beam. The combination of such wave motions at right angles is discussed more fully in the next chapter. Thus a quarter-wave plate can convert linearly polarized light into elliptically or circularly polarized light, or vice versa.

Note that birefringence is constant and is the difference between the two principal refractive indices, whereas retardation, although it depends on the birefringence, also depends on the thickness of the plate.

## 10-6 Applications of polarized light

The applications of polarized light are many and varied, but only a few can be mentioned here.

In sunlight the main source of light is overhead and the most brightly illuminated surfaces are horizontal. The bright surfaces are generally viewed obliquely, near the Brewster angle, so that the reflected light is partially polarized with its electric-vibration direction mainly horizontal. Thus in *polarizing sunglasses* the lenses are of a dichroic material orientated to absorb this horizontal vibration and to transmit only vertical components.

When the angle of reflection is nowhere near the Brewster angle there is no polarization. This situation is of especial interest, for example, to radar operators who are looking for faint spots on an oscilloscope screen. The screen also reflects light which tends to mask these faint spots. To stop these reflections, a circular polarizer plate is used, consisting of a linear polarizer and a quarter-wave plate together, in front of the oscilloscope screen. Light from the room passing through the plate is converted from unpolarized light into circularly polarized light, say, right-circularly polarized. This light is reflected from the oscilloscope screen but now as left-circularly polarized, because the direction of circular polarization is defined with respect to the ray direction. Therefore, on reversing the ray direction, the direction of circular polarization appears reversed. This reflected left-circularly polarized light is then totally absorbed by the circular polarizer plate because they are now of opposite directions. About 99% of the room light that was reflected from the oscilloscope screen is absorbed, whereas nearly half the wanted light from the screen is transmitted. In addition, to ensure that room light is not reflected from the circular polarizer itself, it must be tilted at an angle to the viewing direction.

Two linear polarizers inclined at a variable angle to each other may be used to produce a *variable density filter*. Figure 10-11 shows the transmission planes of a polarizer and analyser—for example, two sheets of Polaroid—at an angle $\theta$ to each other. Let $A$ be the amplitude of the vibration transmitted by the polarizer. This is resolved into two components at the analyser. One component, of amplitude $A_1$, has a vibration

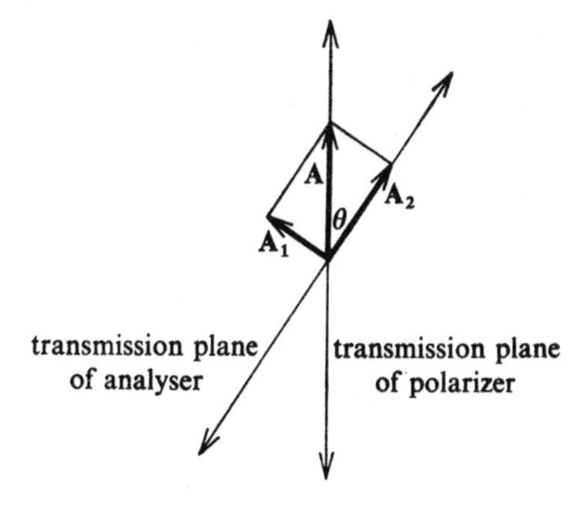


Fig. 10-1

direction at right angles to the transmission plane of the analyser and is therefore absorbed. The other component, of amplitude $A_2$, has its vibration direction parallel to the analyser transmission plane and is therefore transmitted. This transmitted amplitude $A_2$ equals $A \cos \theta$. Therefore the transmitted intensity, which is proportional to the square of the amplitude, varies as the square of the cosine of the angle between the transmission directions of the polarizer and analyser.

A more specialized scientific use of polarized light is connected with *optical activity*. It is known that certain substances—for example, a sugar solution—have the property of rotating the plane of polarization without changing the nature of the linear vibrations. The rotation of the plane of polarization depends on the distance travelled through the solution and the concentration of sugar in the solution. *A polarimeter* is an instrument for measuring this rotation. It contains a glass cell of known length. With the sugar solution in the cell plane-polarized light is shone through it. The amount of rotation of the plane of polarization is measured by a suitably sensitive analyser and the strength of the sugar solution may be determined if the instrument has previously been calibrated.

*Photoelastic analysis* is used in the design of complicated mechanical parts which are to be subjected to strain. A model of the part made from transparent plastic is placed between crossed polarizers. On looking through the assembly only a dark field is seen. The plastic model is now subjected to forces similar to those which would be applied to the actual part under use. Light- and dark-coloured areas are seen in the model as the molecules of the plastic are distorted by the applied stress and the plastic becomes birefringent, acting as a retardation plate. The amount of birefringence introduced at each point in the model is a measure of the strain occurring at that point, and the two axes of the retarder at each point are parallel and perpendicular to the directions of the strain at that point. Thus the points of greatest strain are points of greatest retardation, resulting in greatest miss-match of polarization direction with that of the analyser and in most light being transmitted. The actual measurement of strain in a particular region may be made in several ways. Calibrated retarders may be used to cancel out the retardation introduced by strain in the model, and hence give a measure of the retardation introduced. From subsidiary experiments on models of simpler shape the relation between strain and retardation, that is, the *stress-optic constant*, may be determined. Alternatively, it is possible to determine the retardation from the colour of the fringe at a particular point, since retardation depends on wavelength and the assembly is illuminated by white light. For very large retardations, amounting to many cycles, the number of fringes between the region of interest and a point known to be strain-free is counted and the strain determined by means of the stress-optic constant for the plastic.

Other materials, such as glass, show strain-birefringence. By examining glass between crossed polarizers it is easy to determine whether it has been correctly annealed and hence is strain-free. If it has not, the strain may relieve itself in time by causing the glass to fracture.

## PROBLEMS

**10-1** An incident beam of light is reflected from a liquid of refractive index 1·45 so that the reflected ray is completely polarized. What is the angle of refraction? [34° 36′]

**10-2** What should be the angular elevation of the sun for polarizing sunglasses to be at their most effective over water? [37°]

**10-3** Two sheets of Polaroid are orientated for maximum transmission. What percentage of this maximum intensity will be transmitted when one of the Polaroid sheets is rotated through (a) 30°, (b) 60°? [(a) 75%; (b) 25%]

**10-4** Explain why, on looking at an object through a block of calcite, two images are seen. Explain what is seen if the object is then viewed through two blocks of calcite with one of the blocks being rotated about the line of vision.

**10-5** Two Glan Thompson prisms are arranged in tandem so that the transmitted light has an intensity 0·1 times that of the intensity of the natural light incident onto the first prism. What is the angle between the principal planes of the two prisms? [63° 26′]

**10-6** Show that the minimum thickness of a half-wave retardation plate is given by $\lambda/[2(n_O - n_E)]$. Calculate the thickness, if the plate is quartz with a birefringence of 0·0091, for sodium yellow light of wavelength 5,893 Å. [0·032 mm]

**10-7** Explain how to determine whether a beam of light is (a) unpolarized, (b) plane-polarized, (c) circularly polarized, (d) elliptically polarized, or (e) a mixture of polarized and unpolarized light.

**10-8** A prism is cut from a negative uniaxial crystal so that the direction of the optic axis is parallel to the refracting edge of the prism. For a particular wavelength the angles of minimum deviation for the two resulting rays are 36° and 52° when the prism has a refracting angle of 60°. Calculate the principal refractive indices of the crystal. [$n_O = 1{\cdot}658$; $n_E = 1{\cdot}486$]

# 11 Waves and vibrations

## 11-1 Introduction

Sounds are produced by vibrating bodies. For sounds of low frequency, the vibrations of the body, or source, are clearly visible. At higher frequencies, they are not quite so obvious, but nevertheless can be demonstrated, for example, by bringing a pith-ball pendulum into contact with the source. Similarly, the receiver of the sound can be shown to be vibrating in sympathy with the source. The final receiver is normally the ear. Just as the human eye cannot see ultra-violet light, there are ultrasonic vibrations beyond the range of the ear.

The sound produced by a vibrating body, or for example a vibrating air column, cannot be heard unless the medium between the source and the receiver is capable of conveying the vibrations. This medium is normally air, but it can also be some other gas, a solid, or liquid. In 1705 Hawksbee first demonstrated the necessity for a carrier medium by the classical experiment of the electric bell in the evacuated glass jar. Since both the source and the receiver are known to vibrate, it is reasonable to assume that the carrier medium also vibrates and that some wave-like motion occurs. Two types of wave motion are possible: transverse waves or longitudinal waves.

## 11-2 Transverse waves

In a transverse wave, the vibrating particles of the carrier medium move from side to side, in paths at right angles to the direction of propagation of the wave. For example, waves produced on the surface of water are transverse waves. If a cork is thrown onto the water, it bobs up and down along a vertical line, but does not move in the direction in which the wave is travelling, providing of course that there is no wind. Similarly, if one end of a piece of rope is jerked from side to side, a wave travels along it. The jerking sideways motion can produce a sine wave, if the rope is not too stiff. Each particle of the rope then moves at right angles to the direction of propagation of the wave and executes a simple harmonic motion.

## 11-3 Longitudinal waves

In these, the vibrating particles of the carrier medium move in the same direction as that in which the wave travels, so that the motion produces a continuous series of rarefactions and compressions. It is in this way that sound is transmitted. Figure 11-1 depicts the movement of the prong of a tuning fork. Movement to the right causes a compression in the air, which moves to the right; movement to the left, in turn, causes a rarefaction which again moves to the right. Thus, the vibrating prong propagates a longitudinal wave travelling to the right. At the same time (but not shown) a wave is also propagated to the left. If, however, the prong moves very slowly, no wave motion is produced as the air merely flows around it.

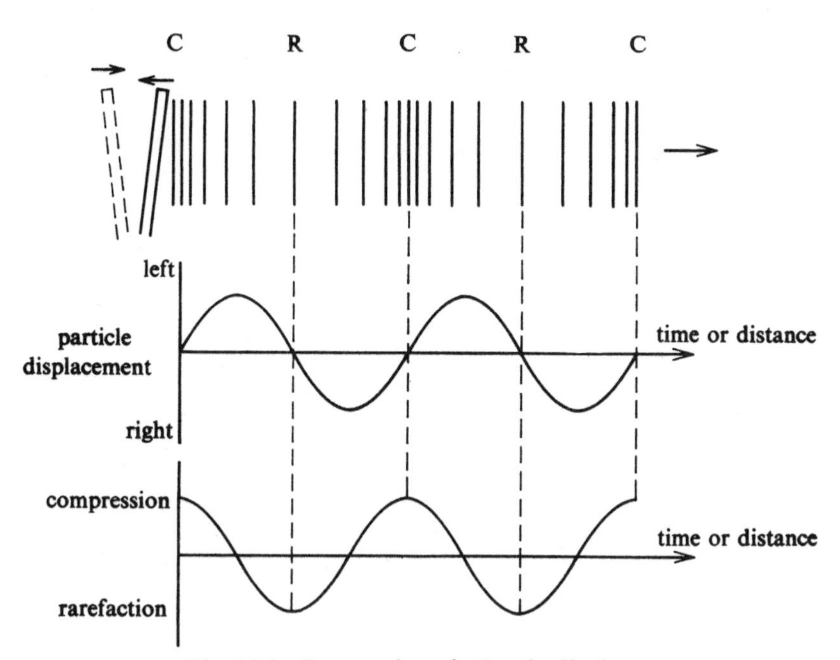


Fig. 11-1 Propagation of a longitudinal wave

## 11-4 Wavelength, frequency, and velocity

In Fig. 11-1, as the prong moves to the right, a compression is sent out. This is followed by another as the prong completes one vibration. The time taken to complete one vibration is the *period*, $T$, and the number of times this happens in one second is termed the *frequency*, $f$, measured in hertz (Hz). The distance between the two compressions, or for that matter the distance between any two successive corresponding parts of the wave, is the wavelength, $\lambda$. If there are $f$ vibrations per second, then a particular compression travels a distance $f\lambda$ in one second. This is the *wave velocity*, $v$, or phase velocity of the wave—that is, $v = f\lambda$—independent of the form of the wave.

$v$ is the wave velocity, but the individual particles that make up the wave motion also have a velocity. For a transverse wave, the particle motion and the wave motion are at right angles. The wave velocity, $v$, is constant and in the direction of propagation, but the *particle velocity* varies with time. If the wave travels a small distance AB in unit time (Fig. 11-2) and the particle travels a distance CB in the same time, then the gradient of the displacement curve is given by CB/AB, particle velocity/wave velocity. This varies with time.

For a longitudinal wave, the particle motion and the wave motion are both in the same direction. Again the wave velocity is constant and the particle velocity varies with time.

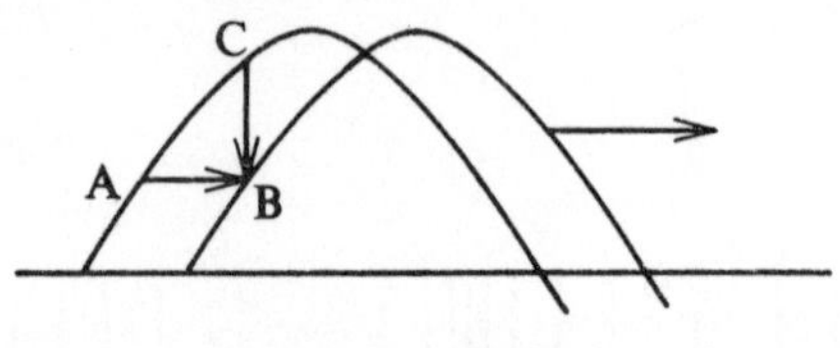


Fig. 11-2

## 11-5 Simple harmonic waves

This is the most fundamental of periodic motions. It is a sinusoidal motion, by which the motion of many kinds of vibrating bodies can be described if the appropriate amplitudes and periods are selected. For example, many machines involve the changing of a circular motion into a motion along a fixed line which is often simple harmonic, or a combination of simple harmonic motions. Similarly, the equations describing the behaviour of an electrical circuit in which there is an alternating current have the same form as those describing the harmonic motion of a vibrating body. It is this type of motion, frequently occurring both in practice and in Nature, that is the most applicable to the study of sound waves. In the case of a sound wave, the wave transfers energy to the carrier medium causing the wave motion to decay, but at this stage we can ignore this complication.

Consider a particle, P, moving along a circular path (Fig. 11-3) with a constant angular velocity $\omega$ radians per second. At some time, $t$, OP makes an angle $\phi$ with the $x$-axis, then

$$\phi = \omega t$$

Also, the projection of OP onto the $y$-axis is given by

$$y = a \sin \omega t \tag{11-1}$$

therefore, the velocity

$$\frac{dy}{dt} = a\omega \cos \omega t$$

and the acceleration

$$\frac{d^2y}{dt^2} = -a\omega^2 \sin \omega t$$

$$= -\omega^2 y$$

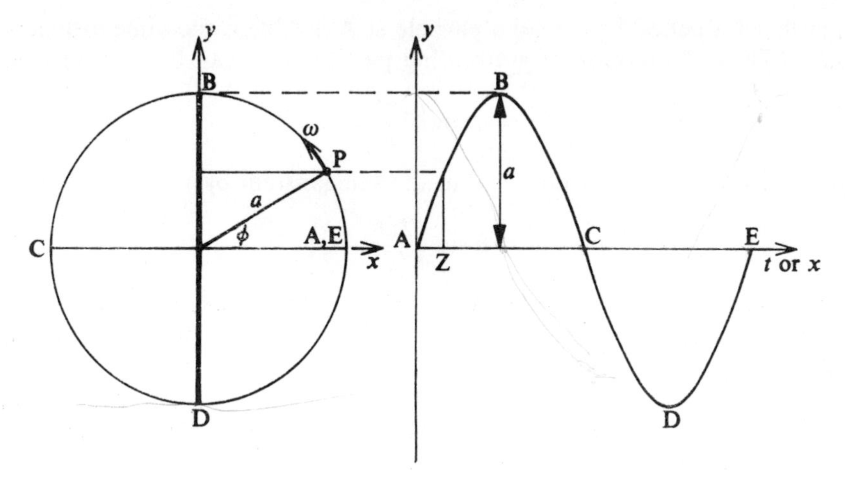


Fig. 11-3 Generation of a simple harmonic wave

That is, the projection of P moves such that its acceleration towards the mean position, O, is always proportional to its distance from O, and the motion is described as simple harmonic.

**Simple harmonic motion** *is defined as a motion in a straight line, under a force which is always directed towards a fixed point in the line and of magnitude such that it is directly proportional to the distance from the fixed point.*

If $T$ is the period, that is, the time for the particle to travel once round the circle, then

$$\omega = \frac{2\pi}{T}$$

$$\therefore\ y = a \sin \frac{2\pi t}{T}$$

This displacement from the $x$-axis, which varies simple harmonically, may be represented graphically by plotting $y$ against the time, $t$, as ordinate (Fig. 11-3). Then the curve, A to E, representing a simple harmonic motion, is generated by one complete revolution of the particle P around the circle. The extreme displacement on either side of the equilibrium position represents the *amplitude, a.*

Let this curve now represent a wave form—that is, the displacements from the mean of the vibrating particles of the carrier medium, where the distance of these particles from the origin of the wave is measured in terms of $x$. Then the distance AE represents one wavelength. Suppose this wave now progresses, such that the same form of the wave exists at any time $t$. If the wave progresses to the right with time, every vibrating particle in the carrier medium is vibrating slightly later than its neighbour on its left. Thus, a particle that produces a displacement at B does so a quarter of a period later than a particle at A produced the same displacement. Thus, the motion of a vibrating particle at A can be described by

$$y = a \sin \left(\frac{2\pi t}{T}\right)$$

and the motion of a particle at B at the same instant by

$$y = a \sin \left(\frac{2\pi t}{T} - \frac{\pi}{2}\right)$$

The minus sign is the result of a convention: the particle at B is *late* with respect to the particle at A—that is, a retardation in phase causes the wave to travel from left to right.

A general equation is now required to describe the motion of a particle at any position, not just at B. If A is the origin and $x$ the distance to a particle whose mean position is Z (Fig. 11-3)

$$\text{AZ} = \frac{x}{\lambda} \text{ of one wavelength}$$

Therefore, the retardation in phase between A and Z is $2\pi x/\lambda$, and the equation of motion of a particle whose mean position is Z is given by

$$y = a \sin 2\pi\left(\frac{t}{T} - \frac{x}{\lambda}\right)$$

$$= a \sin (\omega t \pm \alpha) \qquad (11\text{-}2)$$

This is the general equation for a particle moving under a simple harmonic motion with a phase angle, or simply *phase*, of $(\omega t \pm \alpha)$, where $\omega = 2\pi f = 2\pi/T$; the initial value of the phase (the *phase constant*) $\alpha = 2\pi x/\lambda$; the distance of the vibrating particle from the origin is $x$; and $y$ is its displacement from its mean position. A negative value of $\alpha$ denotes a wave travelling to the right and a positive value represents a wave travelling to the left. The displacement $y$ for a transverse wave is measured in a direction at right angles to the $x$-direction. Where the displacement is in the $x$-direction, as in a longitudinal wave, the symbol $y$ is conveniently replaced by $\xi$.

At any instant, that is, taking $t$ as a constant, the general equation gives the form of the complete wave. For a constant value of $x$, the equation describes the motion of the particle whose mean position is distance $x$ from the origin. Also it may be noted that for a particle at A—that is, $y = 0$

$$\sin 2\pi\left(\frac{t}{T} - \frac{x}{\lambda}\right) = 0$$

$$\therefore \sin \frac{2\pi t}{T} \cos \frac{2\pi x}{\lambda} = \cos \frac{2\pi t}{T} \sin \frac{2\pi x}{\lambda}$$

i.e.,

$$\tan \frac{2\pi t}{T} = \tan \frac{2\pi x}{\lambda}$$

$$\therefore t \propto x$$

Thus, point A of the wave form moves to the right—that is, the wave is progressive as first postulated.

## 11-6 Velocity and acceleration of a particle in a progressive wave

Since $f = 1/T$, where $f$ is the frequency and $T$ is the period, and $\lambda = vT$, where $\lambda$ is the wavelength and $v$ is the wave velocity, the general equation [Eqn (11-2)] for a particle having a simple harmonic motion may be written in the form

$$y = a \sin \omega \left(t - \frac{x}{v}\right) \tag{11-3}$$

where $\omega = 2\pi f$. Then, since for a particular particle $x$ is a constant, the velocity of the particle in the $y$-direction is given by

$$\frac{\partial y}{\partial t} = a\omega \cos \omega \left(t - \frac{x}{v}\right) \tag{11-4}$$

Also, since at a particular instant $t$ is a constant, the gradient of the displacement curve is given by

$$\frac{\partial y}{\partial x} = -\frac{a\omega}{v} \cos \omega \left(t - \frac{x}{v}\right) \tag{11-5}$$

Combining Eqns (11-4) and (11-5)

$$\frac{\partial y}{\partial t} = -v \frac{\partial y}{\partial x} \tag{11-6}$$

i.e., (particle velocity) = − (wave velocity)
× (gradient of displacement curve)

For a longitudinal wave, where the particle is vibrating along the $x$-axis with $\xi$ as its displacement from its mean position, the corresponding equation is

$$\frac{\partial \xi}{\partial t} = -v \frac{\partial \xi}{\partial x} \tag{11-7}$$

where $\partial\xi/\partial x$ is the gradient of the displacement with position.

From Eqn (11-4), the acceleration of a particular particle in the $y$-direction is

$$\frac{\partial^2 y}{\partial t^2} = -a\omega^2 \sin \omega \left(t - \frac{x}{v}\right) \tag{11-8}$$

and from Eqn (11-5), the rate of change of the gradient of the displacement curve is

$$\frac{\partial^2 y}{\partial x^2} = -\frac{a\omega^2}{v^2} \sin \omega \left(t - \frac{x}{v}\right) \tag{11-9}$$

Again combining these equations

$$\frac{\partial^2 y}{\partial t^2} = v^2 \frac{\partial^2 y}{\partial x^2} \qquad (11\text{-}10)$$

i.e., (particle acceleration) = (wave velocity)$^2$
× (rate of change of gradient of
displacement curve)

As before, for a longitudinal wave motion, the corresponding equation is

$$\frac{\partial^2 \xi}{\partial t^2} = v^2 \frac{\partial^2 \xi}{\partial x^2} \qquad (11\text{-}11)$$

## 11-7 Velocity of a sound wave

The velocity of a sound wave depends on the density and the elasticity of the carrier medium. As a sound wave travels through a medium, a series of compressions and rarefactions are produced at each point along the wave path. The pressure, and hence the density, varies locally. The magnitude of the variation depends on the elasticity of the medium. The following terms are used to describe these variations.

**Dilation** ($\Delta$) This is the amount of compression and rarefaction produced by a longitudinal wave. It is defined as the ratio of the increment of volume, $\delta V$, to the original volume, $V_0$,

i.e., $$\Delta = \delta V/V_0 \qquad (11\text{-}12)$$

and $$V = V_0(1 + \Delta)$$

**Condensation** ($s$) This is the amount of compression and rarefaction expressed in terms of density changes. It is defined as the ratio of the increment of density, $\delta\rho$, to the original density, $\rho_0$,

i.e., $$s = \delta\rho/\rho_0 \qquad (11\text{-}13)$$

and $$\rho = \rho_0(1 + s)$$

Also, for a constant mass of gas

$$\rho V = \rho_0 V_0$$

$$\therefore\ (1 + s)(1 + \Delta) = 1$$

$$\therefore\ 1 + \Delta + s + s\Delta = 1$$

i.e., neglecting $s\Delta$ as small $$s = -\Delta \qquad (11\text{-}14)$$

**Bulk modulus of elasticity** ($K$) This is the ratio of stress to strain, that is, the ratio of the force per unit area (the pressure) to the change in volume per unit volume.

$$\therefore\ K = \frac{\delta p}{-\delta V/V_0} = -V_0 \frac{\delta p}{\delta V} \qquad (11\text{-}15)$$

where $\delta p$ is the change in pressure that produces the change $\delta V$ in volume. The negative sign signifies a decrease in volume with increasing pressure. Also

$$K = \frac{\delta p}{s} \tag{11-16}$$

since $$V_0/\delta V = 1/\Delta = -1/s$$

Consider a plane wave progressing in the positive $x$-direction. Then, if the displacement at time $t$ of a particle from its mean position, $x$, is $\xi$, the displacement of another particle from its mean position, $(x + \delta x)$, is $(\xi + \delta\xi)$. If $\partial\xi/\partial x$ is the gradient of displacement with position, the displacement of the second particle is $\left(\xi + \frac{\partial \xi}{\partial x}\,\delta x\right)$ (Fig. 11-4). Thus the

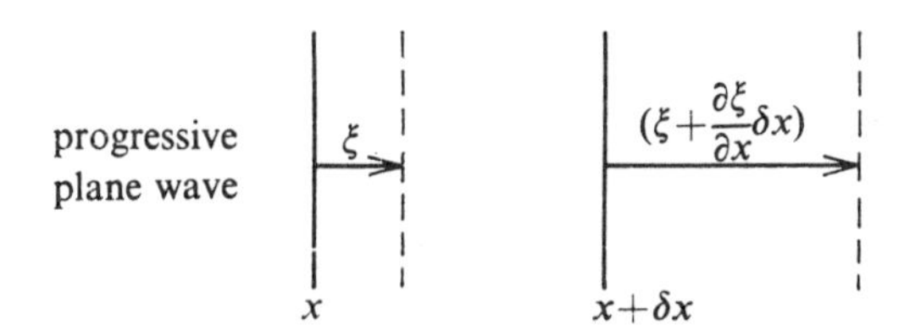


Fig. 11-4 Motion of a progressive plane wave

thickness of the layer bounded by the planes changes from $\delta x$ at time $t = 0$ to $\left(\delta x + \frac{\partial \xi}{\partial x}\,\delta x\right)$ at the later time—that is, to $\delta x\left(1 + \frac{\partial \xi}{\partial x}\right)$. Therefore the volume of the layer for a plane wave of unit cross-sectional area changes from $V_0$ to $V = V_0\left(1 + \frac{\partial \xi}{\partial x}\right) = V_0(1 + \Delta)$, from the definition of 'dilation'.

$$\therefore\ \Delta = \frac{\partial \xi}{\partial x} = -s \tag{11-17}$$

Also, from the definition of the bulk elasticity modulus Eqn (11-16), the excess pressure on the $(x + \delta x)$ plane is

$$\delta p = Ks = -K\frac{\partial \xi}{\partial x} \tag{11-18}$$

and therefore the total pressure is

$$p = p_0 - K\frac{\partial \xi}{\partial x} \tag{11-19}$$

By differentiation, providing $\xi$ is small enough for $K$ to be constant,

$$\frac{\partial p}{\partial x} = -K\frac{\partial^2 \xi}{\partial x^2} \tag{11-20}$$

Also, since $\delta p$ is the extra pressure, or force, on the second plane,

$$\delta p = \text{mass} \times \text{acceleration}$$

$$= -\rho\, \delta x \frac{\partial^2 \xi}{\partial t^2}$$

for unit cross-section of wave, and where the acceleration decreases with increasing force.

$$\therefore \frac{\partial p}{\partial x} = -\rho \frac{\partial^2 \xi}{\partial t^2} \qquad (11\text{-}21)$$

Therefore, from Eqns (11-20) and (11-21)

$$\frac{\partial^2 \xi}{\partial t^2} = \frac{K}{\rho} \frac{\partial^2 \xi}{\partial x^2} \qquad (11\text{-}22)$$

By comparison with Eqn (11-11),

i.e.. $$\frac{\partial^2 \xi}{\partial t^2} = v^2 \frac{\partial^2 \xi}{\partial x^2}$$

$$v = \sqrt{\frac{K}{\rho}}$$

i.e., $$\text{velocity} = \sqrt{\frac{\text{elasticity}}{\text{density}}} \qquad (11\text{-}23)$$

## 11-8 Principle of superposition

The principle of superposition, as first expounded by Huygens for rays of light, applies to all waves of small amplitude. It says that, if two wave motions cross each other, the effect to an outside observer is as if the waves had interfered with each other in the region of crossing, with neither modifying the other beyond this region. For two sound waves of the same frequency, the resultant wave in the region of crossing appears as the vector sum of the amplitudes and phases summed independently. Thus, if two or more waves come together, they are replaced in the region of crossing by some new wave motion that is different from all of the original waves. This must be so because it is impossible for two or more waves travelling together to be independent: each vibrating particle of the carrier medium that makes up the wave motion cannot have two or more velocities or displacements at the same time. Thus, if two waves come together and have the same phase, they enhance each other and produce a new wave of greater amplitude than either of them. The opposite occurs if the waves are out of phase. Sound waves have the same velocity in the region of interference; they can differ only in frequency (and wavelength), phase, and amplitude.

The most important cases of interference in sound waves occur with the waves travelling in the same line or with two waves moving along paths at right angles to each other.

## 11-9 Composition of simple harmonic motions in the same line

Two cases occur; first, for waves of the same frequency and, second, for waves of different frequencies.

**Composition of simple harmonic motions of the same frequency** If the wave motions have the same frequency, they differ only in amplitude and phase. Then for two waves, if AB (Fig. 11-5) describes one wave motion vectorially, both in amplitude and phase, and BC describes another wave similarly, then AC describes the resultant wave in both amplitude and phase. That is, if the equations of the two waves are

$$y_1 = a \sin(\omega t + \alpha) \tag{11-24}$$

and

$$y_2 = b \sin(\omega t + \beta) \tag{11-25}$$

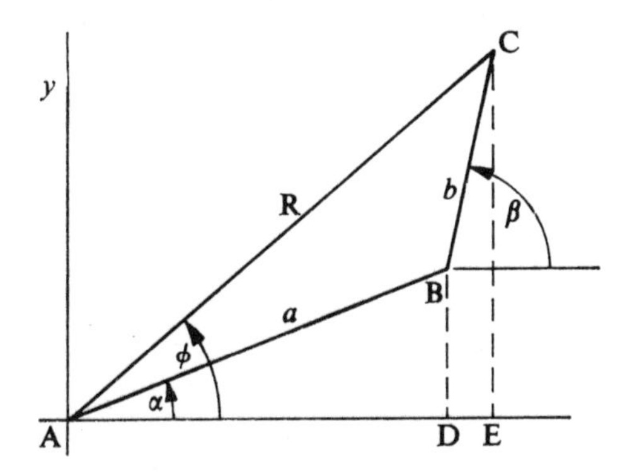


Fig. 11-5 Vector addition of amplitudes and phases

where $a$ and $b$ are the corresponding amplitudes, and $\alpha$ and $\beta$ are the phase constants, $y_1$ and $y_2$ are the displacements, and $\omega = 2\pi f = 2\pi/T$, then the resultant displacement is $y = y_1 + y_2$. From Fig. 11-5

$$\text{AE} = a\cos\alpha + b\cos\beta = R\cos\phi$$

$$\text{EC} = a\sin\alpha + b\sin\beta = R\sin\phi$$

then

$$R^2 = (\text{AE})^2 + (\text{EC})^2$$

$$= a^2 + b^2 + 2ab(\cos\alpha\cos\beta + \sin\alpha\sin\beta)$$

i.e.,

$$R = [a^2 + b^2 + 2ab\cos(\alpha - \beta)]^{1/2} \tag{11-26}$$

and

$$\tan\phi = \frac{\text{EC}}{\text{AE}} = \frac{a\sin\alpha + b\sin\beta}{a\cos\alpha + b\cos\beta} \tag{11-27}$$

from which $\phi$ may be found. The equation of the resultant displacement is then

$$y = R\sin(\omega t + \phi) \tag{11-28}$$

This method may be continued in the same way to sum any number of waves of the same frequency and in the same line.

Alternatively, the waves may be summed graphically by drawing the waves with the correct amplitudes and phase relationships and adding the amplitudes algebraically point by point (Fig. 11-6). This method obtains the form of the resultant wave and is applicable to any number of waves, and to waves of different frequency.

If the two waves have the same frequency but are travelling in opposite directions, *standing waves* are formed. Equations (11-29) and (11-30) represent two such waves having the same frequency and amplitude and travelling in opposite directions.

$$y_1 = a \sin(\omega t - \alpha) \tag{11-29}$$

$$y_2 = a \sin(\omega t + \alpha) \tag{11-30}$$

By addition, for the resultant wave motion

$$y = y_1 + y_2 = 2a \sin \omega t \cos \alpha \tag{11-31}$$

where $\alpha = \omega x/v$ from Eqn (11-3).

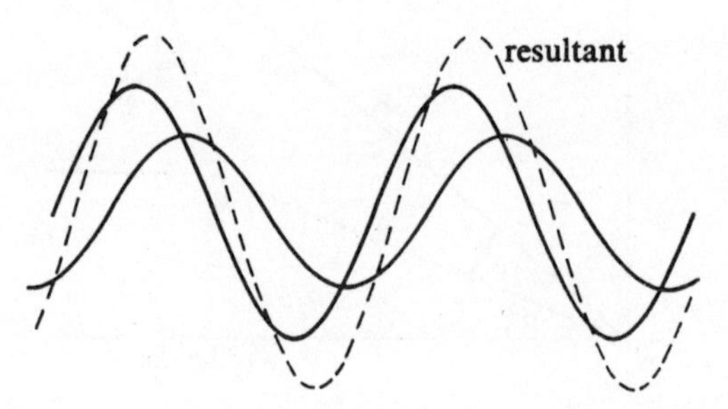


Fig. 11-6 Graphical summation of two wave forms

For any value of $x$—that is, $\alpha$ constant—this represents a simple harmonic motion; and for any value of $t$, the displacement $y$ is zero whenever the term $\cos \alpha$ is zero. This is so whenever

$$\alpha = \frac{\pi}{2}, \frac{3\pi}{2}, \frac{5\pi}{2}, \ldots$$

$$= (2m + 1)\frac{\pi}{2}, \quad \text{where } m = 0, 1, 2, \ldots$$

But $$\alpha = \frac{\omega x}{v} = \frac{2\pi}{T}\frac{x}{v} = \frac{2\pi x}{\lambda}$$

i.e., $$\frac{2\pi x}{\lambda} = (2m + 1)\frac{\pi}{2}$$

$$\therefore \; x = (2m + 1)\frac{\lambda}{4}$$

Thus, a standing wave is formed with the points given by $x = \lambda/4$, $3\lambda/4$, $5\lambda/4, \ldots$ being points of zero displacement, $y$, for all $t$. These points are called *nodes*. The distance between successive nodes is half a wavelength. The displacement, $y$, is a maximum—an *antinode*—when $\cos(\omega x/v) = 1$, that is, when $x = 0, \lambda/2, \lambda, 3\lambda/2, \ldots = m\lambda/2$, giving the maximum displacement as $2a$. The antinodes too are separated by a distance equal to half a wavelength. Thus, the resultant curve now oscillates between an equilibrium position with no particle displacement and a simple harmonic wave of maximum amplitude $2a$ with fixed nodes and which does not progress to left or right. Figure 11-7 shows a standing wave formed from two transverse waves moving in opposite directions.

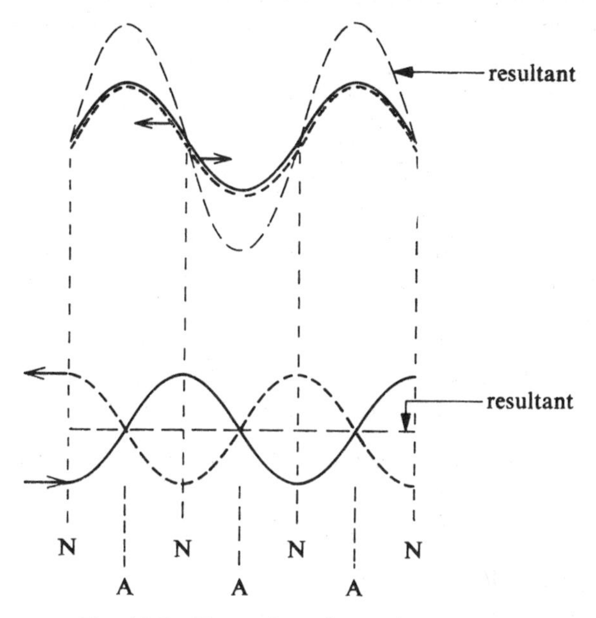


Fig. 11-7 Formation of a stationary wave

If the particle displacements had been in the $x$-direction instead of the $y$-direction, a standing longitudinal wave would have been formed in the same manner.

**Composition of simple harmonic motions of different frequencies** For wave motions of different frequencies, the resultant wave is periodic but not simple harmonic. If two waves of different frequencies, amplitudes, and phases are described by the equations

$$y_1 = a \sin(\omega_1 t + \alpha) \tag{11-32}$$

$$y_2 = b \sin(\omega_2 t + \beta) \tag{11-33}$$

then the resultant wave motion is

$$y = a \sin(\omega_1 t + \alpha) + b \sin(\omega_2 t + \beta) \tag{11-34}$$

The motions may also be summed graphically as before (Fig. 11-6).

One case involving waves of different frequencies is of particular importance. This is when the frequencies are only slightly different, which means that two waves are only rarely in step. Whenever they are in step, their resultant amplitude is a maximum, falling to a minimum when they are completely out of step. Thus the loudness of the sound waxes and wanes, and *beats* are heard. The difference in frequencies of the sounds, which gives the number of times the waves are in step in one second, is equal to the number of beats produced per second. This is an accurate method of tuning two notes to the same frequency, since beats as slow as one in several seconds can be easily detected. That the beat frequency is equal to the difference in frequencies of the sounding bodies may be shown as follows.

If two waves of almost the same frequency have

$$y_1 = a \sin (\omega t + \alpha) \qquad (11\text{-}35)$$

$$y_2 = b \sin [(\omega + \delta\omega)t + \beta]$$
$$= b \sin (\omega t + \delta\omega . t + \beta) \qquad (11\text{-}36)$$

where $\delta\omega = 2\pi\delta f$ and where $\delta f$ is the difference in frequency between the two wave motions, then, since $\delta\omega . t$ is small, $(\delta\omega . t + \beta)$ approximates to the phase constant term in Eqn (11-25) and the Eqns (11-35) and (11-36) become similar to Eqns (11-24) and (11-25). Therefore, from Eqns (11-26) and (11-27), the resultant amplitude is

$$R = [a^2 + b^2 + 2ab \cos (\alpha - \beta')]^{1/2} \qquad (11\text{-}37)$$

where $\beta' = (\delta\omega . t + \beta)$ and the resultant phase angle is $\phi$, where

$$\tan \phi = \frac{a \sin \alpha + b \sin \beta'}{b \cos \alpha + b \cos \beta'} \qquad (11\text{-}38)$$

The resultant wave then approximates to a simple harmonic motion of the form [Eqn (11-28)]

$$y = R \sin (\omega t + \phi)$$

where $R$ and $\phi$ vary slowly as $\beta'$ varies with time. The amplitude $R$ is a maximum whenever $\cos (\alpha - \beta') = 1$, from Eqn (11-37). Then $R$ equals $(a + b)$. It also equals $(a - b)$, a minimum, whenever $\cos (\alpha - \beta') = -1$.

For a maximum amplitude (when $\cos (\alpha - \beta') = 1$) $\alpha - \beta' = 0, 2\pi, \ldots$, that is, $\alpha - \beta - \delta\omega . t = 0, 2\pi, \ldots$ If we take the case when $\alpha = \beta$, then at time $t = 0$ the resultant amplitude is a maximum. This amplitude is a maximum again at a time $t$ later, given by

$$\delta\omega . t = 2\pi$$

but $\delta\omega = 2\pi\, \delta f$ where $\delta f$ is the difference in the frequencies of the two wave motions.

$$\therefore\ 2\pi\, \delta f . t = 2\pi$$

that is, the resultant amplitude is again a maximum after a time $t = 1/\delta f$. Therefore, maximum amplitudes, beats, occur $1/t$ times per second; that is, the beat frequency is $1/t = \delta f$ Hz, where $\delta f$ is the difference in frequency between the original two waves.

## 11-10 Composition of simple harmonic motions at right angles

Again two cases occur; first, for simple harmonic motions of the same frequency, and second, for motions of different frequencies.

**Composition of simple harmonic motions of the same frequency** A simple harmonic motion may be thought of as being that of the projection of a moving point P, as in Section 11-5 and Fig. 11-3. Thus, two simple harmonic motions of the same frequency, and at right angles, may be thought of as due to two points moving around two circles in the same plane with the same angular velocity, with the projection of one being onto the $y$-axis and of the other onto the $x$-axis. The compounded displacements give the resultant motion known as a *Lissajous' figure*.

Lissajous' figures are easily demonstrated by using a cathode ray oscilloscope. Alternating potentials applied to the X- and Y-plates of the oscilloscope move the electron beam under two simple harmonic motions at right angles. Varying the phase difference between the applied potentials produces various Lissajous' figures on the fluorescent screen.

Figure 11-8 represents the resultant displacements due to the in-phase simple harmonic motions generated by projecting the motion of $P_1$ onto

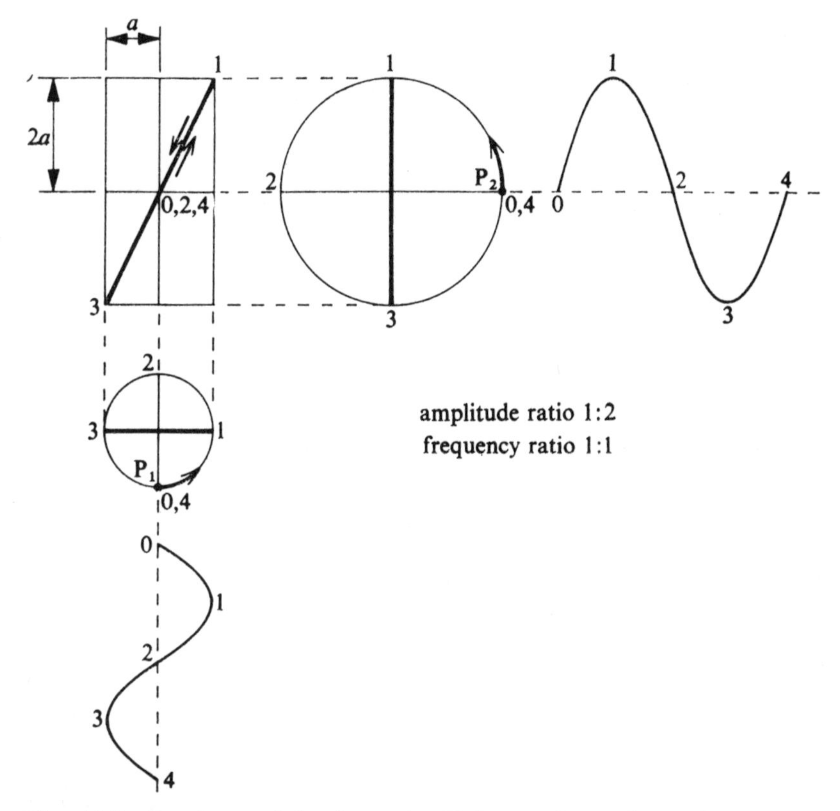


Fig. 11-8 Resultant of in-phase simple harmonic motions of the same frequency

the $x$-axis and of $P_2$ onto the $y$-axis. The radii of the two circles represent the amplitudes of their respective simple harmonic motions.

The phase difference is determined by the respective starting points. Figure 11-9 represents a phase difference $(\beta - \alpha)$ of $\pi/4$, that is, the $y$-motion (due to $P_2$) starts an eighth of a period ahead of the $x$-motion. If the phase difference is $7\pi/4$, a similar ellipse is generated but it rotates in the opposite direction.

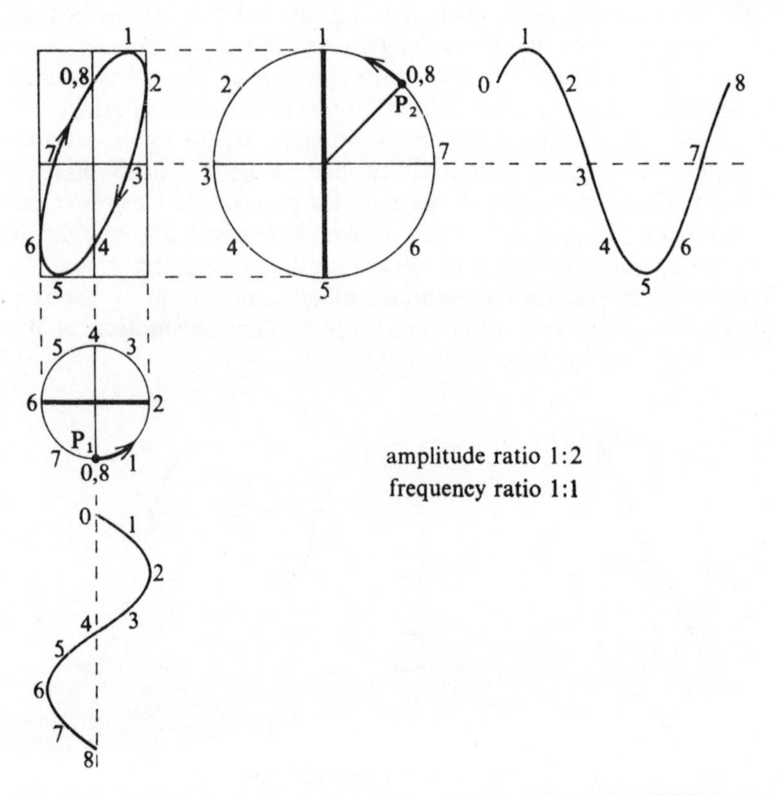


Fig. 11-9 Resultant of simple harmonic motions differing in phase by $\pi/4$

For a phase difference of $\pi/2$, an ellipse rotating in a clockwise direction is generated with the $x$- and $y$-axes as the axes of the ellipse. A similar ellipse is generated for a phase difference of $3\pi/2$, but this time it rotates in an anticlockwise direction. If the amplitudes of the two generating simple harmonic motions are equal, then these ellipses become circles. Similarly, two plane polarized beams of light with their electric vibration directions at right angles will combine to produce elliptically or circularly polarized light.

These motions may also be considered analytically. Thus, let the Eqns (11-39) and (11-40) represent the generating simple harmonic motions, having the same frequency but different amplitudes, $a$ and $b$, and different phase constants, $\alpha$ and $\beta$, and with displacements in the $x$- and the $y$-directions respectively.

$$x = a \sin(\omega t + \alpha) \tag{11-39}$$

$$y = b \sin(\omega t + \beta) \tag{11-40}$$

These equations must be compounded, by the principle of superposition, to find the path of the resultant displacements. Then, expanding the equations,

$$\frac{x}{a} = \sin \omega t \cos \alpha + \cos \omega t \sin \alpha \tag{11-41}$$

$$\frac{y}{b} = \sin \omega t \cos \beta + \cos \omega t \sin \beta \tag{11-42}$$

Multiplying Eqn (11-41) by $\cos \beta$ and Eqn (11-42) by $\cos \alpha$, and subtracting the first equation from the second, gives

$$-\frac{x}{a}\cos \beta + \frac{y}{b}\cos \alpha = \cos \omega t(\sin \beta \cos \alpha - \cos \beta \sin \alpha)$$

$$= \cos \omega t \sin(\beta - \alpha) \tag{11-43}$$

Similarly, multiplying Eqn (11-41) by $\sin \beta$ and Eqn (11-42) by $\sin \alpha$, and subtracting the second equation from the first

$$\frac{x}{a}\sin \beta - \frac{y}{b}\sin \alpha = \sin \omega t(\sin \beta \cos \alpha - \cos \beta \sin \alpha)$$

$$= \sin \omega t \sin(\beta - \alpha) \tag{11-44}$$

By squaring and adding Eqns (11-43) and (11-44) $t$ may be eliminated:

$$\frac{x^2}{a^2} + \frac{y^2}{b^2} - \frac{2xy}{ab}\cos(\beta - \alpha) = \sin^2(\beta - \alpha) \tag{11-45}$$

This is the equation of the resultant path, where $(\beta - \alpha)$ is the difference in phase between the generating simple harmonic motions.

In particular, when $(\beta - \alpha) = 0, 2\pi, 4\pi, \ldots$, the resultant motion has the equation

$$\frac{x^2}{a^2} - \frac{2xy}{ab} + \frac{y^2}{b^2} = 0$$

i.e.,

$$\left(\frac{x}{a} - \frac{y}{b}\right) = 0$$

and

$$x = \frac{a}{b}y$$

This is the equation of a straight line passing through the origin and with slope $a/b$ (Fig. 11-8). If $(\beta - \alpha) = \pi, 3\pi, 5\pi, \ldots$ a straight line is again formed but with slope $-a/b$.

When $(\beta - \alpha) = \pi/2, 3\pi/2, 5\pi/2, \ldots$, the resultant motion becomes

$$\frac{x^2}{a^2} + \frac{y^2}{b^2} = 1$$

an ellipse, with semi-axes $a$ and $b$ coincident with the $x$ and $y$ axes. For generating motions with equal amplitudes, $a = b$, this ellipse becomes a circle.

**Composition of simple harmonic motions of different frequencies** The only case of interest in compounding simple harmonic motions at right angles when the generating motions have different frequencies is when the

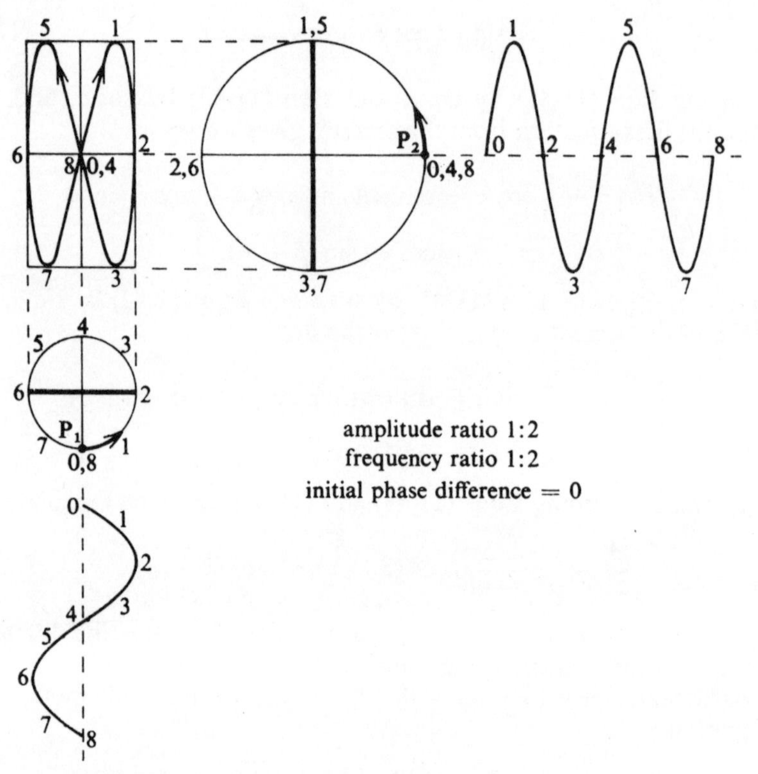


Fig. 11-10 Resultant of in-phase simple harmonic motions with frequency ratio 1:2

two frequencies are commensurate—that is, they are in a simple ratio to each other. For example, Fig. 11-10 represents the resultant motion due to two simple harmonic motions having frequencies in the ratio 1:2 and starting with their initial phases the same. Figure 11-11 represents the resultant of the same two motions, but this time they start off with an initial phase difference corresponding to a quarter of the smaller period. Again the radii of the circles represent the respective amplitudes of the generating simple harmonic motions.

If the frequencies are not exactly commensurate but nearly so, the form of the resultant slowly changes as the phase difference slowly varies in time. This has an important application in the measurement of frequencies, since the rate of change of the Lissajous' figure is the frequency difference involved. A standard known frequency is used by compounding it with the slightly different unknown frequency. In general, when the frequencies are in the ratio 1:1, one loop is formed; when the frequencies are in the ratio 1:2, then two loops are formed. This is still generally true when the frequencies are in the ratio 1: *n*, where *n* loops are formed. Although in theory the resultant motion may be found analytically, in practice this method becomes far too cumbersome.

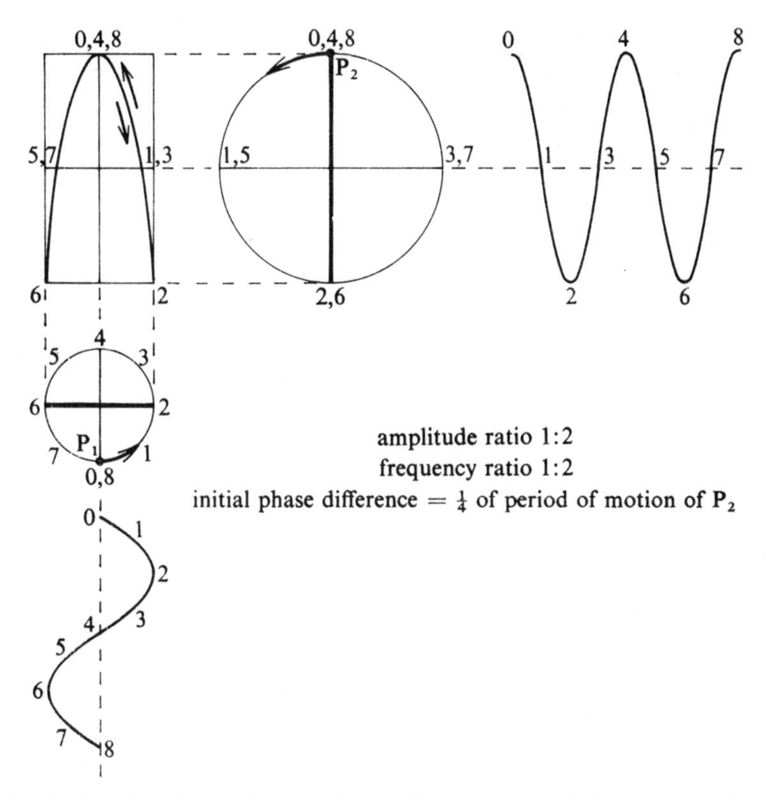


Fig. 11-11 Resultant of simple harmonic motions with frequency ratio 1:2 and with an initial phase difference

## PROBLEMS

**11-1** Explain why the loudest sound is heard at a node of displacement of a standing wave in air, and not at an antinode.

**11-2** Distinguish between progressive and standing waves and show that two trains of equal simple waves moving with equal velocities in opposite directions produce a series of standing waves.

**11-3** Explain the production of beats by the interference of two notes, and how two notes, themselves inaudible, may produce an audible note.

**11-4** Explain how a longitudinal wave may be represented by means of a sine curve, and show the distinction between a compression curve and a displacement curve.

**11-5** A particle moves at constant velocity round a circular path. If the projection of the particle onto a diameter moves with a velocity of 3 m/s when it is 0·3 m from the centre, and 2 m/s when it is 0·6 m from the centre, find the radius of the circular path. [0·609 m]

**11-6** Compound graphically the following pairs of simple harmonic motions in the same straight line: (a) frequencies $3n$ and $n$, amplitudes $a$ and $2a$, initially in phase; (b) frequencies in ratio 6:5, equal amplitudes, initially in phase.

**11-7** Compound graphically the following pairs of simple harmonic motions at right angles: (a) frequencies $3n$ and $n$, amplitudes $a$ and $2a$, initially in phase; (b) $x = a \sin 2\pi nt$, $y = 2a \sin (6\pi nt + \pi/2)$.

**11-8** A standard frequency signal of 250 Hz is applied to the X-plates of an oscilloscope and a second signal of approximately twice this frequency is applied to the Y-plates, producing Lissajous' figures that go through a complete cycle of change once in 10 s. If the frequency of this second signal is then reduced slightly it is observed that the cycle of change now takes place in 5 s. What is the frequency of the second signal before and after its frequency reduction? [499·9 Hz; 499·8 Hz]

# 12 Velocity of sound

## 12-1 Introduction

The velocity of sound depends on the elasticity and density of the medium through which the sound wave is being propagated [Eqn (11-23)]. Thus by a measurement of the sound velocity, information may be obtained that may not readily be obtainable by direct measurements. For example, the relevant elasticity modulus of a gas may be expressed in terms of the ratio of its specific heats at constant pressure and at constant volume (Section 12-2) and from this Lord Rayleigh was able to demonstrate that argon is monatomic.

The velocity of sound is also important in the design of buildings—for example, in the design of auditoria and the elimination of echoes—and as a measure of the velocity of propagation of mechanical disturbances.

## 12-2 Velocity of sound in gases

The velocity of sound through a medium is given by $\sqrt{(\text{elasticity/density})}$. It was Newton who first made a theoretical calculation of the velocity, by applying Boyle's law to this equation. According to Boyle's law the volume occupied by a gas is inversely proportional to its pressure, providing the temperature remains constant. This law is only true for an ideal gas, but is obeyed closely by air at normal temperature and pressure. As a longitudinal sound wave passes a particular point in a medium, a series of compressions and rarefactions succeed each other at that point. For a gas this results in a given mass at that point changing its pressure and volume, $p_1$ and $V_1$, to $p_2$ and $V_2$ such that

$$p_1 V_1 = p_2 V_2 \tag{12-1}$$

i.e.,

$$pV = \text{const.} \tag{12-2}$$

$$\therefore\ p\,\mathrm{d}V + V\,\mathrm{d}p = 0$$

$$p = -V\frac{\mathrm{d}p}{\mathrm{d}V}$$

i.e.,

$$p = -\frac{\mathrm{d}p}{\mathrm{d}V/V} = K \tag{12-3}$$

$dp$ is the change in stress at the point and $dV/V$ is the change in volume per unit volume, that is the strain. Thus Eqn (12-3) is in fact equal to the bulk modulus of elasticity, $K$.

Therefore, from Eqn (11-23)

$$\text{velocity} = \sqrt{\frac{\text{absolute pressure}}{\text{density}}} \tag{12-4}$$

Substituting appropriate values for air at 0°C gives

$$\begin{aligned}\text{velocity} &= \sqrt{\left(\frac{1{\cdot}013 \times 10^5 \text{ N/m}^2}{1{\cdot}29 \text{ kg/m}^3}\right)} \\ &= 280 \text{ m/s} \\ &= 919 \text{ ft/s}\end{aligned}$$

which is about 15% lower than the experimentally observed value.

Some years later Lagrange suggested that the pressure changes taking place as a wave passes are too rapid to be isothermal as assumed by Newton. That is, the heat gains and losses were considered so rapid that there was no conduction to or from the surrounding medium, and therefore the changes should be treated as taking place adiabatically. Later still (in 1816), Laplace modified Newton's equation in accordance with Lagrange's suggestion of adiabatic conditions. It was, however, pointed out by Herzfeld and Rice (1928) that the adiabatic conditions are due to the slowness of the pressure changes, not their rapidity as originally suggested. This is because for ordinary frequencies, say the audible range, the wavelength is too large and the thermal conductivity too small for any appreciable flow of heat.

Equation (12-2) must therefore be modified to

$$pV^\gamma = \text{const.} \tag{12-5}$$

where

$$\begin{aligned}\gamma &= \frac{\text{specific heat of gas at constant pressure}}{\text{specific heat of gas at constant volume}} \\ &= \frac{C_p}{C_V}\end{aligned}$$

Then, differentiating Eqn (12-5),

$$p\gamma V^{\gamma-1}\,dV + V^\gamma\,dp = 0$$

$$\therefore \gamma p = -\frac{V^\gamma\,dp}{V^{\gamma-1}\,dV}$$

$$= -V\frac{dp}{dV}$$

i.e.,

$$\gamma p = K \tag{12-6}$$

That is, $\gamma p$ is now equal to the bulk modulus, $K$.

The velocity is therefore given by

$$\text{velocity} = \sqrt{\frac{\gamma \times \text{absolute pressure}}{\text{density}}} \tag{12-7}$$

that is, for air at 0°C

$$\begin{aligned}\text{velocity} &= \sqrt{\gamma} \times 280\\ &= 332 \text{ m/s}\\ &= 1{,}089 \text{ ft/s}\end{aligned}$$

where $\gamma = 1{\cdot}41$ for air. This is in good agreement with the experimental value (331·45 m/s = 1,087·42 ft/s, at 0°C).

For gases other than air the same equation for the velocity holds. Different velocities are then due to different densities. Then, comparing the velocity in air with that in some other gas at the same pressure

$$v_{\text{air}} = \sqrt{\frac{\gamma p}{\rho_{\text{air}}}}$$

$$v_{\text{gas}} = \sqrt{\frac{\gamma p}{\rho_{\text{gas}}}}$$

i.e.,

$$\frac{v_{\text{air}}}{v_{\text{gas}}} = \sqrt{\frac{\rho_{\text{gas}}}{\rho_{\text{air}}}} \tag{12-8}$$

That is, the velocity of sound in a gas varies inversely as the square root of the density of the gas. Thus, for example, the velocity of sound in hydrogen is 1,269 m/s at 0°C, compared with a velocity in air of 331 m/s. This proportionality however assumes that the ratio of the specific heats, $\gamma$, is constant. This is approximately true since most gases are diatomic; for these $\gamma$ is approximately 1·40. For example, $\gamma$ has the value of 1·403 for air at room temperature and atmospheric pressure, and 1·410 for hydrogen. For monatomic gases such as argon, $\gamma$ is approximately 1·67; for polyatomic gases, such as hydrogen sulphide, $\gamma$ falls to approximately 1·33.

## 12-3 Variations affecting velocity

For a fixed mass of gas that obeys Boyle's law, the pressure varies inversely as the volume. Also, the density of the gas varies inversely as the volume, and hence the pressure and the density of a fixed mass of gas vary in the same ratio. This means [Eqn (12-7)] that the velocity is independent of the pressure, providing Boyle's law holds for the gas.

The effect of temperature changes on velocity is, however, far greater. The pressure of the atmosphere does not depend on temperature, but density does. Thus, if a given mass of gas at absolute temperature $T_0$ and density $\rho_0$ changes by $t$°C to some new temperature $T_t$ and a corresponding density $\rho_t$, then

$$\frac{\rho_0}{\rho_t} = \frac{T_t}{T_0} \tag{12-9}$$

The corresponding velocities of sound are then at $T_0$

$$v_0 = \sqrt{\frac{\gamma p}{\rho_0}} \tag{12-10}$$

and at $T_t$

$$v_t = \sqrt{\frac{\gamma p}{\rho_t}} \tag{12-11}$$

i.e., by Eqn (12-9)

$$v_t = \sqrt{\left(\frac{\gamma p}{\rho_0}\frac{T_t}{T_0}\right)}$$

$$\therefore \; v_t = v_0 \sqrt{\frac{T_t}{T_0}} \tag{12-12}$$

That is, the velocity of sound in air is proportional to the square root of the absolute temperature.

Equation (12-12) may be written in a more convenient form as follows. Let $T_0 = 273°\text{K} = 0°\text{C}$, then the velocity at $t$°C is given by

$$v_t = v_0 \sqrt{\left(\frac{273 + t}{273}\right)}$$

$$= v_0 \sqrt{\left(1 + \frac{t}{273}\right)} \tag{12-13}$$

This gives a change in velocity of approximately 61 cm/s or 2 ft/s for each degree Celsius rise of temperature at ordinary atmospheric temperatures. Thus a velocity of 331 m/s (1,087 ft/s) at 0°C is increased to 343 m/s (1,125 ft/s) at 20°C.

Moisture slightly affects the velocity of sound, but only to the extent that it changes the mean density of the gas. The effect for saturated air is to increase the velocity by about 1 m/s over that in dry air.

If the sound is being propagated through a restricted mass of gas—for example, in a long pipe—the simple expressions for velocity are no longer true. For pipes less than about one metre in diameter, the velocity is reduced by an amount which depends on a number of factors, including the diameter of the tube, the nature of the tube material, its wall thickness, its heat conductivity, and its surface finish, the intensity of the sound, and the way heat is conducted away from the pipe. Helmholtz, Kirchhoff and others have investigated this problem and have shown that for a fixed frequency, the measured velocity $v'$ in a tube of radius $r$ is related to the true 'free' velocity $v$ by the equation

$$v' = v(1 - \mathrm{k}/r) \tag{12-14}$$

where k is a constant for the particular frequency and depends on the physical properties and condition of the tube. Thus, by using two tubes of different radii, but otherwise exactly similar in all other respects, and measuring the corresponding sound velocities $v'$, the tube effect may be eliminated and the true velocity $v$ determined.

## 12-4 Velocity of sound in liquids and solids

Equation (11-23) applies equally well to liquids, solids, and gases, providing the correct elasticity modulus is chosen. Liquids, like gases, require the bulk modulus. Since, however, liquids are almost incompressible, there is no need to distinguish between isothermal and adiabatic changes. Water has a compressibility of 0·000047 per atmosphere at atmospheric pressures. That is, the bulk modulus $K$ is equal to $(1{\cdot}013 \times 10^5\ \mathrm{N/m^2})/0{\cdot}000047 = 2{\cdot}15 \times 10^9\ \mathrm{N/m^2}$, and the velocity of sound through water is given by

$$v = \sqrt{\frac{\text{elasticity}}{\text{density}}}$$

$$= \sqrt{\left(\frac{2{\cdot}15 \times 10^9\ \mathrm{N/m^2}}{1{,}000\ \mathrm{kg/m^3}}\right)}$$

$$= 1{,}466\ \mathrm{m/s}$$

taking the density of water at its maximum value.

This calculated velocity may be compared with the measured value of 1,403 m/s at 0°C and 1,481 m/s at 20°C. In sea water the corresponding velocities are about 1,449 and 1,522 m/s depending on the salinity. The velocity varies with both temperature and pressure, but not in a simple manner. For pure water at a pressure of one atmosphere the variation of velocity $v$ in m/s with temperature $t$ in °C is given by

$$v = 1{,}403 + 5t - 0{\cdot}06t^2 + 0{\cdot}0003t^3 \qquad (12\text{-}15)$$

over the range 0°C to 60°C.

For longitudinal waves travelling along solids in the form of thin bars, Young's modulus of elasticity is involved. For a bar of copper the velocity is given by

$$v = \sqrt{\left(\frac{12{\cdot}2 \times 10^{10}\ \mathrm{N/m^2}}{8{,}900\ \mathrm{kg/m^3}}\right)}$$

$$= 3{,}701\ \mathrm{m/s\ at\ 20°C}$$

However, this depends on the past history of the bar, such as heat treatment, which affects the elastic modulus.

For solids having a large cross-section at right angles to the direction of propagation, Young's modulus is no longer applicable and a combination of the bulk and shear moduli are involved. Then, the so-called *bulk* or *plate* velocity for longitudinal waves is given by

$$v = \sqrt{\left(\frac{K + \frac{4}{3}G}{\rho}\right)} \qquad (12\text{-}16)$$

where $K$ is the bulk modulus and $G$ is the shear or rigidity modulus. The bulk velocity is always greater than the velocity in a thin bar.

For porous solid materials such as acoustic tiles and bricks, the discussion of velocity is more complicated, because it involves not only the elasticity of the solid but also the vibrations in the gas within the voids.

## 12-5 Measurement of velocity of sound in gases and liquids

Direct methods of velocity measurement may be used providing the medium is available in bulk. This generally restricts these methods to water and to air measurements. The earliest determinations were based on the fact that light may be regarded as being propagated instantaneously as compared with sound. Thus by observing the firing of a gun a known distance away the time that elapses between seeing the flash and hearing the bang may be measured with a stopwatch and the velocity calculated. Preferably, the experiment should be repeated immediately with the source and observer in reversed positions so that the velocity is measured in the opposite direction. By taking the mean velocity the effect of wind is eliminated. Similar measurements have been made under water. An explosive charge above the water was made to sound a bell under water. The flash was seen directly by a distant observer and the sound travelling through the water was received by means of a horn with the wide end under water and pointed towards the bell.

Modern methods are similar in principle. Two microphones, separated by a known distance, are arranged in line with the explosive source of sound. This arrangement has the advantage that measurements are not made near to the source of sound, where there is an increase of velocity due to the increased amplitude of the particle displacements due to the explosion. It also has the advantage that the observations are made at two points by the same method and with the same accuracy. Human errors are avoided by timing the interval between the pulses from the two microphones by an electrical method. For example, the signal may activate a pen recorder which marks a paper chart moving with constant velocity. The distance between the two marks on the chart is related to the time taken for the sound to travel between the microphones and hence to the velocity of sound.

Direct measurements have the advantage of simplicity but they are limited to media which are available in bulk. Indirect methods use sound sources of known frequency. For example, the method for two microphones may be modified as in Fig. 12-1(a). The signals from the two microphones A and B are fed separately to the X- and Y-plates of an oscilloscope as shown, resulting in a Lissajous' figure due to the composition of two waves at right angles, as discussed in the previous chapter. One of the microphones is then moved until the Lissajous' figure is a straight line [PQ in Fig. 12-1(a)]. The two signals then have a phase difference of 0°, or 180° (Fig. 11-8). If one of the microphones is now moved a further distance so that the Lissajous' figure is again a straight line, but with the opposite slope—that is, the phase difference has been changed by another 180°—then the distance the microphone was moved through must equal a half wavelength of the sound. So by moving the microphone through a number of straight-line Lissajous' figure conditions, the number of half wavelengths moved through is known. By measuring this distance the mean value of the wavelength is determined, and by knowing the frequency of the sound wave its velocity is determined from

$v = f\lambda$. The fault with this method is that the setting position as the Lissajous' figure changes from an ellipse to a straight line is not a very precise one. It is to reduce this error that it is necessary to go through a number of half-wave positions. Alternatively, if the signals from the two microphones or, what comes to the same thing, one signal direct from the loudspeaker source supply and the other signal from one of the microphones, are fed separately into a two beam oscilloscope as in Fig. 12-1(b), the two wave traces are displayed together. As the distance between the source and microphone changes the relative positions of the traces change.

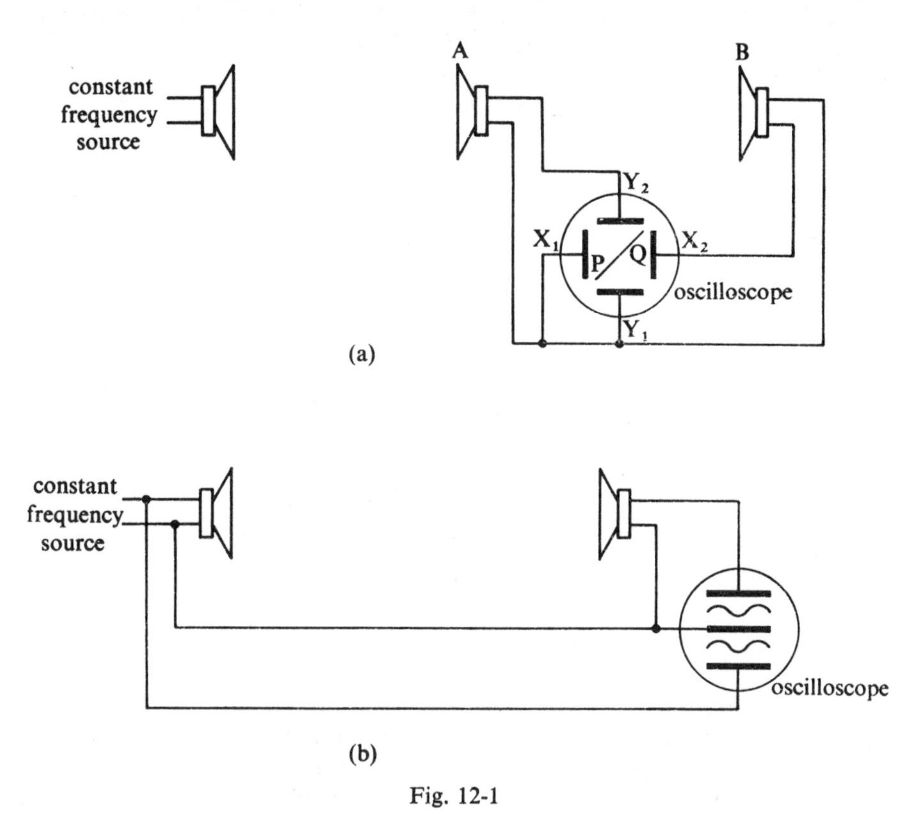


Fig. 12-1

A setting position can therefore be obtained when the two wave traces line up correspondingly. They again line up when the source to microphone distance is changed by one whole wavelength. Providing that the wave form being transmitted is a series of sharp pulses rather than a sine wave, the alignment may be made more precisely than in the Lissajous' figure method.

An earlier but somewhat similar method was first used by T. Hebb. He used two parabolic reflectors of 1·5 m aperture made from plaster of Paris (Fig. 12-2). At the focus of each reflector a microphone is located (A and B). Each microphone is connected through a battery to a separate primary of a transformer with a common secondary, as shown. The source of sound, which in Hebb's experiment was a whistle emitting a frequency of approximately 2,400 Hz, is also at A. Sound from the source at A reaches the microphone at A directly. Since the source at A is at the focus of the first reflector, a parallel beam of sound is reflected towards the second reflector, where it is again reflected to the microphone at B. The intensity of sound heard at the receiver R, connected to the secondary of the transformer, is thus the vector sum of the signals from the microphones at A and B. If now the second reflector, complete with its microphone at B, is shifted back or forth, the relative phases of the signals from the two microphones change. A position may then be found where the resultant sound intensity from the receiver R is a minimum. If the reflector and its microphone are shifted through a distance equal to one wavelength of the sound, the sound intensity at the receiver R again

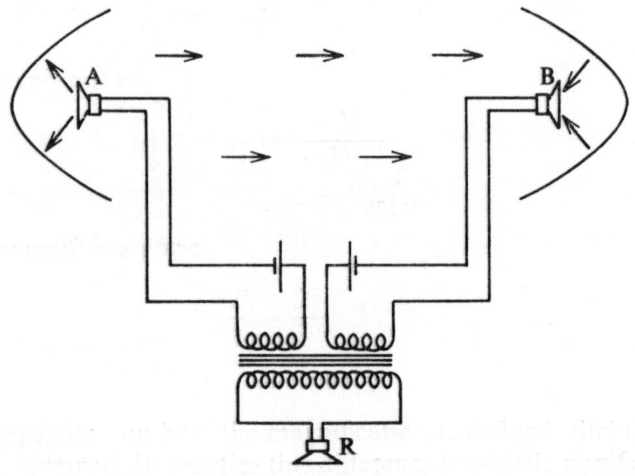


Fig. 12-2 Hebb's method for measuring the velocity of sound

falls to a minimum. Shifting the movable reflector through, say, 101 successive positions of minimum sound, gives the length of 100 wavelengths, about 14 m in air at 0°C using Hebb's source of sound. Assuming a minimum setting position can be located to within one-tenth of a wavelength, the wavelength calculated from a hundred wavelengths is accurate to two parts in a thousand, 0·2%. The velocity may be determined from $v = f\lambda$. The limitation of this method is the accuracy with which the frequency is known. Its advantage is that it can be used in the laboratory, so that there is no wind and the temperature and humidity may be controlled. Although this was originally a method for the measurement of velocity in air, it is adaptable for other gases and liquids and for measurements at different frequencies. The medium is contained in a tube between the two reflectors. Use of modern detectors, for example, an oscilloscope, in place of the telephone-type receiver, improves the accuracy of the measurements and means that the dimensions of the apparatus, especially the aperture of the reflectors, may be reduced. It should be noted that the diameter and other properties of the tube which contains the gas or liquid may affect the velocity (Section 12-3).

Kundt's tube method is interesting, although not very accurate. In its simplest form the apparatus consists of a long rod clamped at its centre point; a disc fixed at one end of the rod acts as a loose-fitting piston in a glass tube sealed at the other end (Fig. 12-3). The tube is movable so that the length of the air column contained in the glass cylinder may be adjusted. A layer of dry lycopodium or similar light powder is sprinkled inside the glass tube. By stroking the rod lengthways with a rosined cloth or holding a piece of solid carbon dioxide ice to one end, the rod is made to vibrate longitudinally, emitting a high-pitched note. The disc fixed to the end of the rod acts as a piston vibrating at this frequency and causes a sound wave to travel through the air in the glass tube. This wave is then reflected at the far end of the tube. The length of the air column is adjusted until a standing wave is set up (Section 11-9). When this happens, the powder in the tube is violently agitated and eventually comes to rest in a series of small piles. These form at the nodes, the positions of permanent rest of the standing wave. The distance between a number of nodes is

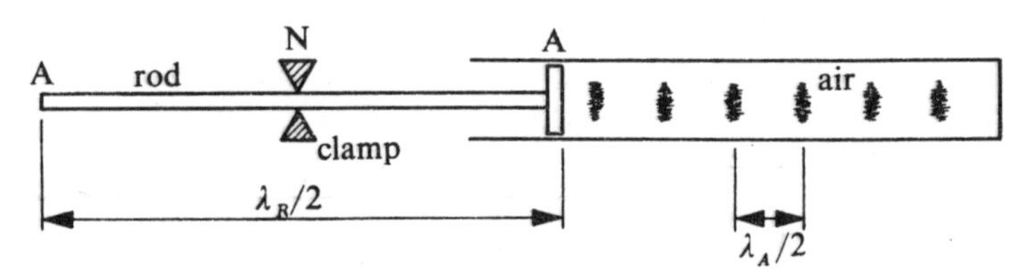


Fig. 12-3 Kundt's tube experiment

measured and hence the mean distance between consecutive nodes, equal to half a wavelength of the sound in air, $\lambda_A/2$, determined. Since the rod is clamped at its centre point, this point is a node for vibrations of the rod. The maximum movements of the vibrating rod occur at its ends, which are therefore antinodes. Thus the length of the rod, the distance between two consecutive antinodes, is equal to half a wavelength of the sound in the rod. The frequency of the sound emitted by the vibrating rod is $f = v_R/\lambda_R$ where $v_R$ is the velocity of sound in the rod and $\lambda_R$ the corresponding wavelength. By calculating the velocity of sound in the rod from $v_R = \sqrt{(E/\rho)}$ [Eqn (11-23)], where $E$ is Young's modulus of elasticity and $\rho$ the density of the material of the rod, and by measuring the wavelengths $\lambda_R$ and $\lambda_A$, the velocity of sound in air, $v_A$, may be determined, since

$$f = \frac{v_R}{\lambda_R} = \frac{v_A}{\lambda_A} \qquad (12\text{-}17)$$

The method may also be used for other gases, but it always has the limitation that the gas is contained in a tube, which affects the velocity.

The velocity, $v_G$, in a gas may be simply calculated from the velocity in air, since the frequency is the same in both instances and

$$\frac{v_A}{v_G} = \frac{\lambda_A}{\lambda_G} \tag{12-18}$$

where $\lambda_A$ and $\lambda_G$ are determined from measurements on the distance between nodes (the dust piles). In a more modern arrangement of this experiment the rod is replaced by a loudspeaker at the end of the glass tube. The other end of the tube is still closed. The loudspeaker is driven by a variable frequency oscillator. The frequency is then adjusted, keeping the length of the gas column constant, until a standing wave is formed, causing the powder to pile up at the nodes as before.

## 12-6 Measurement of velocity of sound in solids

If the velocity of sound in air is known, Kundt's tube method may be used to determine the velocity of sound in a solid in rod form [Eqn (12-17)]. The velocity in a solid in the form of a pipe may be determined by a method due to Biot, who used a pipe 950 m long. A sound pulse at one end of a pipe produces two sound waves: one, with a velocity $v_S$ through the solid wall of the pipe, and the other, with a velocity $v_A$, through the air contained inside the pipe. $v_S$ is greater than $v_A$. The interval between the times when the two waves reach the other end of the tube is

$$t = \frac{l}{v_A} - \frac{l}{v_S} \tag{12-19}$$

where $l$ is the length of the pipe. For an iron pipe 1 km long, the time interval $t$ is approximately $2\frac{1}{2}$ s. The length of the pipe is an obvious limitation, but if the sound pulse is sharp and an oscilloscope is used to measure the time interval, the length of pipe may be reduced to a more reasonable amount.

A better method involves direct measurement of the time taken for a sound pulse to travel along a bar, which may conveniently be less than half a metre in length. The ends of the bar specimen should be parallel to each other. A suitable transducer (Chapter 17), which converts an electric signal into a mechanical movement, is attached to one end of the bar. For example, this may be an electrostrictive or magnetostrictive transducer which changes its dimensions as an alternating electric field is applied, or simply a magnetic movement as used in a loudspeaker. An electronic trigger activates the pulse at regular intervals from 50 to 1,000 Hz. Besides starting the pulse the trigger simultaneously activates the time base control of an oscilloscope which shows a peak on the screen. The pulse travels the length of the bar and is detected by a gramophone-type pick-up that rests lightly against the other end. This signal is fed to the Y-plates of the oscilloscope where it produces a second peak on the screen. From the distance between the peak due to the pulse starting out and the peak due to it reaching the further end of the bar, the time taken for the pulse to travel the length of the bar may be determined. Hence, knowing the length of the bar, the velocity may be determined. If the time-base frequency is synchronized with the pulse repetition frequency, the peaks on the oscilloscope screen remain in the same positions. The oscilloscope time-base may be calibrated from a standard frequency source connected to the Y-plates. Alternatively, the pulse in the bar may be allowed to be reflected at the far end and then detected again at the starting end. This doubles the time of travel and improves the accuracy of measurement.

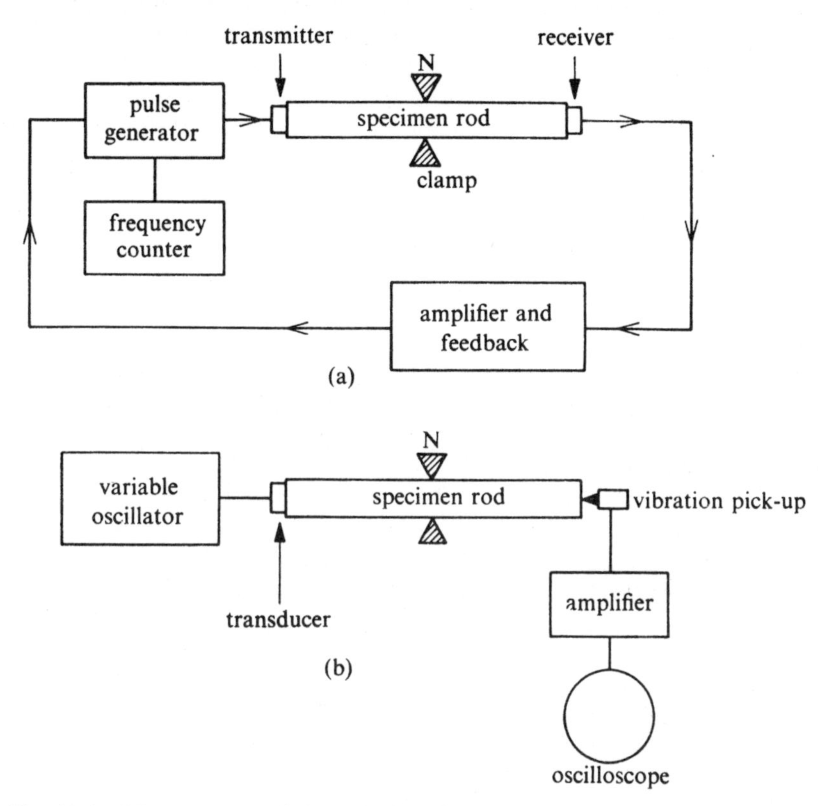


Fig. 12-4 Measurement of the velocity of sound in solids using (a) the 'sing around' and (b) a resonance method

A more accurate pulse method, known as the 'sing around' method, uses two transducers. One is attached to one end of the bar and is the transmitter, whilst the other is attached to the other end and acts as a receiver [Fig. 12-4(a)]. A pulse from the transmitter is received by the receiver which then triggers the next pulse from the transmitter. From the resulting trigger frequency the time of travel through the specimen may be determined. For precise measurement of velocity it is necessary to take account of the time delay in the electrical circuit. This may be done by repeating the measurement with a different length of bar. The constant electrical time delay may then be cancelled out.

In resonance methods the bar is made to vibrate continuously by an alternating signal fed to the transducer at one end. At the other end of the bar, which should be clamped at its mid-point, there is a gramophone-type pick-up [Fig. 12-4(b)], by which the vibration at this end of the bar may be displayed on an oscilloscope screen. The frequency of the driving transducer is increased until the bar resonates. The natural frequency of vibration as detected at the far end of the bar by the pick-up is then the same as the driving frequency of the transducer, and the amplitude of vibration rises to a maximum. Then the longitudinal vibrations of the bar have the same frequency as the signal fed to the transducer and the length of the bar is half a wavelength of the sound; that is, the two ends are antinodes and the centre, where the bar is clamped, is a node, as in the Kundt's tube method. Hence, knowing the frequency for resonance, the velocity may be calculated.

## 12-7 Doppler effect

The Doppler effect is a change in the apparent frequency of a sound wave due to relative motions between the source of sound and the observer. For example, the whistle of an approaching train may appear to be emitting a high frequency note, but as the train passes and travels away from the observer there appears to be a sudden drop in the frequency of the note. There are a number of such possible relative motions:

**Source in motion** Consider first a stationary source S [Fig. 12-5(a)], emitting a note of frequency $f$. Then in one second $f$ waves are emitted, which occupy a total distance $v$ where $v$ is the velocity of sound in the normal stationary situation. Thus the wavelength is $v/f$.

If now the source is moving towards the observer at O, with a velocity of $v_S$, the same number of waves, $f$, travelling with the velocity $v$ are emitted in one second. The velocity is a property of the medium and is independent of the source once the waves have left the source. However these waves now occupy a distance $v - v_S$. That is, the wavelength is now $(v - v_S)/f$. Thus the apparent frequency is given by.

$$f' = \frac{\text{velocity}}{\text{wavelength}} = \frac{v}{(v - v_S)/f}$$

$$= f\left(\frac{v}{v - v_S}\right) \qquad (12\text{-}20)$$

That is, the frequency is increased due to the source approaching the observer.

For the source travelling away from the observer, the $f$ waves emitted in one second are spread out over a distance $v + v_S$. Thus the apparent frequency is

$$f' = f\left(\frac{v}{v + v_S}\right) \qquad (12\text{-}21)$$

That is, there is a reduction in frequency.

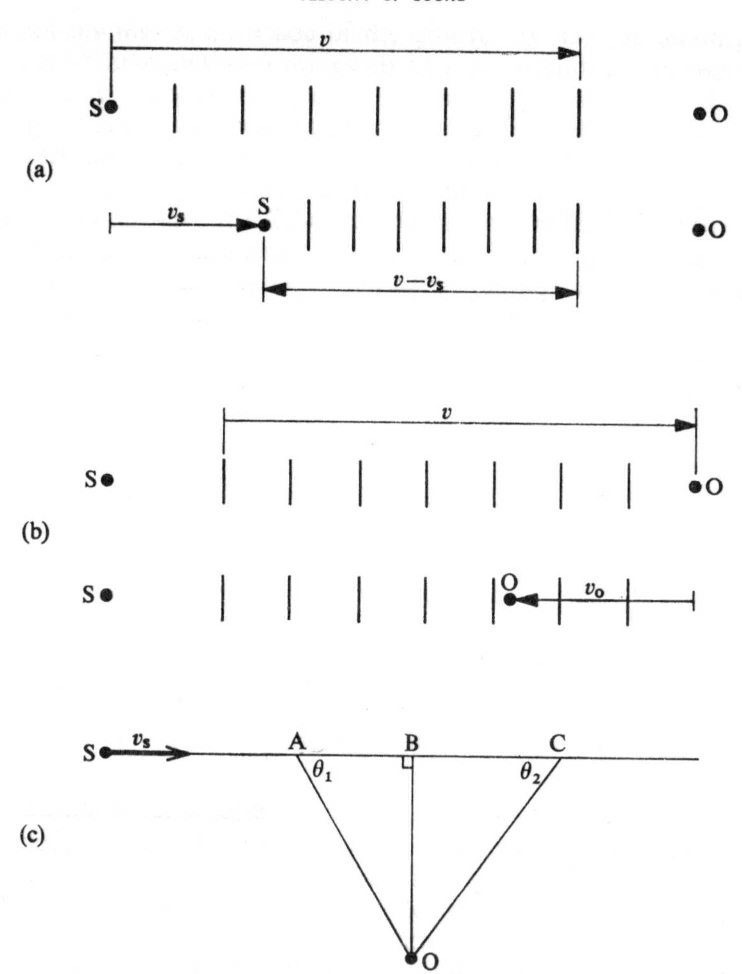


Fig. 12-5 Doppler effect due to (a) source approaching observer, (b) observer approaching source, and (c) source not moving in direction of observer

**Observer in motion** If the observer is at rest, then in one second $f$ waves pass him, occupying a distance $v$. If he is moving towards the source with a velocity $v_O$ [Fig. 12-5(b)] then $f$ waves still occupy a distance $v$. Their wavelength is unchanged and remains $v/f$. However, to the observer the waves appear to be travelling faster and to him they have a relative velocity $v + v_O$. Thus to him the apparent frequency would be

$$f' = \frac{\text{velocity}}{\text{wavelength}} = \frac{v + v_O}{v/f}$$

$$= f\left(\frac{v + v_O}{v}\right) \tag{12-22}$$

The frequency is thus increased for the observer approaching the source.

Conversely, if the observer is moving away from the source, the wave velocity relative to him is $v - v_O$ and the apparent frequency is

$$f' = f\left(\frac{v - v_O}{v}\right) \tag{12-23}$$

That is, there is a decrease in frequency.

**Source and observer in motion** Consider the case when the source and the observer are travelling towards each other. As far as the observer is concerned the true frequency being emitted by the source is $f'$ [Eqn (12-20)]. But, due to the motion of the observer, this frequency $f'$ is changed to $f''$ where by Eqn (12-22)

$$f'' = f'\left(\frac{v + v_O}{v}\right)$$

That is, the apparent frequency to an observer when the observer and the source are approaching each other is

$$f'' = f\left(\frac{v}{v - v_S}\right)\left(\frac{v + v_O}{v}\right)$$

$$= f\left(\frac{v + v_O}{v - v_S}\right) \tag{12-24}$$

If either the source or the observer is travelling away from the other, then the sign of its corresponding velocity must be reversed.

If in addition the medium is in motion, as for example, a wind blowing with a velocity $v_M$ relative to the ground and in a direction from the source to the observer, then the sound waves have a velocity of $v + v_M$. This velocity must then replace $v$ in the previous equations. Conversely, if the medium is moving in the opposite direction, from the observer to the source, the velocity $v$ must be replaced by $v - v_M$.

If the relative motions are not in the line joining the source and observer, the situation is a little more complicated. In Fig. 12-5(c), a source S emitting a note of frequency $f$ is moving along a line ABC not passing through the observer at O, with a velocity $v_S$. At A the component of this velocity along AO is $v_S \cos \theta_1$ and is *towards* the observer. That is, there is an apparent increase of frequency and, from Eqn (12-20), the new frequency is

$$f' = f\left(\frac{v}{v - v_S \cos \theta_1}\right) \tag{12-25}$$

where $v$ is the velocity of sound.

When the source reaches B there is no component of velocity along BO and therefore the observer hears the true frequency $f$. At C the component of velocity along CO is $v_S \cos \theta_2$, but this time directed *away* from the observer. Thus there is an apparent decrease in frequency and, from Eqn (12-21), the new frequency is

$$f' = f\left(\frac{v}{v + v_S \cos \theta_2}\right) \tag{12-26}$$

The Doppler effect applies not only to sound waves but also to light and radio waves. Light emitted from distant stars consists of discrete wavelengths—that is, spectral lines are emitted. It is assumed that vibrating atoms of a particular element emit spectral lines of the same frequencies whether they are here on earth or on some distant star. However, it is observed that some stars emit light of slightly higher or slightly lower frequencies than we would expect for the particular spectral lines. This frequency shift may be explained by the Doppler effect and can be used to calculate the velocity of stars relative to the earth.

The effect is also being used in the tracking of artificial satellites. By measurements on earth of the known frequency emitted by a transmitter in the satellite, its direction and velocity may be determined.

## 12-8 Reflection and refraction of sound

The reflection and refraction of sound waves obey the simple laws developed for optics. Thus for reflection, the angle of incidence equals the angle of reflection, and the incident ray, reflected ray, and the normal to the surface all lie in the same plane. The wavelengths of light waves are usually very small compared with the dimensions of the reflecting surface. However, since sound waves have much longer wavelengths, diffraction effects may predominate. However, if the reflecting surface is sufficiently large, all the simple geometrical treatment of light rays is equally applicable to sound waves, as also is Huygens' principle. Figure 12-6(a) shows how waves radiating from a point source are reflected from a plane surface, still with a spherical wave front, and appear to radiate from an acoustical image behind the reflecting surface. Similarly, waves radiating from a point source situated at the focus of a parabolic mirror are reflected to form a parallel beam with a plane wave front, whilst concave spherical reflecting surfaces can be used to form acoustical images of sources.

Refraction takes place whenever a sound wave crosses a boundary separating two media in which the velocity of sound is different. Again, in agreement with the simple laws of optics, the sine of the angle of incidence bears a constant ratio to the sine of the angle of refraction, and the incident ray, refracted ray, and normal to the surface all lie in the same plane. In practice, it is not so easy to make an acoustic lens as it is to make an acoustic mirror, which need be only a hard surface. However, a balloon filled with a denser gas than air acts as a converging lens. Conversely, when filled with a lighter gas, such as hydrogen, a balloon acts as a diverging lens.

Refraction also occurs in the atmosphere. For example, at night the layer of air near to the earth is usually colder than the air above. Since the velocity of sound increases with temperature, the velocity is greatest higher up. This causes sound to be refracted towards the earth [Fig. 12-6(b)], with a consequent increase in the intensity of sound received by an observer at O. Conversely the sound intensity at ground level is reduced during the daytime, as the layer of air next to the earth is warmest and the sound is consequently refracted away from the earth.

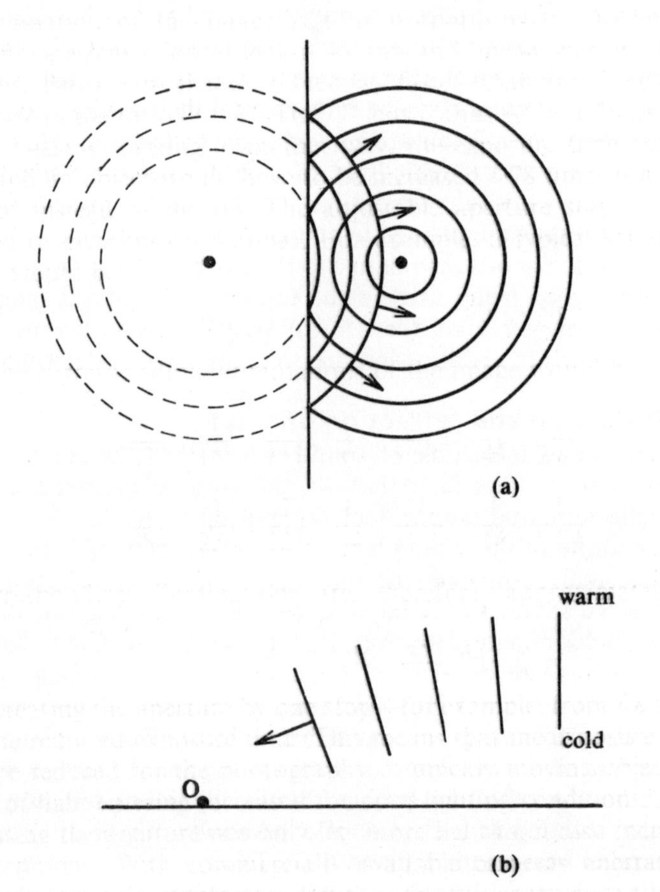


Fig. 12-6 Examples of (a) reflection and (b) refraction of sound

## 12-9 Echoes

The reflection of sound causes the formation of echoes. If a sound is heard both directly from the source and after reflection, and if both sounds reach the observer at the same time, the intensity is increased. If, however, the interval between receiving the sound directly and after reflection is about $\frac{1}{30}$ s the sound seems to be drawn out. If this interval is about $\frac{1}{20}$ s or longer the two sounds are distinguishable and an *echo* is heard. Suppose an observer is standing next to the source of sound and facing a reflecting surface, such as a large wall or a cliff. The observer hears the sound almost instantly from the source and again after reflection. If the shortest time interval for a distinguishable echo is $\frac{1}{20}$ s, the sound must take $\frac{1}{40}$ s to travel to the reflecting surface, and $\frac{1}{40}$ s to travel back. Assuming a velocity of sound in air of 332 m/s, the minimum distance to a reflecting surface for a distinguishable echo is $332 \times \frac{1}{40} = 8{\cdot}3$ m, or about 27 ft. Thus, when a reflecting surface is about 8·3 m away it may just be possible to hear an echo.

## PROBLEMS

**12-1** Calculate the velocity of sound in air at 0°C given the following data: barometric height 76 cmHg, density of mercury 13,600 kg/m$^3$, ratio of specific heats at constant pressure and volume 1·407, density of air 1·29 kg/m$^3$, acceleration due to gravity 9·81 m/s$^2$. [332·5 m/s]

**12-2** If the densities of air and hydrogen are 1·29 kg/m$^3$ and 0·09 kg/m$^3$ respectively at s.t.p., and the ratios of their specific heats, $\gamma$, are 1·40 and 1·41 respectively, calculate the velocity of sound in hydrogen at s.t.p. given that the corresponding velocity for air is 332 m/s. [1,261 m/s]

**12-3** The velocity of sound in air is 332 m/s at 0°C. What is the velocity of sound in air at (a) 20°C, (b) −10°C? [(a) 343·6 m/s; (b) 325·9 m/s]

**12-4** Calculate the velocity of sound in a steel bar, given that its density is 7,700 kg/m$^3$ and Young's modulus of elasticity is $19{\cdot}5 \times 10^{10}$ N/m$^2$. [5,032 m/s]

**12-5** Discuss the requirements for accurate determinations of the velocity of sound in (a) solids, (b) liquids, (c) gases, describing a method for each. To what extent do each of these methods satisfy your requirements for an accurate determination?

**12-6** What are the properties which determine the velocity of sound in a solid, a liquid, and a gas? Explain why Newton's value of the velocity of sound in air differs from the true value.

**12-7** A train travelling at 70 km/h (43·5 mile/h) is sounding its whistle, of frequency 500 Hz. What is the apparent frequency heard by the driver of a train travelling at 100 km/h (62·1 mile/h) in the opposite direction (a) as the trains are approaching, and (b) once they have passed each other? (Velocity of sound = 340 m/s). [(a) 574 Hz; (b) 434 Hz]

**12-8** A source of sound of constant frequency is travelling along a circular path of radius 5 m. If it is making 100 rev/min and the velocity of sound in air is 330 m/s, what is the ratio of the frequencies of the lowest and highest notes heard by a distant observer standing in the plane of the circular path? [1:1·38]

**12-9** The light from a certain star is found to contain the red hydrogen line with a measured wavelength of $6{,}565 \times 10^{-10}$ m. This same line has a wavelength of $6{,}563 \times 10^{-10}$ m when measurements are made on a source in the laboratory. Calculate the speed and the direction of movement of the star relative to the earth (velocity of light = $3 \times 10^8$ m/s). [91·4 km/s, away]

**12-10** After stroking the rod and adjusting the air column in a Kundt's tube experiment, it is found that nine nodal heaps of dust occupy a distance of 77·6 cm. If the rod is of brass, density 8,500 kg/m$^3$, 1 m long and clamped at its midpoint, calculate (a) the velocity of sound in brass, and (b) Young's modulus of brass. With the tube filled with carbon dioxide it is found that the nine nodal heaps then occupy a distance of 58·8 cm. Calculate (c) the velocity of sound in carbon dioxide (velocity of sound in air at temperature of experiment = 340 m/s). [(a) 3,505 m/s; (b) $10{\cdot}4 \times 10^{10}$ N/m$^2$; (c) 258 m/s]

# 13 Transverse vibrations

## 13-1 Velocity of waves in stretched strings

Before discussing the modes of vibration of a stretched string and the formation of standing waves, it is first necessary to examine the velocity of propagation of waves along the string. When a wave travels along a stretched string, it is the form of the wave that travels. Each point on the string is displaced in turn from its mean position. As a section of the string is displaced to one side, the tension in the string causes this displaced section to return to its mean position. In this way the displacement is caused to travel along the string, with a velocity $v$.

Figure 13-1(a) shows a stretched string ABCDEF lying generally in the direction of the $x$-axis and with part of it displaced in the $y$-direction at some instant $t$ in time. Figure 13-1(b) shows a small section CD of the displaced string. If the displacement from the mean position is small, then the tension, $T$, in the string may be regarded as everywhere constant.

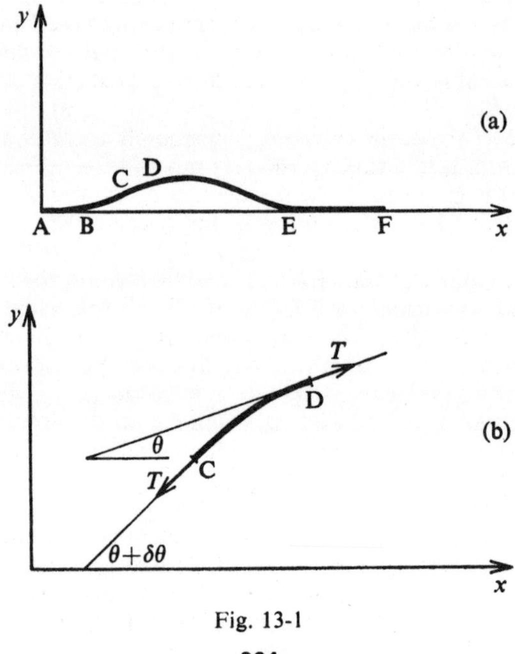


Fig. 13-1

Thus for the element of string CD the tension $T$ acts in the direction as shown in the figure. Resolving these tensions in the $y$-direction,

$$\text{force at C} = T\sin(\theta + \delta\theta) \tag{13-1}$$

$$\text{force at D} = T\sin\theta \tag{13-2}$$

These tensions act in opposite directions, so that the resultant force, $F$, on the section CD, measured along the $y$-axis, is

$$F = T[\sin(\theta + \delta\theta) - \sin\theta] \tag{13-3}$$

$$= T(\sin\theta\cos\delta\theta + \cos\theta\sin\delta\theta - \sin\theta)$$

but since $\delta\theta$ is small, $\cos\delta\theta \rightarrow 1$ and $\sin\delta\theta \rightarrow \delta\theta$

$$\therefore\ F = T\cos\theta.\delta\theta$$

$$= T\,\delta(\sin\theta)$$

If the section CD is sufficiently small, then the resultant force, $\mathrm{d}F$, on the element CD, length $\mathrm{d}x$, is

$$\mathrm{d}F = T\frac{\partial(\sin\theta)}{\partial x}\mathrm{d}x \tag{13-4}$$

If also the mass per unit length of the string is $m$, then the mass of the element CD is $m\,\mathrm{d}x$. The equation of motion of this element moving under a restoring force acting in the $y$-direction is

$$\text{force} = \text{mass} \times \text{acceleration}$$

i.e.,

$$T\frac{\partial(\sin\theta)}{\partial x}\mathrm{d}x = m\,\mathrm{d}x\frac{\partial^2 y}{\partial t^2} \tag{13-5}$$

Also, since the gradient of the curve is small

$$\sin\theta = \tan\theta = \frac{\partial y}{\partial x}$$

$$\therefore\ \frac{\partial(\sin\theta)}{\partial x}\mathrm{d}x = \frac{\partial^2 y}{\partial x^2}\mathrm{d}x \tag{13-6}$$

Therefore from Eqn (13-5)

$$\frac{\partial^2 y}{\partial t^2} = \frac{T}{m}\frac{\partial(\sin\theta)}{\partial x}$$

i.e.,

$$\frac{\partial^2 y}{\partial t^2} = \frac{T}{m}\frac{\partial^2 y}{\partial x^2} \tag{13-7}$$

But by Eqn (11-10),

$$\frac{\partial^2 y}{\partial t^2} = v^2\frac{\partial^2 y}{\partial x^2}$$

which is the equation for the motion of a transverse wave travelling with a velocity $v$. Hence by analogy, the velocity of the wave form travelling along the stretched string is

$$v = \sqrt{\frac{T}{m}} \tag{13-8}$$

If the tension $T$ is expressed in newtons and the mass per unit length in kg/m, then the velocity of the wave form is in m/s. For $T$ in dynes and $m$ in g/cm, velocity $v$ is in cm/s.

## 13-2 Transverse vibrations of stretched strings

If a string stretched between two rigid points is vibrating, then the end points must be nodes. Figure 13-2(a) shows the simplest mode of vibration in which this can occur; there is one antinode between the fixed nodes. This mode of vibration is termed the *fundamental*, or *1st harmonic*. Figure 13-2(b) shows the second simplest mode; this has an extra node

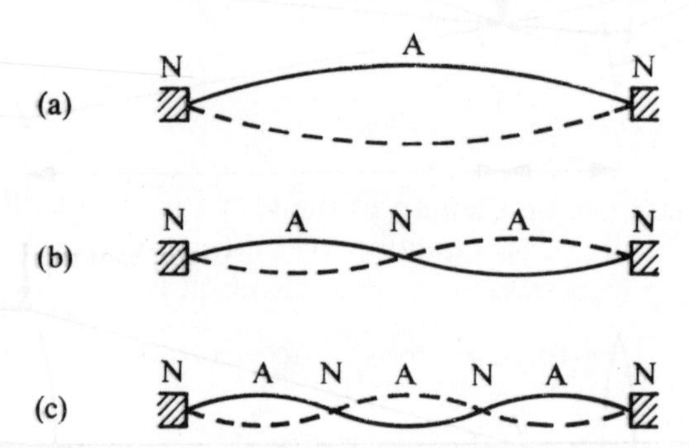


Fig. 13-2 Vibrations of a stretched string (a) fundamental, (b) 1st overtone, and (c) 2nd overtone

at the centre and hence two antinodes are formed between the fixed ends. This mode is termed the *1st overtone*, or *2nd harmonic*. A *harmonic* has a frequency that is a whole-number multiple of the fundamental frequency, whereas an *overtone* is a natural mode with a higher frequency than the fundamental but not generally an exact multiple. In Fig. 13-2(b), the string vibrates in two loops as compared with the one loop of the fundamental. Similarly higher overtones may be formed as in Fig. 13-2(c).

When a string is vibrating in the fundamental mode, the length of the string—that is, the distance between the two fixed nodes—equals half the wavelength of the wave being propagated along the string. Thus, since the frequency $f_1$ of this vibration is equal to *velocity/wavelength*, and the wavelength is equal to $2l$, where $l$ is the length of the vibrating string, from Eqn (13-8) the frequency is

$$f_1 = \frac{1}{2l}\sqrt{\frac{T}{m}} \tag{13-9}$$

Similarly, for the 1st overtone, the total length $l$ of the string is equivalent to a whole wavelength, so that the frequency of vibration is now

$$f_2 = \frac{1}{l}\sqrt{\frac{T}{m}} \tag{13-10}$$

Thus generally

$$f = \frac{n}{2l}\sqrt{\frac{T}{m}} \tag{13-11}$$

where $l$ is the total length of the string and $n$ is the order of the harmonic, being unity for the fundamental. This equation is only strictly true for an ideal string which has negligible stiffness and does not suffer appreciable changes in length whilst vibrating. In a piano the bass wires are wound with copper wire so as to increase their mass per unit length, $m$; otherwise to obtain a sufficiently low frequency it would be necessary to use an impracticably long length or low tension. We cannot simply use a thick wire to increase the mass per unit length as this would cause too great a departure from the condition of negligible stiffness; the method adopted does not interfere with this condition.

Equation (13-11) may be verified experimentally by means of a sonometer. This consists simply of a sounding board or box over which the string or wire is stretched (Fig. 13-3). The wire is held away from the

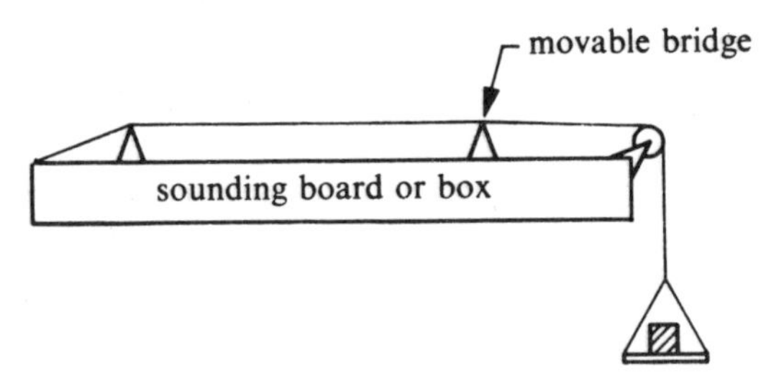


Fig. 13-3 Sonometer

sounding board by knife-edge bridges whose distance apart may be varied in order to vary the vibrating length of the wire. The wire passes over a pulley wheel at one end and is then attached to a scale pan. The pan may be loaded to obtain the tension $T$ in the wire. Attaching the wire to a sounding board means that as the wire vibrates a periodic pull is exerted on the bridges, which flexes the board and forces it to vibrate in unison with the wire. In this way more energy is given to the resulting sound wave in air, since a larger volume of air is forced to vibrate than would be the case due to the wire alone. This makes the sound produced louder but, of course, the vibration energy is used up more quickly and hence the sound dies away more quickly. In addition, when the wire alone is vibrating, rarefactions on one side of the wire and corresponding compressions on the other side partially interfere with each other and lead to a further decrease in the sound intensity. The wire is made to vibrate by plucking or bowing.

Wherever it is plucked cannot be a node and therefore to produce the 1st harmonic the wire should be plucked at its centre point. An even-numbered harmonic cannot be formed with an antinode at the centre.

To show that $f$ is proportional to $1/l$ as in Eqn (13-11) we tune the sonometer to the same musical note as a tuning fork of known frequency, by keeping the tension, $T$, constant and changing the frequency by varying the length, $l$, of the vibrating wire. Slow beats are heard when they are almost in unison. This is repeated for different frequencies and a graph drawn of frequency, $f$, against the reciprocal of the corresponding length, $l$. This graph is a straight line which verifies that

$$f \propto \frac{1}{l} \tag{13-12}$$

This method may also be used to find the unknown frequency of a sounding body. By tuning the sonometer to the unknown frequency, $f_1$, the corresponding length, $l_1$, of the sonometer wire may be found. The sonometer is next tuned to a known sound frequency, $f_2$, with a corresponding wire length, $l_2$. The unknown frequency, $f_1$, is then determined since

$$\frac{f_1}{f_2} = \frac{l_2}{l_1} \tag{13-13}$$

To show that the frequency, $f$, is proportional to $\sqrt{T}$, where $T$ is the tension applied to the wire we use a two-stage method. To verify this directly would require a large number of tuning forks of known frequency or a large number of small weights to adjust the tension. In the two-stage method we apply various tensions to the wire and find the different lengths necessary to tune it to a particular frequency. The particular frequency may be that of another wire of fixed length and tension. By plotting $\sqrt{T}$ against corresponding length, $l$, for constant frequency, a linear relation is found, so that $\sqrt{T} \propto l$. Also by Eqn (13-12), $1/f \propto l$, for constant tension, hence for constant length

$$f \propto \sqrt{T} \tag{13-14}$$

(Note: if $A \times B = C$, then $A \propto C$, $B \propto C$, but $A \propto 1/B$.)

To show that the frequency $f$ is proportional to $1/\sqrt{m}$, where $m$ is the mass per unit length of the wire, we again use a two-stage method. Different wires, at a constant tension, are tuned to a constant frequency by adjusting their lengths. The corresponding lengths show experimentally that $l \propto 1/\sqrt{m}$, at constant frequency. Hence, as before, from a knowledge of Eqn (13-12) it is shown that

$$f \propto \frac{1}{\sqrt{m}} \tag{13-15}$$

and hence combining the experimental results

$$f \propto \frac{1}{l}\sqrt{\frac{T}{m}} \tag{13-16}$$

## 13-3 Reflection of waves in stretched strings

We have shown that a transverse wave form travels along a stretched string with a velocity given by Eqn (13-8) and that a standing wave may be formed with the ends of the string being nodes. Section 11-9 shows that a standing wave is formed when two wave trains having the same frequency and amplitude are travelling in opposite directions. Thus in the case of the stretched strings in Section 13-2, the wave must be reflected when it reaches one end so as to produce a wave travelling in the opposite direction, which may then interfere with the incident wave to produce a standing wave, as in Fig. 11-7.

Figure 13-4 shows how the disturbance on the string is reflected. Suppose a stretched string is fixed at one end to a rigid support or wall

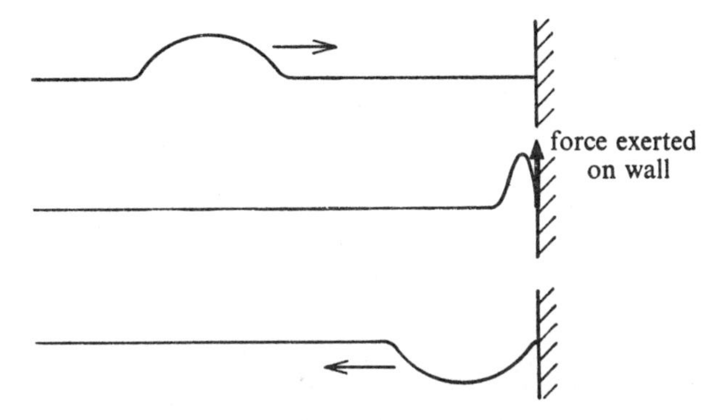


Fig. 13-4 Reflection of a transverse wave

and a single disturbance or pulse is sent along the string as shown in the upper figure. On hitting the wall this pulse exerts an upward force on it. But by Newton's third law the wall in its turn exerts an equal and opposite force on the string, that is, a downward force. Thus due to the action of the wall a displacement is sent out that is opposite to the incoming displacement. In other words, the displacement is reflected with a 180° phase change.

This may also be shown analytically. Let the incident transverse progressive wave be described by the following equation [Eqn (11-3)]

$$y_1 = a \sin \frac{\omega}{v}(vt - x) \tag{13-17}$$

where $y_1$ is the displacement from its mean position at a time $t$ of an element of the string a distance $x$ from an end fixed at $x = 0$, and where $v$ is the velocity of the wave along the string. Equation (13-17) may conveniently be written as

$$y_1 = \mathrm{f}_1(vt - x) \tag{13-18}$$

where f represents the function. Similarly, for the reflected wave, travelling in the opposite direction

$$y_2 = \mathrm{f}_2(vt + x) \tag{13-19}$$

Thus the displacement at the point $x$ along the string is

$$y = y_1 + y_2 = \mathrm{f}_1(vt - x) + \mathrm{f}_2(vt + x) \tag{13-20}$$

Since the string is clamped at the point $x = 0$, there can be no transverse motion at this point

i.e.,
$$y = \mathrm{f}_1(vt - 0) + \mathrm{f}_2(vt + 0) = 0$$

so that

$$\mathrm{f}_2(vt) = -\mathrm{f}_1(vt) \tag{13-21}$$

That is, the two functions have the same form but are of opposite sign. Therefore

$$\mathrm{f}_2(vt + x) = -\mathrm{f}_1(vt + x) \tag{13-22}$$

and hence Eqn (13-20) may be written as

$$y = \mathrm{f}_1(vt - x) - \mathrm{f}_1(vt + x) \tag{13-23}$$

This then represents the resultant displacement from the mean position due to a wave $y_1 = \mathrm{f}_1(vt - x)$ travelling to the right and another, $y_2 = -\mathrm{f}_1(vt + x)$, travelling to the left. Since the functions f are the same but of opposite sign, the incident and reflected waves have the same form but are of opposite displacement, as already shown by Fig. 13-4.

Reflection at a rigid boundary as just discussed is in fact a special case of the more general reflection problem. Consider now two strings of different mass per unit length, $m_1$ and $m_2$ (Fig. 13-5) joined at the point

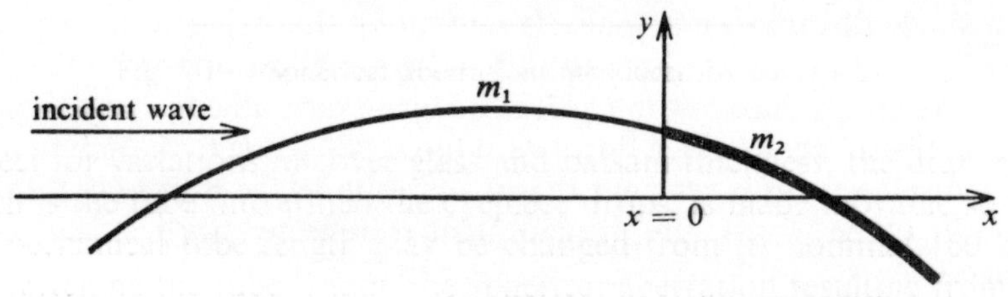


Fig. 13-5

$x = 0$. With a constant tension $T$ throughout the two strings, the general equations for the incident, reflected, and transmitted waves may be written respectively as [from Eqn (11-3)]

$$y_\mathrm{i} = a_\mathrm{i} \sin \omega\left(t - \frac{x}{v_1}\right) \tag{13-24}$$

$$y_\mathrm{r} = a_\mathrm{r} \sin \omega\left(t + \frac{x}{v_1}\right) \tag{13-25}$$

$$y_\mathrm{t} = a_\mathrm{t} \sin \omega\left(t - \frac{x}{v_2}\right) \tag{13-26}$$

That is, the incident and transmitted waves are travelling from left to right, and the reflected wave travelling right to left. The incident and reflected waves are in the first string and have the same velocity $v_1$, whereas the transmitted wave is in the second string, with a velocity $v_2$. The frequency of vibration is unchanged across the junction.

At the junction of the two strings—that is, the point $x = 0$—the displacement in the $y$-direction must be the same for both strings. Therefore, for either side of this junction

$$y_i + y_r = y_t \tag{13-27}$$

i.e.,

$$a_i + a_r = a_t \tag{13-28}$$

since $x = 0$ at the boundary.

Also, at the boundary between the two strings, the transverse forces on the two strings at any instant in time must be the same. Thus equating these forces for the two strings, at $x = 0$,

$$T\left(\frac{\partial y_i}{\partial x} + \frac{\partial y_r}{\partial x}\right) = T\frac{\partial y_t}{\partial x} \tag{13-29}$$

as in Eqn (13-2), where the displacement and hence the angle involved is small.

Differentiating Eqns (13-24), (13-25), and (13-26) gives, at any time $t$,

$$\frac{\partial y_i}{\partial x} = -\frac{a_i}{v_1}\cos\omega\left(t - \frac{x}{v_1}\right) \tag{13-30}$$

$$\frac{\partial y_r}{\partial x} = +\frac{a_r}{v_1}\cos\omega\left(t + \frac{x}{v_1}\right) \tag{13-31}$$

$$\frac{\partial y_t}{\partial x} = -\frac{a_t}{v_2}\cos\omega\left(t - \frac{x}{v_2}\right) \tag{13-32}$$

Therefore, at the boundary $x = 0$,

$$\frac{1}{v_1}(a_i - a_r) = \frac{a_t}{v_2}$$

$$= \frac{1}{v_2}(a_i + a_r) \tag{13-33}$$

Also, since $v_1 = \sqrt{(T/m_1)}$ and $v_2 = \sqrt{(T/m_2)}$, Eqn (13-33) may be written as

$$\sqrt{m_1}(a_i - a_r) = \sqrt{m_2}(a_i + a_r) \tag{13-34}$$

i.e.,

$$a_r(\sqrt{m_1} + \sqrt{m_2}) = a_i(\sqrt{m_1} - \sqrt{m_2}) \tag{13-35}$$

Thus the reflection coefficient $R$, that is the ratio of the reflected to incident wave amplitudes, is given by

$$R = \frac{a_r}{a_i} = \frac{\sqrt{m_1} - \sqrt{m_2}}{\sqrt{m_1} + \sqrt{m_2}} \tag{13-36}$$

Therefore, from Eqn (13-36), if $m_1 > m_2$, the ratio $R$ is positive; that is, at any instant in time the reflected and incident wave displacements are of the same sign. In other words the reflected wave is in phase with the incident wave. If $m_2 > m_1$, the ratio $R$ is negative, meaning that there must be a phase change of 180° on reflection at the boundary for $a_r$ and $a_i$ to have opposite signs at all times. If $m_2 \gg m_1$, as in reflection at a rigid wall, $R$ tends to a value of $-1$ and 100% reflection. The fact that there is a phase change of 180° on reflection at a rigid wall has already been shown by Eqn (13-23) and Fig. 13-4.

## 13-4 Transverse vibrations of loaded strings

Up to now a string has been regarded as having a constant mass per unit length. A different case arises if the stretched string may be regarded as of negligible weight compared to a load fixed to the string at some point.

Consider a mass $M$ fixed to a light string of length $l$ stretched between two fixed points. Let the mass be at a distance $x$ from one end of the string, as in Fig. 13-6. If the mass $M$ is displaced only a small amount from

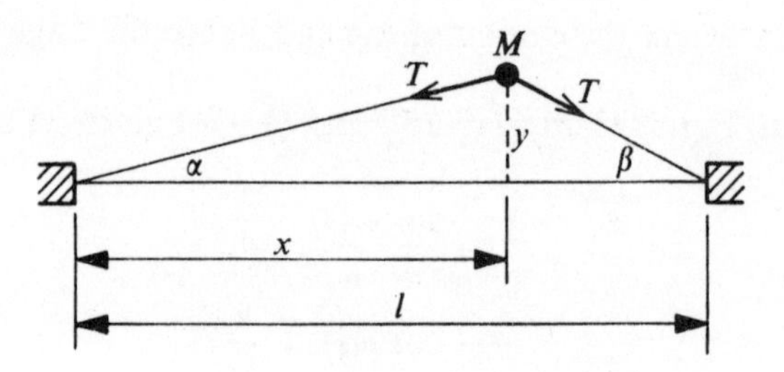


Fig. 13-6 Vibrations of a loaded string

the mean position—that is, $y$ is small compared to $l$—then it may be considered that the tension $T$ in the string and the length of the string remain constant. The resultant force in the $y$-direction, neglecting the effect of gravity, and acting in a direction so as to reduce the displacement, is thus equal to $-(T\sin\alpha + T\sin\beta)$,

i.e.,

$$\text{resultant restoring force} = -\left(T\frac{y}{x} + T\frac{y}{l-x}\right)$$

$$= -Ty\left(\frac{1}{x} + \frac{1}{l-x}\right)$$

$$= \text{mass} \times \text{acceleration}$$

$$= M\frac{d^2y}{dt^2}$$

i.e.,

$$\frac{Mx(l-x)}{Tl}\frac{d^2y}{dt^2} + y = 0$$

That is, the motion is simple harmonic, since the restoring force is proportional to the displacement, and hence the period is equal to

$$2\pi\sqrt{\frac{Mx(l-x)}{Tl}}$$

The frequency of vibration is thus

$$f = \frac{1}{2\pi}\sqrt{\frac{Tl}{Mx(l-x)}}$$

By adjusting the position of the mass $M$ along the string the fundamental frequency of vibration may be changed, since $f^2$ varies inversely as $x(l-x)$. If the length $x$ is in a simple ratio to the length $(l-x)$, for example $n:m$, then the mass $M$ is situated at a node and $(n+m)$ loops are formed in the vibrating string. That is, the $(n+m)$th harmonic [or $(n+m-1)$th overtone] is formed.

## 13-5 Transverse vibrations of rods

As its diameter to length ratio increases a wire becomes a rod. The effect of tension decreases and stiffness becomes of overriding importance. As in the case of a bent beam the restoring force is a function of Young's modulus of elasticity. When one end of a horizontal rod is held rigid and the other end bent downwards, the under side is under compression and the upper under tension, thus producing a restoring couple. On releasing the end, the rod recovers, overshoots, and so vibrates. The actual mode of vibration is not a simple form and is generally of little practical interest; it varies depending on how the rod is held, for example, whether it is clamped in the middle, at one end, or at both ends.

It is found that the frequency of vibration of the rod varies inversely as the square of its length, and is proportional to the thickness of the rod but independent of its width. Figure 13-7 shows the first three modes of vibration of a thin rod clamped at one end. The 1st overtone has a frequency of over six times that of the fundamental, with the 2nd and 3rd overtones having frequencies of $17\frac{1}{2}$ and $34\frac{1}{2}$ times the fundamental. That is, the frequencies do not form a harmonic series like the vibrating string, which formed a harmonic series with frequencies in the ratio 1:2:3:4... times that of the fundamental.

If the vibrating rod is free at the ends and is gripped in the middle, then when sounding its fundamental it has an antinode at each end and a node at the centre. If this rod is now bent into a U-shape, a tuning fork is formed (Fig. 13-8). When a tuning fork is sounded internal friction rapidly causes the overtones to die away leaving the fundamental sounding as a pure tone. As the prongs of the fork vibrate from side to side, they bend into arcs so that the centre of mass moves up and down. Thus a vertical motion is imparted to the stem. If the stem is held against a surface area, such as a table top, the surface is forced to vibrate at the same frequency as the tuning fork. The surface, in turn, causes the surrounding air to

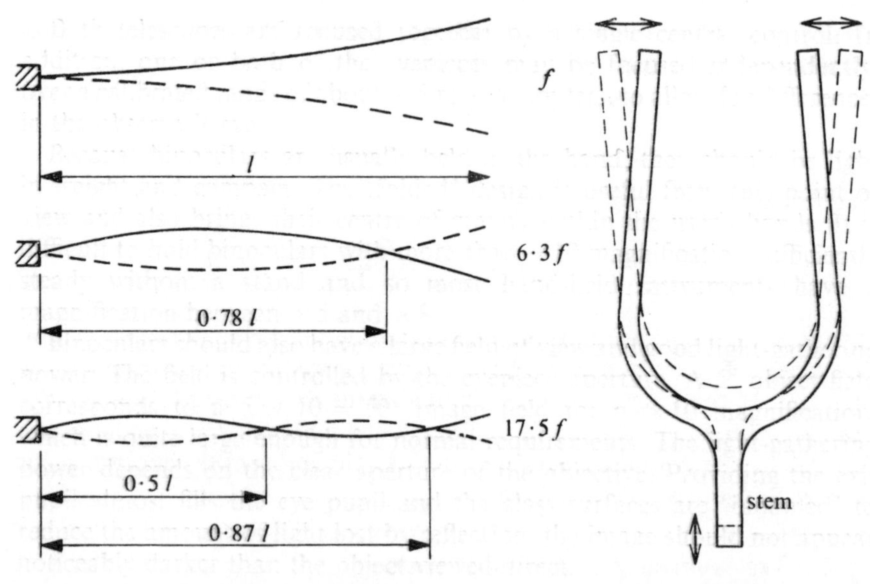


Fig. 13-7 Vibrations of a thin rod

Fig. 13-8 Vibrating tuning fork

vibrate. Since a larger mass of air is in contact with the surface than with the prongs alone, the sound volume is increased, but will of course die away more quickly.

The frequency of vibration of the tuning fork may be adjusted: the frequency may be raised by shortening the length of the prongs thus reducing the moment of inertia without affecting the stiffness; conversely, the frequency may be lowered by reducing the elastic stiffness, without altering the moment of inertia, by removing metal from the base of the U and the top of the stem regions.

Temperature affects the dimensions of a fork and, more significantly, the Young's modulus of the fork material; together these result in a temperature coefficient of frequency of about $1{\cdot}2 \times 10^{-4}$ per degree Centigrade change at room temperature for a steel fork.

## 13-6 Vibrations of diaphragms and plates

A diaphragm may be considered to be a thin membrane possessing perfect flexibility so that its vibrating motion depends almost entirely on the applied tension. A good example is a drum skin which consists of a circular membrane clamped at its outer edge, which therefore must be a node. The general pattern of the other nodes is somewhat complicated, but usually the nodal lines form a series of radial lines and concentric circles with the centre as origin. As with vibrating rods, the overtones are inharmonic, but in a tuned drum, such as a kettle drum, the undesirable overtones may be suppressed by striking the drum at a point about a quarter of the way across, giving a sound that is mainly fundamental.

If the diaphragm is thick enough to be appreciably rigid, it becomes a thin plate. The controlling factor for vibration then becomes its elastic properties and the effect of tension diminishes to become zero for a thick plate. The modes of vibration for plates was first investigated by Chladni (1787). He clamped plates at centres of symmetry and made them vibrate by bowing them at the edge. When vibrating plates are touched lightly with a finger at various points, the points become nodes. The nodal lines are made visible by lightly sprinkling the vibrating plates with fine sand, which distributes itself along the nodal lines. If a very fine powder is used —as, for example, lycopodium powder—the effect is reversed and the powder tends to collect at the antinodes. This is due to the air currents set up in the neighbourhood of the vibrating plate. A more convenient way of getting the plates to vibrate is to use a piece of solid carbon dioxide, which sublimes at $-78°C$; on bringing this up to the warmer metal of the plate a rapid localized build up of gas is produced, which pushes the metal away. The plate therefore vibrates at its natural frequency, receiving an impulse each time it comes in contact with the solid $CO_2$. In this method the plate does not have to be rigidly clamped at some point. The nodal pattern is generally complicated and varied (Fig. 13-9). The mathematical

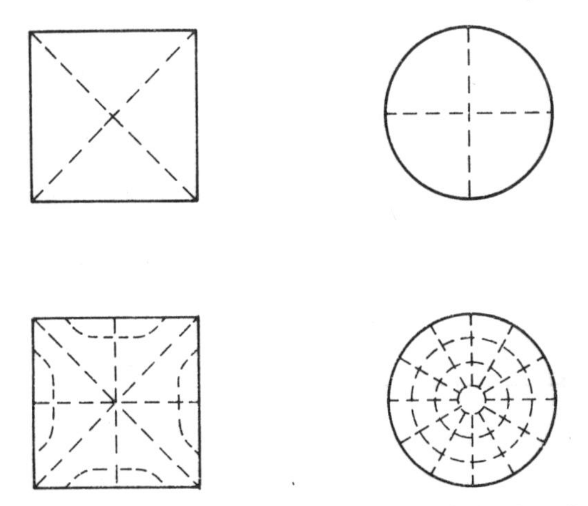

Fig. 13-9 Examples of Chladni's figures on square and circular vibrating plates

analysis is also complicated and often impossible. The study of such vibrations is important, for example, in the design of microphones, which usually employ circular diaphragms, turbine discs, and many mechanical devices, since positions of permanent antinodes can become points for fatigue failure.

Just as a tuning fork may be regarded as a curved rod, so a bell may be regarded as a curved circular plate. The vibration is very complex, although it does tend to have some similarities to the vibrations of a disc. A vibrating bell has nodal lines that radiate out from the support and nodal circles concentric to the support. For a perfectly symmetrical bell the radial nodes are indeterminate in position but, in practice, there is sufficient asymmetry to produce nodal lines in constant positions. An additional complexity is a vibration along the circumference. This tangential motion is the same as that produced in a wine glass by wiping a damp finger rapidly around the rim and causing it to 'sing'. The special properties of a bell are due to its shape and the fact that it has a thickened rim. When a bell is struck a 'strike note' is first heard, which is difficult to account for in theory. This note then changes to the normal pitch of the bell. In addition, there is a 'hum note' which tends to persist after the others.

## PROBLEMS

**13-1** Derive the equation for the velocity with which a lateral disturbance runs along a stretched string.

**13-2** Explain why the frequency of a string vibrating transversely depends on the tension, while that of a string vibrating longitudinally is not greatly affected by changes in tension.

**13-3** Give a general explanation of the reflection of waves. When water waves are running obliquely towards the shore, they tend to become more nearly parallel to the shore as they advance; explain this.

**13-4** A vibrating tuning fork of frequency 256 Hz is sounding in unison with a wire, 50 cm long and tensioned by a load of 10 kg. What must (a) the load, (b) the length be changed to for the wire to vibrate at 260 Hz?

[(a) 10·31 kg; (b) 49·23 cm]

**13-5** A string is attached to a weight such that when the length of the string is altered, the weight is adjusted so as to be always equal to 400 times the weight of the string. Show that the frequency of vibration of the string varies inversely as the square root of its length. If the length of the string is 1 m, calculate its fundamental frequency of vibration. (Acceleration due to gravity = 9·81 $m/s^2$.)

[31·3 Hz]

**13-6** Why is a tuning fork made with two prongs? Would it still function normally if one prong was removed? Why?

**13-7** A tuning fork and a stretched wire of length 71 cm are sounding in unison. If the length of the wire is increased by 1 cm without any change in tension, 5 beats per second are heard. What is the frequency of the tuning fork?

[360 Hz]

**13-8** Calculate the ratio of the fundamental frequencies of vibrating stretched steel and brass wires of the same length and diameter when the tension on the steel wire is twice that on the brass wire. The densities of steel and brass are 7,700 and 8,500 $kg/m^3$ respectively. [1:0·67]

**13-9** One end of a stretched string is fastened to one of the prongs of a vibrating tuning fork. Compare the vibrations excited in the string when the prong is moving (a) parallel, (b) transversely to the string.

[frequencies (a):(b) = 1:2]

**13-10** Two wires are identical except for their lengths, which are in the ratio 9:10. The wires are tensioned by two weights A and B respectively. When the weights are exchanged, the number of beats heard when both wires are sounding their fundamentals is increased by a factor of three. What is the ratio of the weights? [A:B = 1:1·111]

# 14 Forced vibrations and resonance

## 14-1 Free and forced vibrations

When a simple harmonic force is applied to a body, the body is made to vibrate simple harmonically in step with the driving force. Such a *forced vibration* generally has only a small amplitude. When the driving force is removed the vibrating body settles down and vibrates at one of its natural frequencies, or a combination of several of them. This is called a *free vibration*. There is a slow decay in the amplitude, termed *damping*, which is due to the energy supplied to produce the original displacement being used up in overcoming the friction of the vibrating body.

If, however, the simple harmonic driving force has a frequency equal to the natural frequency of free vibration of the body, the resulting forced vibration has a very large amplitude and *resonance* is said to occur. The amplitude tends to increase but is limited by the physical dimensions of the system and friction. For example, consider a thin thread attached to the bob of a heavy pendulum. The pendulum bob may be displaced by pulling the thread horizontally, but too hard a pull breaks the thread. Thus only a small displacement is possible by a direct pull. When the thread is released the pendulum swings back. If we pull on the thread only when the pendulum is swinging towards us and release it when the pendulum is swinging away, its amplitude of swing steadily increases. This is an example of resonance because the applied force, the pull, has a frequency which is equal to the natural frequency of vibration of the pendulum. The amplitude at resonance is limited by the length of the pendulum and by the friction at the support and of the air.

Figure 14-1 shows a heavy pendulum of length $L$ and a very light pendulum of length $l$ swinging from the bob of the heavy pendulum. If the light pendulum is of negligible mass compared to the heavy pendulum, then displacing and releasing the heavy pendulum causes it to perform a free simple harmonic vibration. This drives the light pendulum simple harmonically so that it vibrates with the same frequency, but with an amplitude depending on its natural frequency of vibration; that is, the light pendulum executes a forced vibration. If $l$ is less than $L$, as in Fig. 14-1(a), then the natural frequency of the light pendulum is greater than that applied by the heavy pendulum. The light pendulum therefore has a forced vibration of small amplitude and of the same phase as the applied force. If the two pendulums are of the same length, $l = L$, then the natural

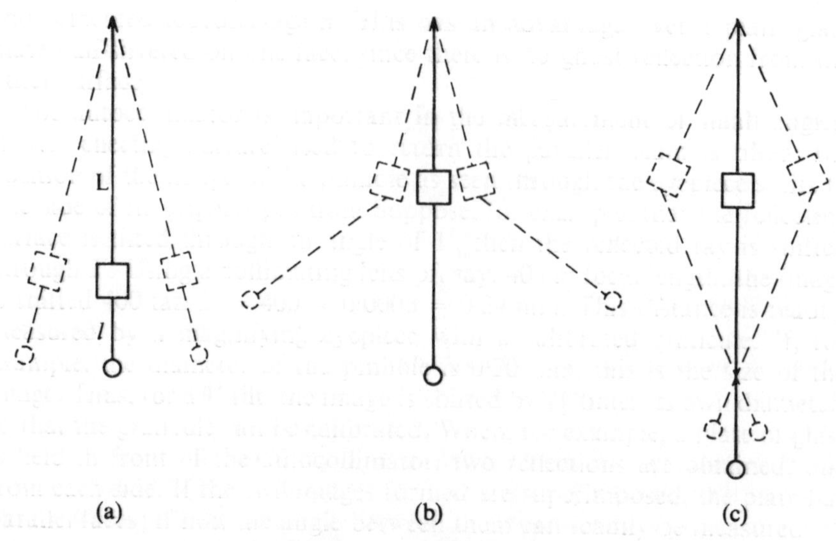


Fig. 14-1 Forced vibrations

frequency of the light pendulum is equal to that of the driving force and resonance occurs [Fig. 14-1(b)]. The amplitude of the light pendulum now becomes great, with the two vibrations still in phase. When the length, *l*, of the light pendulum is greater than the length, *L*, of the heavy pendulum as in Fig. 14-1(c), the natural frequency of the light pendulum is less than that of the driving pendulum and its forced vibration is opposite in phase. When the length of the light pendulum is only a little greater than that of the heavy pendulum—that is, still near to the resonance condition—the amplitude of the light pendulum is still great. But as the length *l* of the light pendulum increases and its natural frequency of vibration decreases, its amplitude falls until, when the driving frequency is about $1\frac{1}{2}$ times that of the natural frequency of the light pendulum, the two amplitudes are equal but opposite in phase. Finally, when the driving frequency is about 2 or 3 times the natural frequency of the light pendulum, the amplitude of the light pendulum becomes zero.

The implications of forced vibrations and resonance are many. All bodies may be made to vibrate and all have a natural frequency of vibration. When they are forced to vibrate at their natural frequency, the vibrational amplitude becomes large. For example, when crossing a stream using a flexible plank it is best to use short quick steps. Otherwise the frequency of the steps may approximate to the low natural frequency of vibration of the plank. Resonance would then occur causing a large oscillation of the plank. When resonance occurs in machines or other vibrating systems, it can cause annoying sounds, deterioration in the strength of the metal due to fatigue, or even complete failure due to breakage. If a wireless set is tuned by means of a variable capacitor until the natural frequency of the circuit is the same as that of the incoming signal, resonance occurs in the electrical circuit.

## 14-2 Analytical treatment of vibrations

The analytical treatment of vibrations is not simple and involves the solution of differential equations which are beyond the scope of this book. For this reason, only a brief outline of the treatment can be given here.

Section 11-5 showed that a simple harmonic motion may be described by

$$\frac{d^2y}{dt^2} + \omega^2 y = 0 \qquad (14\text{-}1)$$

that is, the equation of a particle moving in a straight line such that its acceleration is always towards a fixed point on that line and proportional to the distance $y$ from the point. The solution of this differential equation was shown to be [Eqn (11-2)]

$$y = a \sin(\omega t - \alpha) \qquad (14\text{-}2)$$

where $\omega = 2\pi f$ and $\alpha$ is the phase constant.

However, such a simple harmonic motion does not continue in practice but decays due to frictional losses, for example, air resistance. It may be assumed that this resistance to motion is proportional to the velocity of the body in motion. Hence the equation for a damped simple harmonic motion is

$$\frac{d^2y}{dt^2} + 2k\frac{dy}{dt} + \omega^2 y = 0 \qquad (14\text{-}3)$$

where $dy/dt$ is the velocity of motion at a time $t$ and $2k$ is the resisting force per unit mass per unit velocity. This second-order differential equation may be solved by assuming a solution of the form $y = \text{A}\,e^{mt}$ where A is a constant. On substituting in Eqn (14-3)

$$(m^2 + 2km + \omega^2)\text{A}\,e^{mt} = 0 \qquad (14\text{-}4)$$

For this to be a valid solution

$$m^2 + 2km + \omega^2 = 0 \qquad (14\text{-}5)$$

The general solution of the differential equation is given by the sum of the two separate solutions corresponding to the roots $m_1$ and $m_2$ of Eqn (14-5). That is, the general solution to Eqn (14-3) is

$$y = \text{A}_1\,e^{m_1 t} + \text{A}_2\,e^{m_2 t} \qquad (14\text{-}6)$$

where $m_1$ and $m_2$ are given by

$$\frac{-2k \pm \sqrt{4k^2 - 4\omega^2}}{2}$$

i.e.,

$$m_1, m_2 = -k \pm \sqrt{k^2 - \omega^2} \qquad (14\text{-}7)$$

and where $\text{A}_1$ and $\text{A}_2$ are constants whose values depend upon the defining conditions of the particular motion. Therefore Eqn (14-6) may be written as

$$y = \text{A}_1\,e^{[-k+\sqrt{k^2-\omega^2}]t} + \text{A}_2\,e^{[-k-\sqrt{k^2-\omega^2}]t}$$
$$= e^{-kt}[\text{A}_1\,e^{\sqrt{k^2-\omega^2}.t} + \text{A}_2\,e^{-\sqrt{k^2-\omega^2}.t}] \qquad (14\text{-}8)$$

The actual meaning of this solution depends on the values of $k$ and $\omega$. In particular, if $k^2 > \omega^2$ in Eqn (14-8), the exponential indices are all negative, so that $dy/dt$ is always negative. This then represents a return of $y$ from its maximum displacement to zero, with time $t \to \infty$, without $y$ ever changing sign. This is termed a *dead-beat motion.*

If $k^2 < \omega^2$, the indices are imaginary and Eqn (14-8) becomes

$$y = e^{-kt}(A_1 e^{j\sqrt{\omega^2 - k^2}.t} + A_2 e^{-j\sqrt{\omega^2 - k^2}.t}) \qquad (14\text{-}9)$$

where $j = \sqrt{-1}$.

Then also, since $\cos\theta + j\sin\theta = e^{j\theta}$ and $\cos\theta - j\sin\theta = e^{-j\theta}$,

$$y = e^{-kt}[(A_1 + A_2)\cos\sqrt{\omega^2 - k^2}.t + j(A_1 - A_2)\sin\sqrt{\omega^2 - k^2}.t]$$
$$= e^{-kt}(B\cos\sqrt{\omega^2 - k^2}.t + C\sin\sqrt{\omega^2 - k^2}.t)$$

This may be written in the form

$$y = A e^{-kt}\sin(\sqrt{\omega^2 - k^2}.t - \beta) \qquad (14\text{-}10)$$

where A and $\beta$ are constants and where A involves both $A_1$ and $A_2$ and $\tan\beta = -B/C$. This represents a simple harmonic motion whose amplitude is decreasing exponentially with time. $e^{-kt}$ is termed the *damping coefficient*. The frequency of the vibration is now given by $\sqrt{\omega^2 - k^2}/2\pi$, but this differs little from the original frequency $\omega/2\pi$ if $k$, the factor due to damping, is small.

If now $k^2 = \omega^2$ in Eqn (14-8), *critical damping* occurs. Substituting $k^2 = \omega^2$ in Eqn (14-8) gives $y = (A_1 + A_2)e^{-kt}$, which although a solution of the differential equation (14-3) is certainly not the most general one. The general solution of this equation, for $k^2 = \omega^2$, is

$$y = (At + B)e^{-kt} \qquad (14\text{-}11)$$

where A and B are two independent arbitrary constants. This is the equation of a non-oscillatory motion and is termed critical damping because oscillation only just does not occur. Alternatively, Eqn (14-8) shows that when $k$ is large, the term $\exp(-k - \sqrt{k^2 - \omega^2})t$ tends to zero (since $e^{-\infty} = 0$). Hence the $\exp(-k + \sqrt{k^2 - \omega^2})t$ term is predominant for $k$ large but still finite. This term has its greatest numerical value when $k^2 = \omega^2$, corresponding to the most rapid decay. That is, the body is brought to rest in the shortest time.

Equation (14-3) represents a damped free vibration. If now this vibration is disturbed by a periodic force proportional to $\sin qt$, where $q$ is a function of the period, then the damped forced vibration equation may be written [Eqn (14-3)] as

$$\frac{d^2y}{dt^2} + 2k\frac{dy}{dt} + \omega^2 y = p\sin qt \qquad (14\text{-}12)$$

Deleting the second term—that is making $k = 0$—makes this the equation of an undamped forced vibration. The solution of the differential equation

(14-12) is somewhat difficult, but is obtained by adding a particular integral to the solution of the equation

$$\frac{d^2y}{dt^2} + 2k\frac{dy}{dt} + \omega^2 y = 0$$

For $k^2 < \omega^2$, this solution has already been shown to be [Eqn (14-10)],

$$y = A\,e^{-kt}\sin(\sqrt{\omega^2 - k^2}.t - \beta)$$

It may also be shown that the required particular integral is of the form

$$y = B\sin(qt - \gamma)$$

giving as the complete solution to the damped forced vibration of Eqn (14-12)

$$y = A\,e^{-kt}\sin(\sqrt{\omega^2 - k^2}.t - \beta) + B\sin(qt - \gamma) \qquad (14\text{-}13)$$

where A, B, $\beta$, $\gamma$ are constants of integration depending on the defining conditions of the original motion. This shows that the final motion may be considered as made up of two components. The first term represents a free vibration with its own natural frequency and decaying at a rate depending upon $k$, which is the same rate of decay as the unforced system. This component of the resulting forced vibration is termed a *transient*, and soon dies away leaving the second term only of Eqn (14-13). The persisting forced vibrations then have the equation of motion

$$y = B\sin(qt - \gamma) \qquad (14\text{-}14)$$

This is an undamped simple harmonic motion—that is, it has a constant amplitude—with a frequency equal to $q/2\pi$, the frequency of the periodic driving force. Due to the two frequencies involved in Eqn (14-13), beats may be heard initially, until the damped free oscillation dies away leaving the steady-state condition.

Resonance occurs when the natural undamped frequency of the vibrating system has the same value as the forcing frequency. In this condition the energy of the forced vibration is at its maximum. This type of resonance is also termed *velocity resonance* and corresponds to the condition for current resonance in an electrical circuit.

## 14-3 Pipe resonance

If a source of vibrations—for example, a tuning fork—is held over the end of a pipe, the air in the pipe is set into forced vibration. If the length of the pipe is adjusted so that the natural frequency of vibration of the air column is the same as that of the driving force—that is, the frequency of the tuning fork—then resonance occurs and a loud sound booms out from the pipe.

Figure 14-2 shows a vibrating tuning fork held over the end of a pipe, which is closed at the other end. For simplicity, consider one prong only of the tuning fork. As this moves from position A to position B the air is compressed and this compression travels down the pipe as a longitudinal

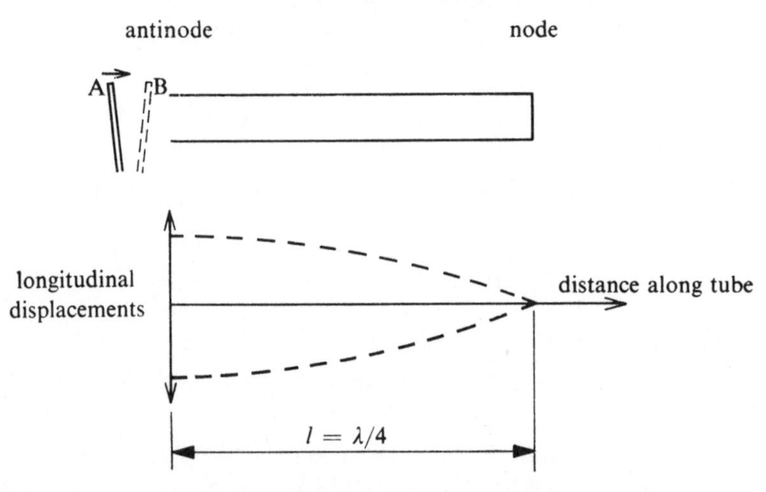


Fig. 14-2 Resonance in a closed pipe

wave. The compression is then reflected at the closed end, still as a compression, and travels back again. This compression tends to overshoot, leaving a rarefaction at the open end, which in its turn now travels down the pipe. If the time taken for the compression to travel down the pipe and back again, to form a rarefaction at the open end, is the same as the time taken for the prong of the tuning fork to travel from A to B, then this rarefaction coincides with the rarefaction due to the prong of the tuning fork starting to travel from B to A. The vibrations are then in step and resonance occurs. Thus for resonance, the prong travels from A to B and back to A in the time taken by a compression and the following rarefaction to travel down the pipe and back again. In other words four times the length of the pipe is equivalent to one wavelength of the sound, due to one complete vibration of the tuning fork. The air at the closed end is unable to vibrate so this point must always be a node, with the air here having the maximum variations in pressure. At the open end, the air is free to vibrate and consequently this is an antinode. Figure 14-2 also shows the particle displacement curve, but it must be remembered that this vibration is longitudinal not transverse, although for convenience it is drawn in the same way.

By lengthening the pipe [Fig. 14-3(a)] other lengths for resonance at the same frequency may be obtained. The closed end is always a node and the open end an antinode. Alternatively, by using forks of higher frequencies, other conditions of resonance may be obtained with the same length pipe [Fig. 14-3(b)]. The air in the first pipe is vibrating at the fundamental or 1st harmonic, whilst in the second tube the first overtone is formed. Since the frequency is proportional to $1/\lambda$, the frequency for this 1st overtone is three times that of the fundamental—that is, the 1st overtone corresponds to the 3rd harmonic. Similarly for the third pipe, the 2nd overtone corresponds to the 5th harmonic. The frequencies of the driving tuning forks must therefore be in the ratio of $1:3:5\ldots$. Compare this

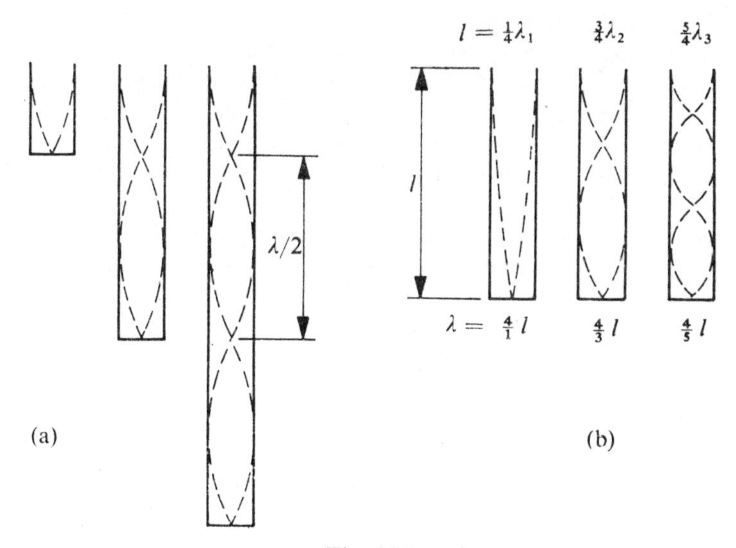


Fig. 14-3

with the frequencies of the overtones of a stretched string which are in the ratio of 1:2:3 . . . times that of the fundamental.

If the pipe is open at both ends, the longitudinal wave is reflected so that an antinode is now at both ends (Fig. 14-4). As with a closed tube, resonance occurs. As the prong of the tuning fork moves from A to B a compression is sent down the pipe. At the far end of the pipe, due to overshooting, this compression is reflected as a rarefaction, which in turn travels back along the pipe to be reflected as a compression again at the near end. For resonance to occur, the vibrations must be in step, so that the time for the prong to travel from A to B and back to A must be the same as the time taken for the compression to travel the length of the pipe

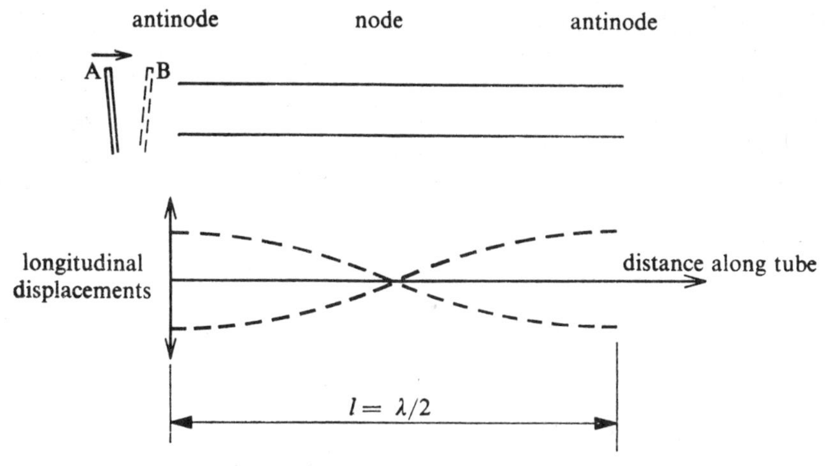


Fig. 14-4 Resonance in an open tube

and the reflected rarefaction to come back; that is, twice the length of the pipe corresponds to one wavelength of the sound.

Again as with the closed pipe, other lengths of pipe give the same conditions for resonance, with an antinode at each open end and with the time taken for the disturbance to travel down the pipe and back again being the same as that for the fork to make a whole number of complete vibrations. Figure 14-5(a) shows the lengths for resonance at the same frequency. As before, overtones may also be produced for a fixed length of pipe [Fig. 14-5(b)]. Since the vibration frequency is proportional to $1/\lambda$, the resonant frequencies are in the ratio 1:2:3... [Fig. 14-5(b)]; that is, the 1st overtone corresponds to the 2nd harmonic, and so on, as

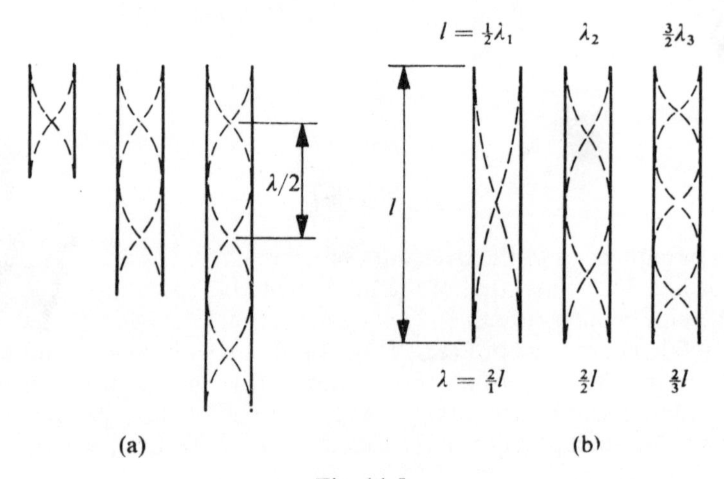


Fig. 14-5

with the vibrations of a stretched string. Thus a pipe open at both ends can resonate with any frequency that is a whole number multiple of its fundamental, whilst a pipe closed at one end only resonates with the odd number multiples. Vibrating air columns in an open and in a closed pipe, each with the same fundamental frequency, sound different, since overtones are always present at the same time as the fundamental is sounding, and the sound from the closed pipe lacks the even harmonics.

So far we have assumed that the plane of reflection at an open end of the pipe is actually at the end of the pipe. However, the reflection in fact takes place just outside the mouth of the pipe so that the pipe must be considered as being $0{\cdot}3d$ longer than it really is, where $d$ is the diameter of the pipe. Thus in Fig. 14-2 a quarter wavelength is in fact equal to the measured length of the tube, $l'$, plus $0{\cdot}3d$,

i.e., $$\lambda/4 = l = l' + 0{\cdot}3d$$

Similarly, for the pipe open at both ends, the *end correction* of $0{\cdot}3d$ applies to each end, so that in Fig. 14-4

$$\lambda/2 = l = l' + 0{\cdot}6d$$

where $l'$ is again the measured length. By a suitable bell-shaped design of the pipe mouth, it is possible for the end correction to be made zero so that the antinode is in the plane of the end. The overtones would then form a harmonic series in practice as well as in theory. This is important in the design of brass wind instruments.

It has also been found that there is a stationary layer of air about 1 mm thick adjoining the pipe wall. The actual thickness of this layer varies with the physical conditions and it has little effect except in pipes with small diameters. For pipes less than about 1-cm diameter the layer causes considerable damping.

Resonant pipes may be used to measure the velocity of sound in air or other gas, since at resonance the wavelength may be determined and the frequency of the tuning fork is known. Hence the velocity may be calculated from $v = f\lambda$. The end correction can be avoided; for example, for a closed pipe the difference between the first and second resonant positions is equal to half a wavelength [Fig. 14-3(a)] and if the distance between these two positions is measured there is no need for an end correction.

## 14-4 Cavity resonance

If water is added slowly to a bottle and a sounding tuning fork held over the neck, an air volume is reached at which resonance occurs. When the bottle is tilted, the shape of the cavity changes but the air volume remains constant. Since the resonance does not change, it must depend on the volume of the cavity, not its shape. It may be shown experimentally that the natural frequency of vibration of a cavity increases with the area of the mouth and decreases with increase of volume. Helmholtz constructed a number of cavity resonators in the form of hollow brass spheres. As well as the mouth to let the sound in, these had a smaller aperture which could be held to the ear so that any resonance within the cavity could be clearly heard. With a cavity resonator of this type, it is possible to detect a particular frequency in a mixed sound, to the exclusion of all other frequencies. In this way complex sounds may be analysed.

In its simplest form such a cavity may be considered as a fixed mass of gas within the neck, with the air within the cavity acting as a spring. If the effective length of the neck is $l$ and the area of the aperture is $A$ (Fig. 14-6), the effective mass of the 'piston' of air in the neck is $Al\rho$, where $\rho$ is the mean density of the air. If now this 'piston' moves through

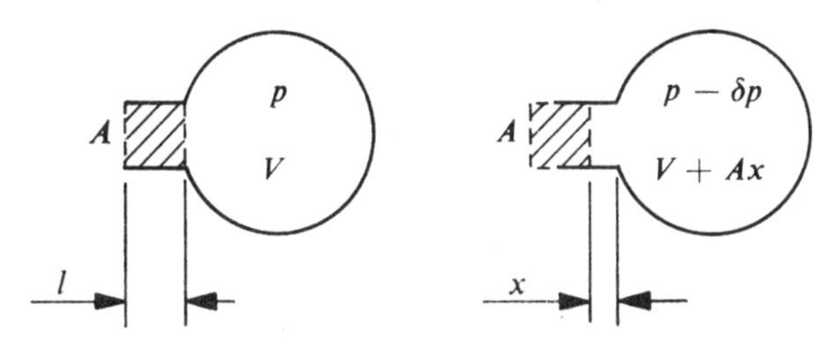


Fig. 14-6 Cavity resonator

a distance $x$, the volume of the cavity changes from $V$ to $V + Ax$. Since the change is adiabatic,

$$pV^{\gamma} = (p - \delta p)(V + Ax)^{\gamma} \tag{14-15}$$

where $p$ was the initial pressure in the cavity. Expanding Eqn (14-15) by the binomial theorem,

$$pV^{\gamma} = pV^{\gamma} + \gamma pV^{\gamma-1}Ax + \frac{\gamma(\gamma-1)}{2!}pV^{\gamma-2}A^2x^2 + \cdots$$

$$- \delta pV^{\gamma} - \gamma\delta pV^{\gamma-1}Ax - \frac{\gamma(\gamma-1)}{2!}\delta pV^{\gamma-2}A^2x^2 - \cdots$$

Hence, neglecting small terms involving powers of $x$, or $x\,\delta p$, and re-arranging

$$\delta pV^{\gamma} = \gamma pV^{\gamma-1}Ax$$

i.e.,

$$\delta p = \frac{\gamma pAx}{V} \tag{14-16}$$

The actual force on the 'piston', due to the pressure difference $\delta p$, is $A\,\delta p$, hence

$$\text{force} = A\,\delta p = \text{mass} \times \text{acceleration}$$

$$= -m\frac{d^2x}{dt^2}$$

$$= -Al\rho\frac{d^2x}{dt^2} \tag{14-17}$$

where the negative sign indicates a restoring force. Therefore, combining Eqns (14-16) and (14-17),

$$\frac{d^2x}{dt^2} + \frac{\gamma pA}{V\rho l}x = 0 \tag{14-18}$$

This is the equation of a simple harmonic motion, and therefore the frequency is given by

$$f = \frac{1}{2\pi}\sqrt{\frac{\gamma pA}{V\rho l}} \tag{14-19}$$

Also the velocity of sound in a gas is given by (Eqn 12-7)

$$v = \sqrt{\frac{\gamma p}{\rho}}$$

Hence the natural frequency of vibration of the cavity is

$$f = \frac{v}{2\pi}\sqrt{\frac{A/l}{V}} \tag{14-20}$$

where $V$ is the volume of the cavity and $A/l$ is termed the conductivity of the neck. This theory is not exact since it is complicated by the end corrections in $l$, but it is sufficiently correct to show that for a cavity resonator

$$f^2V = \text{const.} \tag{14-21}$$

When a cork is removed rapidly from a bottle, the pop heard is due to adiabatic oscillations being set up in the neck, with the bottle acting as a cavity resonator.

## 14-5 Bar resonance

A long bar or rod in longitudinal oscillation vibrates in a similar manner to the air inside a pipe open at both ends. A bar that is not constrained and can vibrate freely has maximum movement at its ends which therefore form antinodes. Between these antinodes one or more nodes occur. The wavelengths in the solid (not air) are therefore $2l, 2l/2, 2l/3, \ldots$, where $l$ is the length of the bar [compare this with the air column in Fig. 14-5(b)]. The velocity of sound in a bar is given by $v = \sqrt{(E/\rho)}$ [Eqn (11-23) and Section 12-4] where $E$ is Young's modulus of elasticity for the bar material and $\rho$ is its density. The natural frequencies of vibration of the bar are therefore

$$f = \frac{n}{2l}\sqrt{\frac{E}{\rho}} \tag{14-22}$$

where $n = 1, 2, 3, \ldots$ Thus, the overtones form a complete harmonic series. If the bar is constrained at its middle point this must always be a node. Such a bar cannot vibrate with the frequencies of the even-numbered harmonics, that is, $n \neq 2, 4, \ldots$ in Eqn (14-22).

It is interesting to compare the natural frequencies of transverse and longitudinal vibrations in the same bar. For transverse vibrations the bar must be thin and under tension and may therefore be treated as a stretched wire. Let the natural frequencies be $f_t$ and $f_l$ for the transverse and longitudinal vibrations respectively, and let the length of the thin bar (wire) be $l$. Under a tension $T$ this length increases by $l'$, to $(l + l')$. By Eqn (13-9) the frequency of the transverse vibration is

$$f_t = \frac{1}{2l}\sqrt{\frac{T}{m}}$$

$$= \frac{1}{2l}\sqrt{\frac{T}{A\rho}} \tag{14-23}$$

where $m$ is the mass per unit length, $A$ is the cross-sectional area, and $\rho$ is the density of the wire. But Young's modulus, by definition, is

$$E = \frac{\text{stress}}{\text{strain}} = \frac{T/A}{l'/l} \tag{14-24}$$

and therefore Eqn (14-23) may be rewritten as

$$f_t = \frac{1}{2l}\sqrt{\frac{El'}{\rho l}}$$

$$= f_l\sqrt{\frac{l'}{l}}$$

by Eqn (14-22). Hence $f_t$ becomes equal to $f_l$ only if $l = l'$—that is, only when the wire is stretched to twice its normal length. This is normally an impossible condition in practice. Therefore the natural frequency due to longitudinal vibrations in a rod is much greater than that due to transverse vibrations.

## PROBLEMS

**14-1** Discuss the practical importance of resonance, illustrating your answer with examples from different branches of engineering and physics.

**14-2** Give a brief explanation of resonance. One tuning fork is placed near a cavity whose resonant frequency is the same as that of the tuning fork. A second similar fork is placed at a distance and the two forks are bowed equally. What difference will there be in the rate at which the vibrations decay, and why is there this difference?

**14-3** A vertical tube 1 m long and 3 cm in diameter is initially full of water. If a tuning fork of frequency 256 Hz is made to sound continuously over the top of the tube and the water is allowed to run out slowly, find the distances from the top of the tube at which resonances occur (velocity of sound = 340 m/s). [33·2 cm, 98·7 cm]

**14-4** Five beats per second are heard when two organ pipes are sounding together in air at 10°C. How many beats would be heard if the temperature rose to 25°C? (Velocity of sound in air at 0°C = 331 m/s.) [5·13 beats/s]

**14-5** A tube is closed at one end. Resonance with a particular tuning fork is first found to occur when the tube is 29 cm long and then again when it is 92 cm long. What is the diameter of the tube? [8·3 cm]

**14-6** A steel wire, diameter 1 mm, is stretched by a 10-kg load. If Young's modulus for the wire is $19{\cdot}5 \times 10^{10}$ N/m$^2$, find the ratio of the fundamental frequencies of transverse to longitudinal vibrations of the wire (acceleration due to gravity = 9·81 m/s$^2$). [1:39·5]

**14-7** A fixed frequency source of 256 Hz and an organ pipe at 15°C and of a slightly lower frequency are sounding together, producing 5 beats/s. On changing the temperature of the organ pipe the number of beats heard again becomes 5 beats/s. What is the new temperature of the organ pipe? [38·4°C]

**14-8** Show that if the note given out by an organ pipe is to remain approximately constant for all normal temperatures, the coefficient of cubical expansion of the material from which the pipe is made should be about one and a half times greater than that of a gas.

**14-9** A wire of length 40 cm and weighing 0·8 g is stretched by a load of 6 kg. What is the shortest length of 5-cm diameter pipe, open at both ends, for resonance to occur between the air column and the wire vibrating at its fundamental note? (Acceleration due to gravity = 9·80 m/s$^2$; velocity of sound in air = 340 m/s.) [76·3 cm]

**14-10** The air column contained in a brass tube closed at one end resonates at 600 Hz when at 0°C. What will be its resonant frequency at 50°C, if the coefficient of linear expansion of brass is $19 \times 10^{-6}$ per °C, neglecting the end correction? [652·0 Hz]

# 15 Intensity of sound

## 15-1 Introduction

Noise, loudness, and intensity are all terms that are related but nevertheless have very different meanings. *Noise* is all sounds that are within the hearing range but are undesired. These are the sounds to which the attention is not directed and whose reception is resisted by the hearer. Thus noise causes fatigue, masks desired sounds, and distracts the listener's attention. Noise is made up of a very large number of sound frequencies and may therefore be termed 'unpitched' sound. If the frequency spectrum is continuous and the amplitude density is constant throughout the frequency range, the noise is termed 'white' noise, by analogy with white light. Loud noises may cause hearing damage whilst a less loud noise may cause nervous disturbances, which in some occupations could obviously have dangerous consequences. There is thus a need to be able to assess the degree of loudness.

*Loudness* is a purely subjective sensation, involving the response of the ear to the sound; *intensity*, on the other hand, can be defined scientifically and involves a definite physical quantity. A sound may be placed on a loudness scale from quiet to loud, but its position is purely a psychological choice, although clearly loudness and intensity are related. If a sound is more intense, it is judged to be louder, but there is no one-to-one correspondence. An observer may judge that one sound is louder than another but will have difficulty in saying by how much. If the two sounds are pure tones of different frequencies, then he may even have difficulty in choosing which is the loudest.

Loudness and intensity may be regarded as related in the same way that current is related to its reading on an ammeter. Analogously, an electric current may be precisely measured and causes a deflection on a galvanometer. The current is a precise quantity, as is intensity, but the galvanometer deflection, although it increases as the current increases, depends on the galvanometer itself in the same way that loudness depends on the observer's ear.

Pitch and frequency are related like loudness and intensity. Pitch is the subjective sensation and frequency is objective. Frequency is capable of precise definition and measurement, in vibrations per second. Pitch is related to frequency but also depends on intensity, especially at high and low frequencies. It is a musical term and is expressed as that frequency of a pure tone of equal loudness which is judged by the average normal ear to have the same position in a musical scale. A musical sound is complex, containing sounds of different frequencies as well as harmonics. There is however usually a dominant note, or notes, which has a particular frequency and therefore characterizes the pitch. For pure tones the pitch and frequency are thus the same. It is found that two low-intensity sounds with frequency ratios 1:2 are judged by a normal observer correctly as octaves of each other, but if one sound has its intensity raised greatly, without changing its frequency, it sounds flat, showing that judgement of pitch is not wholly determined by frequency.

Before the discussion of loudness and noise measurement can be carried further it is necessary to consider the meaning of intensity in greater detail.

## 15-2 Intensity

The propagation of sound involves the transmission of energy in the form of a wave. The intensity of a sound is defined as the flow of this energy across unit area in unit time. This is the energy contained in a column of length $v$, the velocity of the sound; that is, the product of the total energy density of the sound wave and the velocity $v$. The total energy density of the wave is the sum of the kinetic and potential energies due to the wave motion. Expressions for these separate energies may be derived, or alternatively the total energy may be calculated directly, as follows.

Consider the work done by a piston in producing a longitudinal sinusoidal wave in air. Let this piston have unit cross-sectional area and be located at $x = 0$, so that the waves travel in the positive $x$-direction with a velocity $v$. By Eqn (11-2) the wave produced by a small displacement $\xi$ of the piston in the $x$-direction is

$$\xi = a \sin(\omega t - \alpha)$$

where $\omega = 2\pi f$ and $\alpha = 2\pi x/\lambda$

i.e.,
$$\xi = a \sin 2\pi\left(ft - \frac{x}{\lambda}\right) \tag{15-1}$$

By Eqn (11-16), the excess pressure due to the movement of the piston is

$$\delta p = Ks \tag{15-2}$$

where $K$ is the bulk modulus of elasticity and $s$ the condensation as defined by Eqn (11-13). Also at the piston, that is, at $x = 0$, from the definition of $s$ in Section 11-7, at any instant $t$,

$$s = -\frac{\delta V}{V_0} = -\left(\frac{\partial \xi}{\partial x}\right)_{x=0}$$

$$= \frac{2\pi a}{\lambda}\cos 2\pi ft \tag{15-3}$$

In addition, the bulk modulus

$$K = \gamma p_0$$

by Eqn (12-6).

Hence, the excess pressure on the piston [Eqn (15-2)] is

$$\delta p = \frac{2\pi a \gamma p_0}{\lambda} \cos 2\pi f t \tag{15-4}$$

The velocity of the moving piston is $(\partial\xi/\partial t)_{x=0}$ so that the actual distance moved by it in a time $\mathrm{d}t$ is equal to

$$\left(\frac{\partial \xi}{\partial t}\right)_{x=0} \mathrm{d}t = 2\pi a f \cos 2\pi f t \, \mathrm{d}t \tag{15-5}$$

Thus the work done by the piston of unit cross-sectional area, from the product of force and distance, is

$$\mathrm{d}W = \frac{4\pi^2 a^2 f \gamma p_0}{\lambda} \cos^2 2\pi f t \, \mathrm{d}t \tag{15-6}$$

Therefore the work done in $n$ complete oscillations of the piston—that is, in $n$ periods of the resulting wave—is

$$W = \frac{4\pi^2 a^2 f \gamma p_0}{\lambda} \int_0^{nT} \cos^2 2\pi f t \, \mathrm{d}t \tag{15-7}$$

$$= \frac{4\pi^2 a^2 f \gamma p_0}{\lambda} \int_0^{nT} \frac{1 + \cos 4\pi f t}{2} \, \mathrm{d}t$$

$$= \frac{4\pi^2 a^2 f \gamma p_0}{\lambda} \left[\frac{t}{2} + \frac{\sin 4\pi f t}{8\pi f}\right]_0^{nT}$$

$$= \frac{4\pi^2 a^2 f \gamma p_0}{\lambda} \frac{nT}{2} \tag{15-8}$$

Putting

$$\gamma p_0 = v^2 \rho_0$$

that is, Eqn (12-7), then the work done by the piston in the time $nT$ is

$$W = \frac{2\pi^2 a^2 f v^2 \rho_0 n T}{\lambda}$$

$$= \frac{2n\pi^2 \rho_0 a^2 v^2}{\lambda} \tag{15-9}$$

This energy is all contained in the wave train. Hence dividing by the time $nT$, i.e. $n\lambda/v$, gives the intensity, $I$ (or energy flow per unit time per unit area normal to the direction of propagation), also termed the 'energy flux':

$$I = \frac{2\pi^2 a^2 v^3 \rho_0}{\lambda^2} \tag{15-10}$$

This may be written in various ways, for example,

$$I = 2\pi^2 \rho_0 a^2 f^2 v \tag{15-11}$$

It will be noted that the *intensity* is proportional to the *square of the amplitude*; this is true for all types of wave motion.

This amount of energy, $I$, must also be the same as that contained in a column of unit cross-section and of length $v$, the velocity of sound. That is, the intensity, $I$, is the product of the 'mean energy density' $2\pi^2\rho_0 a^2 f^2$ and the velocity $v$. It will be seen that the particle velocity $(\partial\xi/\partial t)_{x=0}$ found by differentiating Eqn (15-1) has a maximum value of $2\pi af = v'$. Hence, the mean energy density is also equal to $\frac{1}{2}\rho_0 v'^2$, so that

$$I = \rho_0 v \frac{v'^2}{2} \qquad (15\text{-}12)$$

This equation is of the same form as that for electrical power, where $\rho_0 v$ replaces resistance and is here termed the 'characteristic impedance' of the medium, and $v'^2/2$ takes the place of the current squared, that is, current is represented by the r.m.s. particle velocity amplitude.

Equation (15-12) may be more usefully written in terms of pressure. By Eqn (15-2) [Eqn (11-16)] the change in pressure within the longitudinal wave is

$$\delta p = Ks$$

but $K = \gamma p_0 = v^2\rho_0$ [Eqn (12-7)] and $s = (\rho - \rho_0)/\rho_0$ by definition [Eqn (11-13)] and therefore

$$\delta p = v^2(\rho - \rho_0)$$

This may be written as

$$p' = v^2(\rho' - \rho_0) \qquad (15\text{-}13)$$

for the maximum pressure change within the wave.

Also, since the maximum value of the particle velocity $\partial\xi/\partial t$ is $v' = 2\pi af$ by differentiating Eqn (15-1) with respect to time $t$, and since the maximum value of the condensation is $s = -\partial\xi/\partial x = 2\pi a/\lambda$, from Eqn (11-17) and by differentiating Eqn (15-1) with respect to $x$,

$$v' = \lambda fs = vs$$

$$= v\left(\frac{\rho' - \rho_0}{\rho_0}\right) \qquad (15\text{-}14)$$

so that [from Eqn (15-13)]

$$v' = \frac{p'}{v\rho_0} \qquad (15\text{-}15)$$

Thus the intensity may be expressed as [Eqn (15-12)],

$$I = \frac{p'^2}{2\rho_0 v} \qquad (15\text{-}16)$$

where $p'$ is the pressure amplitude, $\rho_0$ is the average density of the air, and $v$ is the velocity of the sound wave.

Intensity may be expressed in units of watts/square metre or, if in terms of pressure variation, in units of newtons/square metre.

## 15-3 Intensity and loudness scales

Weber was the first to investigate the connection between a stimulus, such as light or sound, and the corresponding sensation, vision or hearing. He showed that the change in the intensity of the stimulus, $\delta I$, necessary to produce the least perceptible change in sensation, $\delta S$, is proportional to the pre-existing intensity of the stimulus,

i.e.,
$$\delta S = K(\delta I/I) \qquad (15\text{-}17)$$

where K is a constant. Fechner integrated Weber's law, assuming $\delta S$ and $\delta I$ were true differentials, to obtain the Weber–Fechner law

$$S = K \log I \qquad (15\text{-}18)$$

In his original work Weber used weights to produce a physical sensation but the law is generally applicable to any sensation involving the nervous system. However, for extremes of intensity and frequency the law is only approximately true. One has only to sit in a quiet room for a while and then to start hearing very faint sounds, such as the creaking of floorboards, to appreciate the general truth of the law, which is that the sensitivity of the ear increases as the intensity of the surrounding sound level decreases.

It will be seen that it is the *fractional* change in the stimulus intensity which is important. The smallest detectable change in this fraction, $\delta I/I$, has in practice been found to be approximately 0·1. Consequently, the Weber–Fechner law has led to the definition of a unit, the *bel* (named in honour of Alexander Graham Bell), which is a unit for the *comparison* of intensities. If $I_1$ and $I_2$ are the intensities of two sources, then the ratio of these intensities is expressed as

$$N = \log_{10} (I_1/I_2) \text{ bels} \qquad (15\text{-}19)$$

It is found more convenient in practice to use decibels (dB) which are tenths of bels, on the same logarithmic scale, so that one decibel is equivalent to a change of 1·26 times ($\text{antilog}_{10}$ 0·1). Thus the intensity ratio of Eqn (15-19) may be expressed as

$$n = 10 \log_{10} (I_1/I_2) \text{ decibels} \qquad (15\text{-}20)$$

This forms a useful scale since the decibel is about the smallest intensity change that can normally be detected by the human ear and because the range of human hearing is large and is therefore more conveniently dealt with by a logarithmic rather than an arithmetic scale. Starting with a sound that can only just be heard, a tenfold increase in intensity is equal to one bel, and a further tenfold increase is equal to another bel. If thirteen such tenfold increases are made the sound produced, thirteen bels, would cause pain to a hearer.

Since all modern sound detectors respond to pressure changes, it is convenient to express the intensity ratio in terms of pressure changes. Since, by Eqn (15-16), the intensity is proportional to the square of the corresponding pressure amplitude, the intensity ratio of Eqn (15-20) may be expressed as a *sound pressure level*,

i.e., $$\text{SPL} = 20 \log_{10} (p_1/p_2) \text{ dB} \tag{15-21}$$

By measuring the pressure changes of sound intensities that are at the threshold of hearing, it has been found that the normal listener is most sensitive to sounds with a frequency of about 2,000 Hz. At this frequency a change in pressure of less than $10^{-3}$ microbar can just be detected. One microbar is equivalent to a pressure of 0·1 N/m$^2$ or 1 dyn/cm$^2$, which is approximately a millionth of the atmospheric pressure. The upper limit of excess pressure, at the threshold of feeling, is about $10^3$ microbar. That is, at medium frequencies, the ear is capable of handling sounds involving pressure amplitude changes over a range of a million to one.

As we have already said, only intensity ratios are measured, not absolute values. To put the intensity on an absolute scale, known as the *intensity level*, a scale has been devised which measures intensity relative to an arbitrary fixed low reference intensity. The reference chosen has an intensity which is approximately at the average threshold of hearing and is defined as having an intensity, for plane waves, of $10^{-12}$ W/m$^2$ = $10^{-9}$ erg/s/cm$^2$. This is approximately equivalent to an r.m.s. pressure variation in the wave of 0·00002 N/m$^2$ = 0·0002 dyn/cm$^2$. Then for a particular sound, of intensity $I$ and r.m.s. pressure amplitude $p$,

$$\text{intensity level in dB} = 10 \log_{10} (I/I_0) \tag{15-22}$$
$$= 20 \log_{10} (p/p_0) \tag{15-23}$$

where $I_0$ and $p_0$ are the reference values referred to above. Figure 15-1 gives an approximate indication of intensity levels of some common sounds. An intensity level of 120 dB, corresponding to an intensity of 1 W/m$^2$, is about the maximum that the ear can tolerate.

An intensity level defined in the above manner must not be confused with a *loudness level*. Loudness is a purely subjective quantity and depends on the pitch. Therefore in the measurement of loudness, the reference level of loudness is not only taken as that at the threshold of hearing as in defining an intensity level, but as that at the threshold of hearing for a particular pitch. This particular pitch, or standard tone, is defined as that of a plane sinusoidal sound wave coming from a position directly in front of the observer and having a frequency of 1,000 Hz. The reference level of the loudness is taken as that corresponding to an r.m.s. sound pressure of 0·00002 N/m$^2$, as before. Thus to measure a loudness level, the standard 1,000 Hz tone is increased in intensity until it sounds as loud as the sound under test, as judged by an observer using both ears, with the standard tone and the sound under measurement being heard alternately. If the intensity of the standard tone has to be raised by, say, $n$ decibels above its reference level for the sound under test to be judged equally loud, then the sound under test is said to have an 'equivalent loudness' of *n phons*.

Thus the intensity level of a sound is the number of decibels above the reference intensity, whereas the loudness level is the intensity level (with

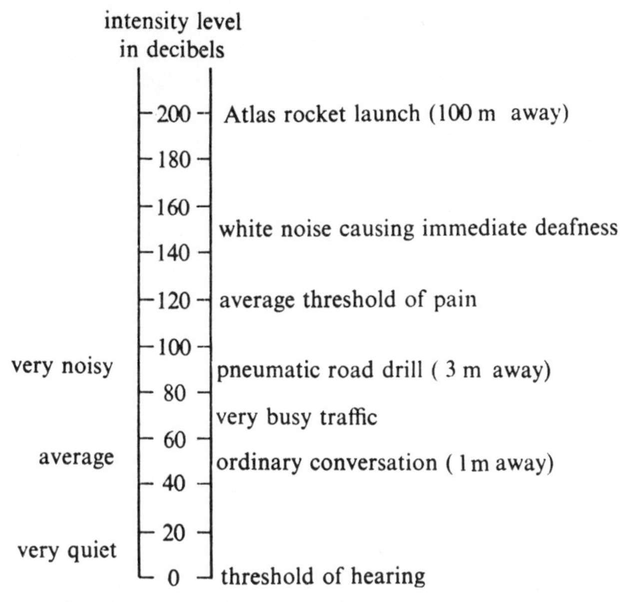


Fig. 15-1 Intensity levels of some common sounds

decibels expressed as phons) of the equally loud reference tone alternately heard at the same place.

**Example 15-1** A 600 Hz tone has an intensity of $5 \times 10^{-7}$ W/m². On comparing with the 1,000 Hz standard tone, it is found that the intensity of the standard tone must be increased from the reference level of $1{\cdot}0 \times 10^{-12}$ W/m² to $2 \times 10^{-9}$ W/m² for the sounds to appear equally loud. What are the intensity and loudness levels of the 600 Hz tone?

Intensity level

$$10 \log_{10} \left(\frac{I}{I_0}\right) = 10 \log_{10} \frac{5 \times 10^{-7}}{1 \times 10^{-12}}$$
$$= 57{\cdot}0 \text{ dB}$$

Loudness level

$$10 \log_{10} \left(\frac{I}{I_0}\right) = 10 \log_{10} \frac{2 \times 10^{-9}}{1 \times 10^{-12}}$$
$$= 33{\cdot}0 \text{ phons}$$

A disadvantage of measuring loudness in terms of phons is that it is difficult to appreciate relative values. To get over this difficulty a *sone* scale has been devised. In this scale a loudness level of 40 phons is defined as having a loudness of 1 sone. This is a linear scale, so $n$ sones appear $n$ times as loud as one sone. In practice, a smaller unit, the *millisone* = 0·001 sone, is often used. Thus a sound judged by an average listener to be four times as loud as a pure 1,000 Hz tone whose intensity level is 40 dB

—that is, a loudness level of 40 phons—would have a loudness of four sones.

This scale is based upon the fact that for the average listener, two sounds differing in intensity by 10 dB have a loudness ratio of approximately two to one. Then, since one sone is defined as a loudness level of 40 phons, the number of sones ($S$) in terms of the number of phons ($P$) is

$$S = 2^{(P-40)/10} \tag{15-24}$$

or in logarithmic form

$$\log_{10} S = 0{\cdot}030P - 1{\cdot}20 \tag{15-25}$$

## 15-4 Intensity measurements

Intensity can conveniently be determined by measurement of the pressure variation in the sound wave and substitution in Eqn (15-23). The pressure variation is easily determined from the output of a suitably calibrated microphone so that the problem becomes one of calibrating the microphone.

At low frequencies the microphone can be calibrated with the *pistonphone*; this is simply a close-fitting piston moving in a cavity, at the other end of which the microphone is mounted. Movement of the piston subjects the microphone to pressure oscillations of known magnitude. A double-piston arrangement, which reduces backlash from bearings, gives more accuracy.

The *electrostatic actuator* is similar in principle to the pistonphone but, instead of a simple mechanically driven piston, the electrostatic force between charged metal plates is used to produce oscillations of the microphone diaphragm. The inertia of the piston and the phase differences occurring in the cavity make the pistonphone unsuitable for high frequency use. The electrostatic actuator, however, may be used well into the ultrasonic range. A metal conductor plate is arranged parallel and close to the microphone diaphragm, which acts as the other conductor plate. This forms a capacitor. Knowing the voltage applied across the capacitor and the thickness of the air gap, the force—that is, the pressure—on the microphone diaphragm may be calculated. The actuator plate may be slotted to reduce air damping. The difficulty of measuring the air gap limits the accuracy of this method.

Probably the best calibration technique involves the *reciprocity method*, but a detailed discussion of this is beyond the scope of this book. The reciprocity principle, well established in electrical circuit theory, is applied to acoustics by using microphones which may also be used as sources of sound, for example, capacitor microphones. Such a microphone may be called a reversible *transducer*, since it is capable of receiving power from one system and transferring it to another, and vice versa. By connecting two of these microphones to a cavity and using one microphone as the source of sound and the other as the receiver, it is possible to calculate the product of the microphones' sensitivities in terms of the volume of the cavity, ambient pressure, driving current, and the output voltage. By next subjecting the two microphones in turn to the same cavity pressure, which is obtained by using a third sound source in the cavity, the ratio of the microphones' sensitivities may be obtained. Knowing the product and ratio of the sensitivities, the absolute sensitivity for each of the two microphones may be obtained in millivolts/microbar.

## 15-5 Noise Abatement Act

The Noise Abatement Act, 1960, makes provisions for the control of noise and vibration. It makes some noises and vibrations a statutory nuisance and therefore brings them under the Public Health Act, 1936. The Noise Abatement Act lays down that three or more persons living near to the source of sound nuisance are sufficient to file a complaint. However, it is a defence in the case of noises or vibrations caused in the course of a trade or business to show that the best practical means have been taken to prevent or reduce the nuisance. A ruling is also made on the use of loudspeakers in the street. It is now an offence, punishable on a summary conviction by a maximum fine of £10, to use a loudspeaker in the street at any time for the purpose of advertising any entertainment, trade, or business, or to use it for any other purpose between the hours of nine in the evening and eight the next morning. Exceptions are made for police, ambulance, fire brigade, local authorities, and certain other specialized purposes, including mechanical loudspeakers not relaying words and attached to vehicles carrying perishable goods. In this last category the use is restricted to the hours between noon and seven o'clock in the evening.

The Act however does not extend to noise or vibration caused by aircraft. Under the Airports Authority Act, 1965, the Minister concerned has the power to take such measures as necessary to limit airport noise and vibration. Local Authorities still have the power to make by-laws which increase the coverage made by the Act, but may not reduce the coverage.

## 15-6 Inverse square law

The inverse square law given for the illumination of a surface (Section 7-3) is equally applicable to sound. Consider a source, S, of sound radiating equally in all directions (Fig. 15-2). If equal and constant amounts of sound energy radiate in unit time from this source, then at any instant the energy contained in a spherical shell of radius $r_1$ and unit thickness is

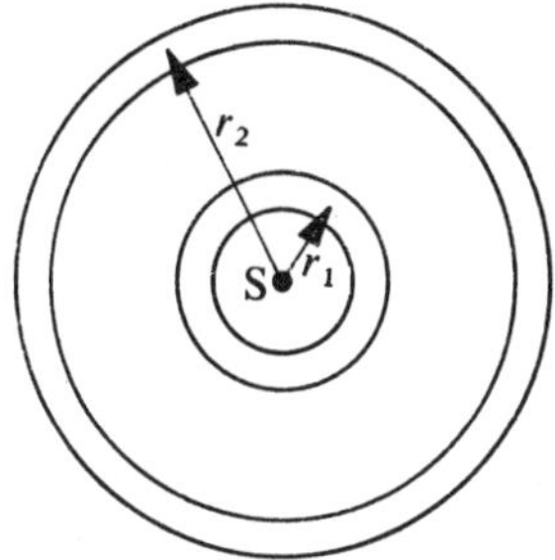


Fig. 15-2 Illustrating the inverse square law

$$\text{(volume of shell)} \times \text{(energy per unit volume)} = 4\pi r_1^2 I_1 \quad (15\text{-}26)$$

where $I_1$ is the intensity of the sound at a distance $r_1$ from the source S. Similarly, for a spherical shell of radius $r_2$

$$\text{energy} = 4\pi r_2^2 I_2 \quad (15\text{-}27)$$

But a constant amount of energy is being radiated in unit time and since the shells are of equal thickness

$$4\pi r_1^2 I_1 = 4\pi r_2^2 I_2$$

i.e.,

$$\frac{I_1}{I_2} = \frac{r_2^2}{r_1^2} \tag{15-28}$$

That is, the intensity of the sound varies inversely as the square of the distance from the source of sound.

**Example 15-2** A hall is 30 m long and 7 m high. If the ceiling reflects 75% of the incident sound, compare the intensity of sound reaching the end of the hall directly from a source at the other end of the hall, with that reflected from the ceiling.

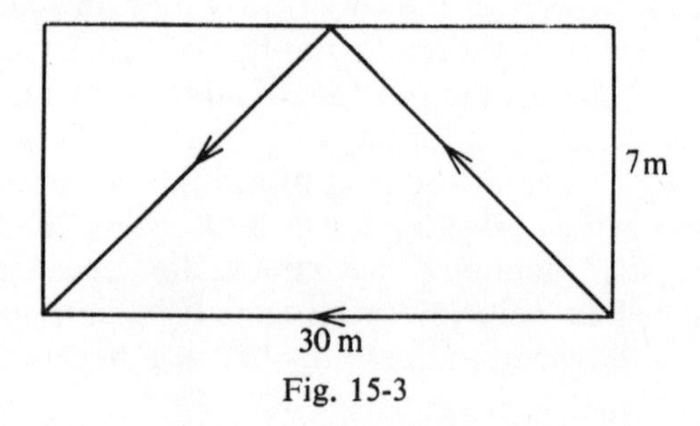


Fig. 15-3

If $W$ is the energy of the source, then the energy reaching the other end of the hall by a direct path (Fig. 15-3)

$$= W/30^2$$

Similarly, for the reflected path, the energy reaching the other end

$$= \frac{75}{100} \cdot \frac{W}{4(7^2 + 15^2)}$$

Hence the difference in the intensities at the rear of the hall, due to the different paths

$$= 10 \log_{10} \left[ \frac{W/30^2}{75W/400(7^2 + 15^2)} \right]$$

$$= 10 \log_{10} 1{\cdot}624$$

$$= 2{\cdot}11 \text{ dB}$$

That is, the sound intensity at the back of the hall due to the direct sound is 2·11 dB greater than that due to the reflected sound.

The intensity $I$ of a sound wave is the average rate at which energy is carried by the wave across unit area perpendicular to the direction of propagation. In other words, it is the average power across this unit area. Thus the total power across a surface is the product of the sound intensity at that surface and the surface area. For example, consider a loudspeaker emitting sound into a room such that the sound intensity is $10^{-4}$ W/m$^2$ at a distance of 4 m from the speaker, and assume that the intensity is the same all over a hemispherical surface centred at the speaker. The area of this hemisphere is approximately 100 m$^2$ and hence the acoustic power of the loudspeaker is $10^{-4} \times 100 = 0{\cdot}01$ W. Since the efficiency of a loudspeaker—that is, the ratio of the acoustic power to the input power—is only a few per cent, the power into the speaker has to be about one watt.

## 15-7 Microphones

Variations in an acoustic wave produce changes in the electrical properties of a microphone enabling it to convert the sound signal into an electrical one. This electrical output can then be amplified.

The *carbon microphone* contains small granules of carbon which touch each other and the metal containing cup (Fig. 15-4) to form a large number

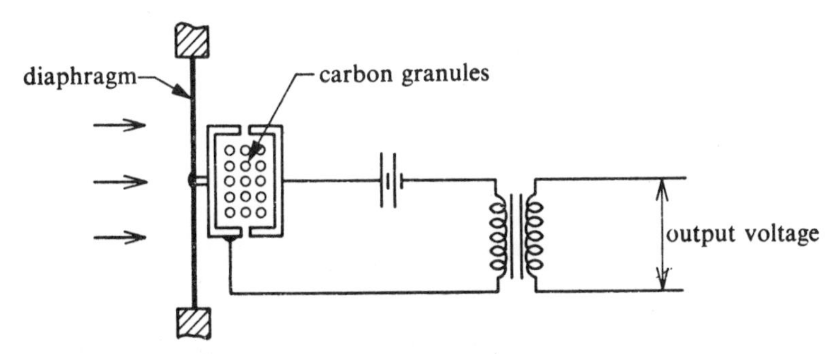


Fig. 15-4 Arrangement of a carbon microphone

of electrical contacts. Displacements of the diaphragm by pressure variations in the sound wave cause changes in the pressure of the electrical contacts and hence change the electrical resistance. If the microphone is incorporated in a closed electrical circuit with a steady source of e.m.f., the change in resistance produces a change in current. A transformer included in the circuit amplifies the current variations. This type of microphone is both cheap, robust, and highly sensitive and so is very commonly used. However, it has the disadvantages that it produces a background noise or hiss, that the granules are liable to pack, and that its frequency response is somewhat irregular—that is, its sensitivity is not constant with frequency.

The *capacitor microphone* consists of a thin metal diaphragm held under tension and separated from a fixed metal plate by a parallel air gap about 0·025 mm thick (Fig. 15-5). This forms a parallel-plate capacitor whose capacitance varies with the movement of the diaphragm. The thinner the air gap the greater is the microphone's sensitivity. This type of instrument has the advantage that, since the capacitance is inversely proportional to the distance between the plates, the output from the microphone may be made directly proportional to the pressure amplitude on the diaphragm for diaphragm displacements that are small fractions of the average plate separation. Consequently this type of microphone has a very good frequency response. There is an electrical disadvantage, however: due to the very high internal impedance of the microphone, careful design and construction are needed for the amplifier, which is often built as an integral part of the microphone assembly. Capacitor microphones also have the disadvantages that moisture may form between the plates and cause failure and they are fragile. On the other hand, their frequency response is good and there is an absence of background noise. Some problems arise due to the resonant frequency of the diaphragm itself. If the tension is increased to raise the resonant frequency above the audio range, there is

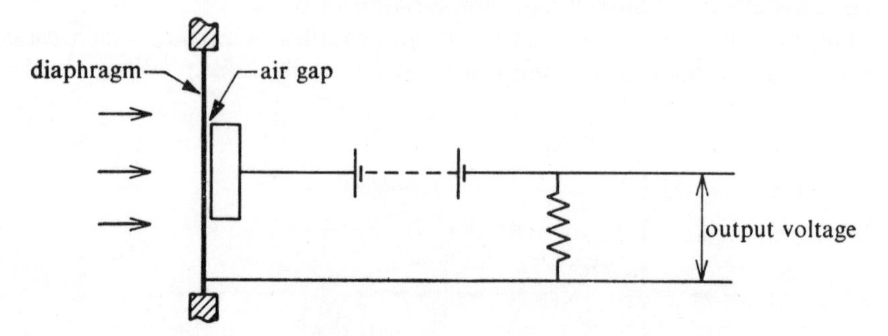


Fig. 15-5 Arrangement of a capacitor microphone

a loss in sensitivity and the diaphragm, which is generally about 0·04 mm thick, may rupture. However, the viscosity of the layer of air between the diaphragm and the fixed plate loads the diaphragm and tends to raise its resonant frequency. Also, the amplifier has to be mounted very close to the capacitor microphone; this disadvantage led to the development of the moving-coil and ribbon microphones.

The *ribbon microphone* has the added advantage of being very directional. The microphones already described made use of the pressure variations in the acoustic wave deflecting a diaphragm, whereas in the ribbon microphone there is no diaphragm but a ribbon conductor that is able to move in a magnetic field. This means that the output voltage is proportional to the velocity of motion. A thin metallic ribbon, usually crinkled aluminium foil, is mounted between the pole pieces of a magnet as in Fig. 15-6. An incident sound wave causes the ribbon to move, inducing a voltage between A and B. If both sides of the ribbon are open to the sound field the microphone operates as a velocity microphone, but at high frequencies the wavelength of the sound is of the same order as the dimensions of the system so that the front of the ribbon shields the back from the sound waves. The instrument then behaves as a pressure microphone. The directional property is due to the ribbon being constrained to move in one direction only and it is independent of frequency. This type of microphone has good frequency response, no background noise, and is

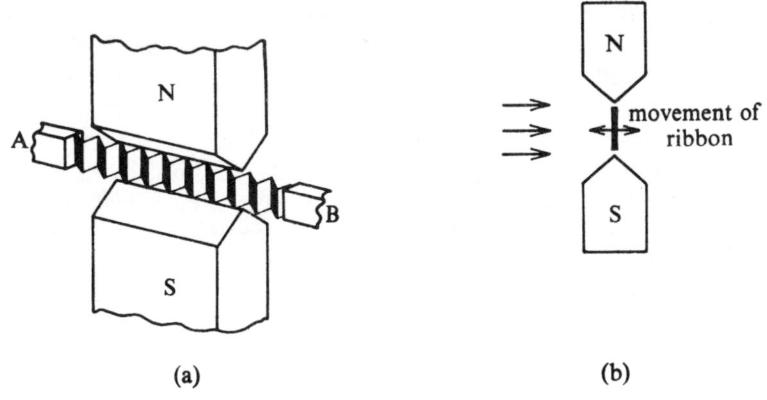


Fig. 15-6 (a) Arrangement of ribbon microphone and (b) section through pole piece

robust. Its disadvantages are its relatively low output and a tendency for the ribbon to flutter in the wind, which can prove very troublesome in outdoor use.

The *moving-coil microphone* consists of a rigid diaphragm with an elastic circumferential suspension so that the diaphragm may move as a piston. A coil mounted on the diaphragm (Fig. 15-7) is free to move in a radial magnetic field. Variations in pressure within the sound wave cause piston-like movements of the diaphragm and its attached coil. As it moves through the magnetic field a current is induced in the coil; this may be amplified in the usual manner, by coupling with a transformer to an electronic amplifier. This type of microphone has a good frequency response and no background noise. It also has good sensitivity and is robust. When used for directional purposes, it has the disadvantage that some frequency discrimination occurs.

The *crystal microphone* makes use of piezoelectric crystals (Section 17-6). In crystals such as Rochelle salt ($KNaC_4H_4O_6 . 4H_2O$) a mechanical strain

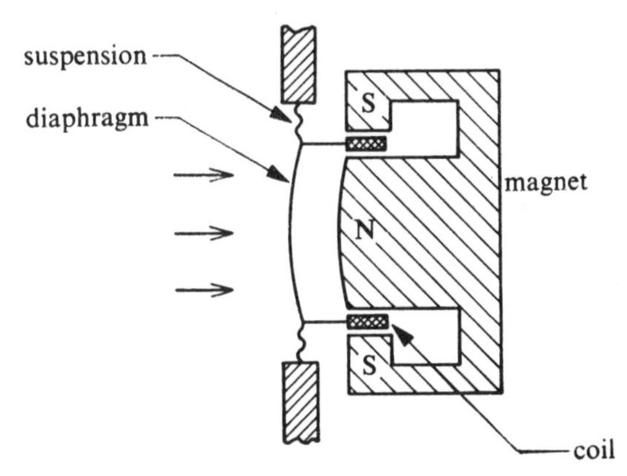


Fig. 15-7 Section showing arrangement of a moving-coil microphone

is associated with an electrical polarization—that is, with the crystal cut along a certain axis, one face becomes positively charged and the other negatively charged when the crystal is strained (Fig. 15-8). In a crystal microphone it is usual for two such crystals to form a sandwiched assembly, termed a *bimorph*. As the assembly is deformed by the sound wave, one crystal is compressed and the other put under tension, as in a bimetal strip. Each crystal is plated on both sides for ease of electrical connection. Both series and parallel connections are used; the series connection gives the larger output voltage and the parallel connection gives the lower internal impedance. To avoid resonances in the range of the working frequencies of the microphone it is necessary to have very thin crystal plates.

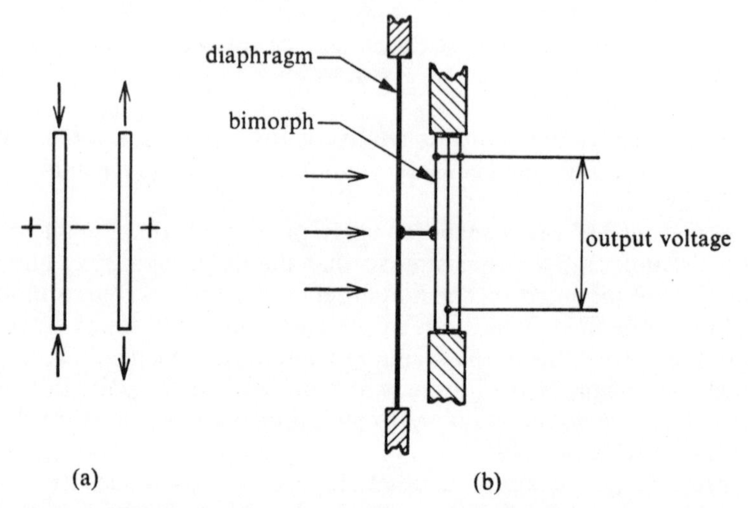


Fig. 15-8 (a) Effect of strain on piezoelectric crystal plates and (b) utilization of a bimorph in a crystal microphone

Alternatively, two such bimorph elements may be arranged coaxially and parallel. The spacer ring between the elements gives an air-tight seal so that both crystal units are distorted by the sound wave. This arrangement has the advantage that if the bimorphs are connected in parallel, their separate outputs are additive when distorted by an incident sound wave but tend to cancel when the distortion is caused by mechanical shock.

Crystal microphones have a very good frequency response, are robust and have no background noise, making them superior to the best carbon microphones. On the other hand they require a carefully designed and constructed amplifier and directional types show frequency discrimination with direction.

## 15-8 Loudspeakers

The most common type is the *moving-coil* loudspeaker. This has the same working principle as the moving-coil microphone. The driving force is provided by a current-carrying coil mounted to move freely in a radial magnetic field (Fig. 15-9). This coil is fixed to a stiffness element, called a *spider*, which centres the coil in the magnet's air gap but allows it to move in and out. To the spider is fixed the diaphragm or *cone*, which is normally of paper. This cone has a corrugated rim to provide elasticity along its attached edge. Ideally, the cone should behave as a rigid piston; however, to minimize inertia the cone and coil assembly are made as light as possible, which results in some loss in rigidity of the cone. The design has to compromise between weight and rigidity. Lack of rigidity leads to resonance of the cone and consequent distortion of the sound. The stiffness

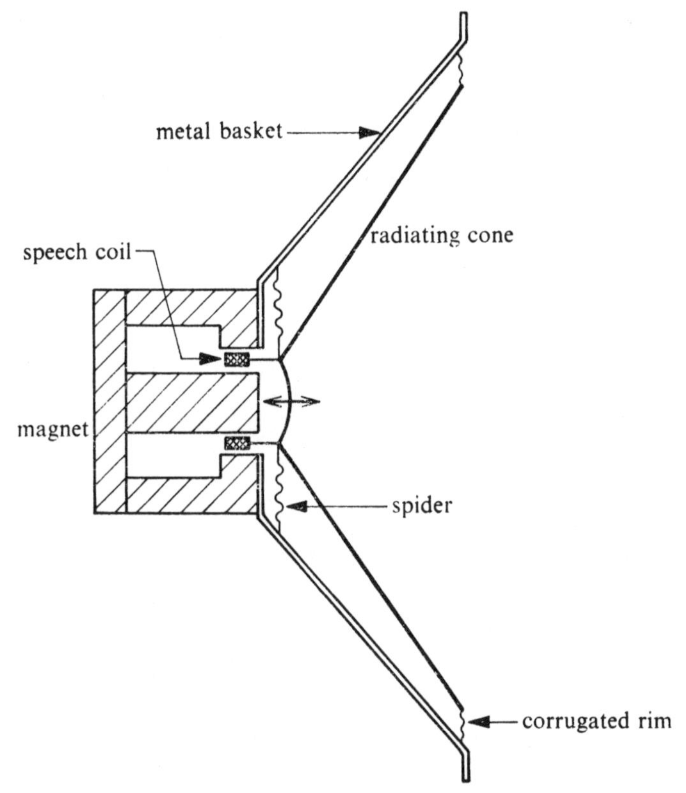


Fig. 15-9 Loudspeaker assembly

of the suspension, the air resistance, and the mass of the cone and coil, also affect the resonant frequency, which should be as low as possible. Generally, from this low resonant frequency up to about 1,000 Hz, the response is fairly uniform, but above about 1,000 Hz the cone begins to show complex modes of vibration which change the response. This varying response may be compensated for to some extent electronically and by a suitable choice of cone material up to about 7,000 or 8,000 Hz. To give high-fidelity reproduction over the full audio frequency range, however, it is necessary to use several loudspeakers, each designed to give an approximately uniform response over a limited range of frequencies. For high frequencies the diameter of the loudspeaker must be small for the simple reason that a large cone with large inertia cannot be driven fast enough.

At low frequencies, problems arise in the mounting of the loudspeaker. As well as radiating to the front, the cone, behaving as a piston, also radiates to the rear. These rear waves have an opposite phase to and interfere with the front waves, giving poor radiation characteristics. One solution is to mount the loudspeaker in an infinite baffle, such as a hole in a wall, but this has obvious practical limitations. A finite baffle with as large an area as possible is satisfactory for high frequencies, but experiments show that, when the size of the baffle is about the same as the wavelength of sound, interference between the front and rear waves again becomes a problem. For example, a frequency of 100 Hz has a wavelength of approximately 3·5 m, leading to an impracticably large baffle. It is, therefore, necessary to mount the loudspeaker in a baffle in the form of a box so that no rear waves can reach the front. Other difficulties then arise. The box enclosure now becomes a resonant volume. This may be overcome by lining the inside with absorbent material to absorb the higher frequencies and by making an opening in the front of the box. A cavity resonator is then formed which by a suitable choice of dimensions may be made to have a natural frequency of vibration below that of the frequencies to be reproduced. The material of the box must also be chosen to give a very low natural frequency of resonance. Such a box or cabinet is termed a *bass reflex cabinet*.

Another type of loudspeaker which has not yet reached its ultimate design and development stage is the '*ion*' *loudspeaker*. This has been developed by D. M. Tombs and utilizes the electrostatic 'wind' associated with high-voltage discharge points. Two point electrodes are used (Fig. 15-10), one of which is surrounded by a suitably positioned ring electrode. The two point electrodes each produce an electric wind, which by an adjustment of electrode positions and voltages may be made equal and opposite so that they cancel. Then, by applying an alternating voltage to the ring electrode a pulsation of the air may be produced making the arrangement act as a sound generator. As yet the acoustic output obtained is only small but the frequency response is much more uniform than that of a cone-type loudspeaker.

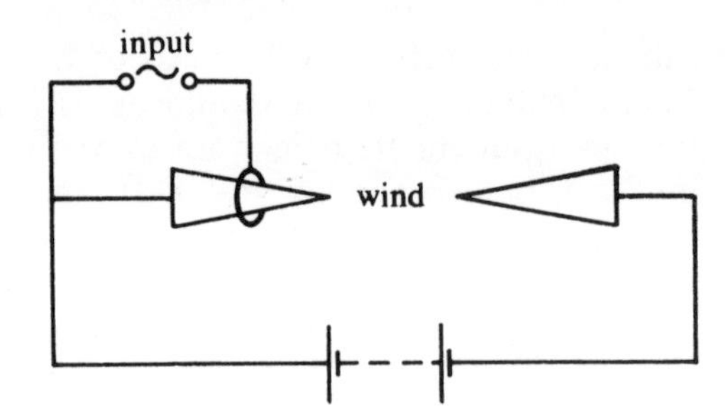


Fig. 15-10 'Ion' loudspeaker

## PROBLEMS

**15-1** Show that the amplitude of a sound wave varies inversely as its distance from the sound source.

**15-2** Show that the kinetic energy of a vibrating body of mass $m$ is proportional to the square of the amplitude of vibration.

A body of mass 1 g is executing a simple harmonic motion of amplitude 2 cm and frequency 50 Hz. Calculate the kinetic energy of the body when at the middle of an oscillation. [k.e. = $2\pi^2 m f^2 a^2$ = 0·0197 J]

**15-3** Define decibel and phon. How may the intensity of a sound be measured?

**15-4** Define 'intensity' of sound. Show that the intensity of a sound due to a wave is proportional to the square of its amplitude of vibration.

**15-5** (a) The power in watts supplied to a loudspeaker is tripled. What is this power ratio in decibels? (b) The pressure amplitude in a sound wave is tripled. By what factor is the intensity increased? [(a) 4·77 dB; (b) 9]

**15-6** What is the intensity level of a sound wave in air (a) when its intensity is $10^{-5}$ W/m$^2$, and (b) when its pressure amplitude is 0·4 N/m$^2$? [(a) 70 dB; (b) 86 dB]

**15-7** A loudspeaker produces an intensity level of 70 dB at a point 3 m away. If the power input to the speaker is doubled, calculate the new intensity level at the same point. [73 dB]

**15-8** The intensity level at a particular distance from a small sound source is 40 dB. What would be the intensity level due to 15 such sources at the same distance? [51·76 dB]

# 16 Acoustics of buildings

## 16-1 Introduction

An auditorium should preferably have certain acoustic properties. Firstly, everybody in the audience must be able to hear clearly. In other words the sound must not only be loud enough but successive sounds must not overlap, causing a confused jumble of sound. Secondly, the quality of the sound must not be changed, that is, one frequency must not be absorbed or its intensities enhanced more than any other. In other buildings the requirements may be different. For example, in a noisy factory as much sound as possible should be absorbed to reduce the general level of noise; in a block of flats as little sound as possible should be transmitted from one flat to another.

The acoustical quality of a building or room depends on a number of interdependent factors. In this chapter these factors will be discussed separately under the following headings: (a) reverberation, (b) absorption, (c) amplification and loudness, (d) resonance, and (e) extraneous noises.

## 16-2 Reverberation

The intensity of a sound at a particular point depends on the amplitude of the sound waves at that point. If the sound is in an enclosure, such as a room, hall, or theatre, the original sound is reflected back and forth between the walls, ceiling, and floor. Thus by superposition of sound waves the amplitude and hence the intensity at a particular point is increased over that occurring without reflection. Thus, the sound dies away less quickly inside an enclosure than in the open air. This prolonging of the sound is termed *reverberation.* If the sound is prolonged too much then each sound may not have died away by the time the next sound arrives. Thus in speech the syllables would overlap, making the words unintelligible.

The *reverberation time* has been defined as the time for the sound intensity to fall to one-millionth of its original value, that is, a drop of 60 dB. For speech, it is found in practice that a reverberation time of about 1 s for a small room, rising to about 2 s for a large room or hall, proves most acceptable to both speaker and listener. Less reverberation causes the speaker to have to work harder to produce adequate loudness, whilst more leads to unintelligibility for the listener. For music the requirements are slightly different. For chamber music, where detail is required to be heard, a short reverberation time is best. For large orchestras and organs a longer time is preferable. The most acceptable time also depends on the volume of the hall. Experience has shown that this time is given by

$$t = (0{\cdot}0118V^{1/3} + 0{\cdot}107)\mathrm{k} \qquad (16\text{-}1)$$

where $t$ is the reverberation time in seconds, $V$ is the volume of the hall in cubic metres, and k is a constant having the values 4, 5, and 6 for speech, orchestra, and choir respectively. If the volume of the hall is in cubic feet, Eqn (16-1) becomes

$$t = (0{\cdot}0036V^{1/3} + 0{\cdot}107)\mathrm{k}$$

For broadcasting studios the reverberation time should be about two-thirds of that normally regarded as acceptable. Here there are two reverberation times involved, that of the studio itself and that of the room in which the broadcast is received and reproduced. In a studio the reverberation time can be artificially lengthened and thus a 'dead' studio—that is, one with a very short reverberation time—may be thought to be most useful. In practice this is not always so, as performers find such studios oppressive and musicians find it a strain to keep in time. To increase the reverberation time, sound from a studio microphone is reproduced by a loudspeaker in a highly reverberant bare room. The sound is then picked up by another microphone and its output added to that of the original studio microphone. Alternatively, a magnetic recording device may be used. The sound is recorded on a magnetic rotating drum and reproduced by detectors around the drum, thus introducing suitable time delays.

## 16-3 Absorption

In a normal room a sound may be reflected several hundred times before becoming inaudible. This means that the nature of the wall coverings, for example, has a great effect on the reverberation time. A more absorbent surface uses up more sound energy at each reflection, giving less reflections and a shorter reverberation time. Sabine was the first to investigate the effect of absorption on reverberation. Whilst Professor of Physics at Harvard University, Sabine made a study of a room in which the reverberation time was about 5½ s. Using an organ pipe of fixed frequency as a source of sound he measured the reverberation time after silencing the pipe. (This may be done nowadays by displaying the amplified output from a microphone on an oscilloscope and observing the signal decay.) Sabine then measured the reverberation times with different numbers of identical cushions spread about the room and found that the time measured depended almost entirely on the number of cushions and not their positions. By finding how many cushions were needed to reduce the reverberation time by a known amount he was able to calculate the area of cushions that was equivalent to the area of the walls, floor, and ceiling of the bare room. By plotting the reciprocal of the reverberation time against the cushion area he showed that

$$At = \text{const.} \qquad (16\text{-}2)$$

where $A$ is the total area of exposed cushions plus the area of the exposed walls, floor, and ceiling expressed in equivalent cushion area, and $t$ is the reverberation time. The value of the constant depends on the particular room and sound frequency.

Obviously an area of cushion is not a particularly satisfactory standard of absorption. Sabine chose as a standard of absorption one square foot of open window, termed an open window unit or sabin. This is a perfect absorber since all sound incident onto an open window is absorbed and none is reflected back into the room. By repeating his measurement of reverberation time, first with cushions in the room and then with the cushions removed but with the window opened an amount to give the same reverberation time, Sabine was able to find the area of open window equivalent to unit area of cushion. This area, always a fraction, is termed the *absorption coefficient*. The absorption of a surface is thus the product of its area and absorption coefficient. If the area of the surface is expressed in square feet, then the absorption is expressed in open window units or sabins. The SI* unit of absorption, with the area expressed in square metres, is equal to 10·8 sabins. Since the absorption coefficient varies with frequency, values normally quoted are for 500 Hz unless otherwise specified.

By making measurements in different rooms, Sabine went on to show that the reverberation time depends on the size of the room, such that

$$t = \frac{0{\cdot}16V}{S} \tag{16-3}$$

where $t$ is the reverberation time in seconds, $V$ the volume of the room in cubic metres, and $S$ is the absorption—that is, $S = \sum aA$ where $A$ is the area of the surface in square metres and $a$ its corresponding absorption coefficient. If the volume is expressed in cubic feet and the absorption in sabins, then Eqn (16-3) becomes

$$t = \frac{0{\cdot}05V}{S}$$

Although Sabine's formula [Eqn (16-3)] may be derived theoretically and is in agreement with experiment, it is incorrect for extreme conditions. For a room in which the surfaces are perfect absorbers, the absorption coefficient is unity and hence the reverberation time as given by Eqn (16-3) has a finite value. This is incorrect, since if the surfaces are perfect absorbers there can be no reflections, hence the reverberation time must correctly be zero. Sabine's theoretical derivation assumed that the sound died away continuously. A different approach by Eyring, in which he assumed that sound energy is lost only at each reflection, produced the result that the reverberation time

$$t = \frac{0{\cdot}16V}{-A \log_e (1 - a)} \tag{16-4}$$

* SI-Système International d'Unités. See BS 3763: 1964 *The International System (SI) Units*, and BSI-PD5686: 1967 *The use of SI Units.*

where $A$ is the surface area. But

$$\log_e (1 - a) = -\left(a + \frac{a^2}{2} + \frac{a^3}{3} + \cdots\right)$$

$$\simeq -a$$

when $a$ is small. Thus from Eqn (16-4)

$$t = \frac{0{\cdot}16V}{Aa}$$

$$= \frac{0{\cdot}16V}{S}$$

Thus Sabine's formula as given by Eqn (16-3) is just a special case, for small values of absorption coefficient, of Eyring's more general formula. Also Eyring's formula gives the correct value $t = 0$ for the reverberation time, when the surfaces are perfect absorbers, and is therefore the preferable formula for absorption coefficients approaching unity. It also has the practical advantage that less absorbing material may be used to produce a 'dead' room—that is, one with a very short reverberation time—than is predicted by the Sabine formula.

These formulae are sufficient to give the basic design of a concert hall. When the absorption coefficient is less than about 0·2, Eyring's formula may be replaced by Sabine's. Knowing the required volume and reverberation time, it is thus possible to determine the required total absorption and hence the most suitable building and decorating materials, but these equations are not sufficient to indicate a suitable distribution of the absorbing and reflecting surfaces. It should be remembered that decorating—for example, painting over cork tiles—as well as curtaining and carpeting can considerably change the designed characteristics. Similarly the presence of an audience has a great effect. Sabine's work suggested that on average an adult person was equivalent to 0·44 SI units of absorption, or 4·7 sabin, although later work has suggested that this figure may be a little high. Plush upholstery is commonly used in theatres in order that an empty seat may roughly approximate to the absorption of that of a person. Table 16-1 gives approximate values of the absorption due to some common materials and objects.

When the hall is used for music, there is also an optimum number of of musical instruments. On average, for each 18·5 SI units, or 200 sabin, of absorption one musical instrument provides the right amount of sound energy to give a satisfactory 'volume' of sound.

**Example 16-1** A hall of floor area $15 \times 30$ m and height 6 m seats 500 people, 250 on upholstered seats and the remainder on wooden chairs. What is (a) the optimum reverberation time for orchestral music, (b) the absorption to be provided by the walls, floor, and ceiling when the hall is fully occupied, (c) the reverberation time if only the upholstered seats are

Table 16-1 Approximate absorption at 500 Hz of some common materials and objects

| *Material* | *Absorption coefficient* |
|---|---|
| Brick | 0·03 |
| Plaster | 0·03 |
| Wood, 3-ply panelling | 0·02 |
| Wood, polished block | 0·06 |
| Carpet, 1 cm thick | 0·25 |
| Carpet on 1 cm felt underlay | 0·5 |
| Curtains, medium weight | 0·3 |
| Curtains, heavy weight | 0·7 |
| Fibre board, 1 cm thick on battens | 0·3 |
| Perforated acoustic tiles on battens | 0·8 |
| Glass wool, 5 cm thick on battens | 0·8 |

| *Object* | *Absorption* | |
|---|---|---|
| | SI units | sabin |
| Audience, per person | 0·44 | 4·7 |
| Seats, upholstered | 0·17 | 1·8 |
| Seats, wood | 0·02 | 0·2 |

occupied, and (d) the most suitable size of orchestra. (e) Would 3-ply panelling on the walls and a plaster ceiling be suitable materials for this hall? The absorption figures of Table 16-1 may be assumed.

(a) From Eqn (16-1), where k = 5 for orchestral music, the optimum reverberation time is

$$\begin{aligned} t &= 5(0{\cdot}0118V^{1/3} + 0{\cdot}107) \\ &= 0{\cdot}059(15 \times 30 \times 6)^{1/3} + 0{\cdot}535 \\ &= 1{\cdot}36 \text{ s} \end{aligned}$$

(b) Using Sabine's equation for SI units

$$t = \frac{0{\cdot}16V}{S}$$

i.e.,
$$1{\cdot}36 = \frac{0{\cdot}16(15 \times 30 \times 6)}{S}$$

Hence the required total absorption is

$$S = 318 \text{ SI units}$$

Absorption due to audience

$$= 500 \times 0{\cdot}44$$
$$= 220 \text{ SI units}$$

Therefore the required absorption to be provided by the walls, floor, and ceiling is

$$318 - 220 = 98 \text{ SI units}$$

(c) When the hall is only half-full, the absorption provided by the audience is now only $(250 \times 0{\cdot}44) = 110$ SI units. The vacant wooden seats however also give an absorption of $(250 \times 0{\cdot}02) = 5$ SI units, so that the total absorption of the hall is now

$$98 + 110 + 5 = 213 \text{ SI units}$$

Hence the reverberation time, given by Sabine's formula, is now

$$t = \frac{0{\cdot}16(15 \times 30 \times 6)}{213}$$
$$= 2{\cdot}03 \text{ s}$$

Under these conditions this hall is possibly more suited to a choir than to an orchestra.

(d) The required total absorption for the hall is 318 SI units. Therefore, assuming on average one musical instrument to every 18·5 SI units of absorption, the optimum number of instruments in the orchestra

$$= \frac{318}{18{\cdot}5}$$
$$= 17$$

The members of the orchestra would of course also affect the reverberation time slightly due to the slight change in the total absorption.

(e) The absorption due to the 3-ply panelling on the walls is

$$0{\cdot}02(2 \times 15 \times 6 + 2 \times 30 \times 6) = 10{\cdot}8 \text{ SI units}$$

and to the plaster ceiling

$$0{\cdot}03(15 \times 30) = 13{\cdot}5 \text{ SI units}$$

giving a total absorption of 24·3 SI units.

The required absorption of the floor, walls, and ceiling is 98 SI units. Thus, 3-ply panelling and plaster would be unsatisfactory alone and it would be necessary to make use of panels of acoustic tiles on the walls and ceiling and maybe carpeting on the floor area clear of the seating to obtain the required acoustical properties.

Little has been said yet about the types of absorption materials. These are important, since they must cover large areas and in general must be considered as part of the decoration. Besides the obvious materials, such as curtains and carpets, other materials, usually in the form of large tiles and panels, are used to cover large areas such as walls and ceilings. It is not so much the actual material that gives an absorber its properties, but how it is used. Thus a thin sheet works best as an absorber of short wavelengths, that is, high frequencies. If now there is an air space between the sheet and a solid backing, there is better absorption of the lower frequencies. A common type of absorber is in the form of a cavity resonator. This is made from a sheet, possibly of fibre board, with holes drilled through to form the necks of resonators. Behind the sheet is a space filled with glass wool and then a rigid backing, usually the room wall (Fig. 16-1). A cavity alone has an optimum absorption for a particular frequency, but with the glass wool packing the damping is increased and a broad range of frequencies is absorbed. In addition, the glass fibres themselves absorb at

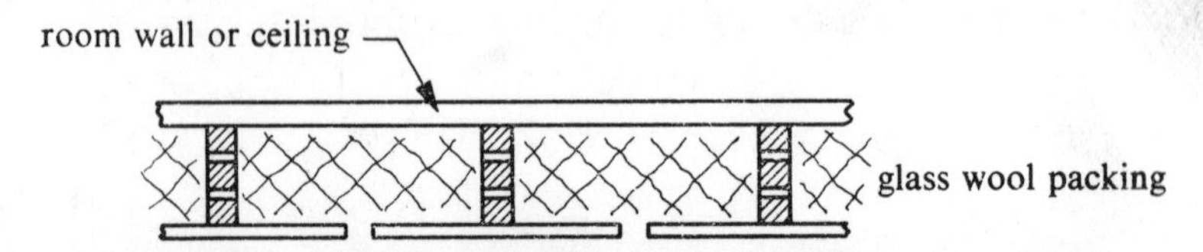


Fig. 16-1 Section through a cavity-type absorber

higher frequencies. Porous and foam-texture materials are also used, because their pores act as cavity resonators. With some types of foam-texture materials, however, the pores are not connected and these materials are then less effective as absorbers.

## 16-4 Amplification and loudness

When the reverberation time in an auditorium is reduced, the sound level is lowered. Before sound amplifying systems were widely used, it was necessary to balance the disadvantages of reverberation against the need for an adequate sound level. Today, auditoria are designed to have an optimum reverberation time for their dimensions and suitable absorbing materials are chosen to achieve this; the required sound level is then achieved by electronic amplification. One problem arising from this involves the positioning of the loudspeakers. The best result is achieved if the loudspeakers are placed above the performers; the direct sound from the performers and the amplified sound from the loudspeakers then travel approximately the same distance to the members of the audience. If, for example, a loudspeaker is placed down the hall, say, a distance of 30 m from the performer, then a member of the audience seated near the loudspeaker hears the amplified and the direct sound separately, in this instance with a time interval between them of approximately a tenth of a second. This causes a confusion of sound unless the amplified sound is delayed by means of a magnetic tape recording, so that it is emitted to coincide with the arrival of the direct sound. Care must also be exercised in the choice of amplifier, since most amplifiers give greater amplification to low frequencies than to high. Also, if the fabric of the hall is such as to give a selective absorption or enhancement of certain frequencies, it is desirable that the amplification system should be able to compensate for this.

The loudspeakers must, of course, beam their sound towards the back of the hall, otherwise a person seated at the front of the audience and close to the performer will be troubled by double sound. However, the sound should not be too directional, since an even distribution of sound is required. This is usually achieved by pillars and boxes which give irregular reflecting surfaces. Similarly, heavy-relief decoration plays its part, although it should have a depth of about one third of a metre to be effective. Thus the Victorian music hall was by chance of good acoustic design in this respect.

Large reflecting surfaces placed above and behind the performers help to direct the sound towards the audience. Similarly, a clear floor area before the front row of the audience is useful in reflecting sound upwards and towards balconies, although this is not always economically practicable. A flat ceiling may also be useful in giving an even distribution of sound whereas a curved ceiling may or may not prove an advantage (Fig. 16-2). Generally, a concave surface should be avoided because it

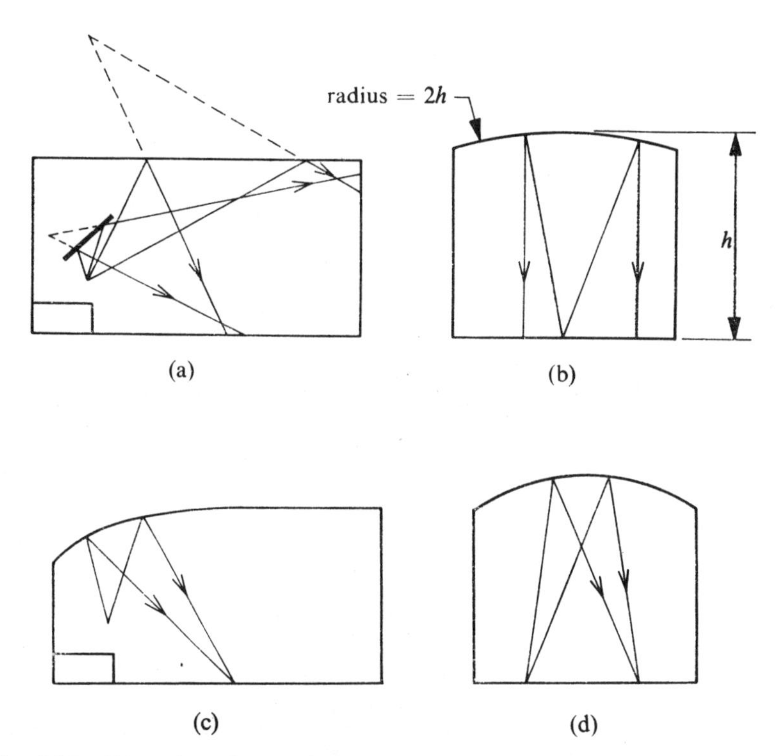


Fig. 16-2 Examples of (a), (b) desirable and (c), (d) undesirable focusing effects in auditoria

often leads to undesirable focusing. The phenomenon of 'creeping' occurs in rooms with a circular plan in which sounds directed tangentially to a wall are reflected around the circumference of the room producing a 'whispering gallery'. A flat surface usually contributes towards even distribution, except when it forms the back wall of a hall. It is then possible for a person in the front of the audience to hear the sound reflected from the back wall which, having travelled twice the length of the hall, may be heard an appreciable time after the direct sound. For a person more than about 12 m from the rear wall this can produce a confusing echo. For this reason many halls are designed to have almost non-existent rear walls by suitably raking the seats, which of course also improves viewing.

Raking of seats may introduce a further complication. The gangways now become flights of shallow steps and the risers act as separate reflecting surfaces in the form of an echelon (Fig. 16-3). An incident sound may be

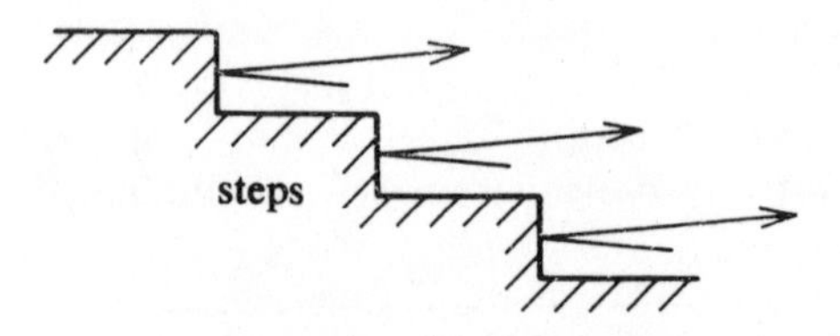


Fig. 16-3 Echelon effect

reflected at each step, with each consecutive echo occurring after a constant time interval. For example, with a 0·25 m tread the path difference between consecutive echoes is 0·5 m, giving a corresponding time interval of 0·5/340 = 1/680 s between echoes, assuming a velocity of sound of 340 m/s. Thus a note of frequency 680 Hz is produced. Obviously, this may be prevented by carpeting, undercutting the steps to remove the flat vertical reflecting surface, or having varying tread widths.

## 16-5 Resonance

In the same way that stretched strings and air columns have natural resonant frequencies, so do rooms and halls. Due to reflections between walls, standing waves are set up. Thus as an orchestra produces its range of notes it will momentarily excite resonances.

Resonant frequencies occur at

$$f = \frac{v}{2}\sqrt{\left[\left(\frac{n_1}{l}\right)^2 + \left(\frac{n_2}{b}\right)^2 + \left(\frac{n_3}{h}\right)^2\right]} \tag{16-5}$$

where $v$ is the velocity of sound in air, $l$, $b$, and $h$ represent the length, breadth, and height of the room, and the $n$'s are any integers, including zero. Such frequencies are also termed normal frequencies and with each there is a corresponding normal mode of vibration—that is, a corresponding geometric arrangement of nodes and antinodes in the room.

Generally for a large room the more energetic lower frequencies for resonance are well below the audible range. However, when a note at a resonant frequency is sustained—for example, by an organ—the resonance builds up and the result is felt rather than heard and may also cause effects such as rattling of windows.

Besides standing waves between parallel walls, other modes of vibration are possible due to more complex reflections (Fig. 16-4). The numerous nodes and antinodes produced cause an uneven distribution of sound. If two or more normal modes formed by different reflection paths have the same resonant frequency—that is, different combinations of the $n$ values in Eqn (16-5)—they are termed *degenerate modes.* More energy is absorbed with these modes than at other frequencies. This means that this vibration lingers unduly long and is energetically favoured at the expense

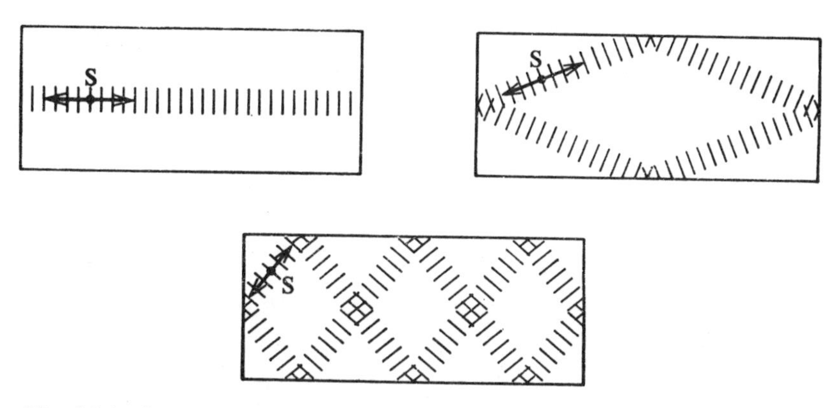


Fig. 16-4 Some possible normal modes of vibration for a rectangular room with a sound source S

of other frequencies, with disastrous effect on the acoustics. In an irregularly shaped room fewer degenerate or near degenerate modes form and consequently there are fewer annoying resonances. This has been known empirically for a long time and has resulted in the usual irregular shape of concert halls and theatres.

Resonance may also occur in the fabric of a room—for example, in wooden wall panelling. This may be beneficial if it helps to compensate for too short a reverberation time, but it is a difficult factor to incorporate into an initial design.

## 16-6 Extraneous noises

Extraneous noises may be either air-borne or structure-borne, with the former being the more difficult to prevent. The obvious solution is to choose a quiet site but this is rarely possible in practice. For example, a theatre is usually required to be built in a town and adjoining a road. However, it is often convenient to shield the acoustic area from road noise by building a foyer at the front and incorporating corridors leading to boxes around the sides. Most sound enters through openings, and, therefore, windows should be eliminated and doors made close fitting. Even the crack under a door lets a surprising amount of noise enter. For example, if the crack under the door is a thousandth part of the total area of the door, then closing the door reduces the sound energy entering to a thousandth of the original, that is, a drop of only 30 dB in sound level is attained and that is assuming that the door itself is a perfect insulator. Therefore, double doors are commonly used.

Sealing the hall to reduce air-borne noise introduces problems of ventilation. The ventilating ducts themselves carry noise originating in the ventilating room and noise between rooms served by the ducts. To reduce noise from the ventilating room, large slow-moving fans and blowers are used, which must be free of rumble, and a large amount of absorbing material is used in the room. Sound may also be transmitted through the walls of the ducts. This structure-borne noise is more easily dealt with by simply breaking the conduction path. Thus, for example, in a steel-framed building, insulating layers of asbestos or lead sheeting may be used at overlap points and bolts sheathed with bitumen-impregnated asbestos. Similarly, insulating gaskets may be used at the junctions of water pipes and lengths of flexible ducting used in ventilation systems.

Footfalls on floors above—for example, in blocks of flats—may be reduced by the use of suspended ceilings (Fig. 16-5) so that there is no

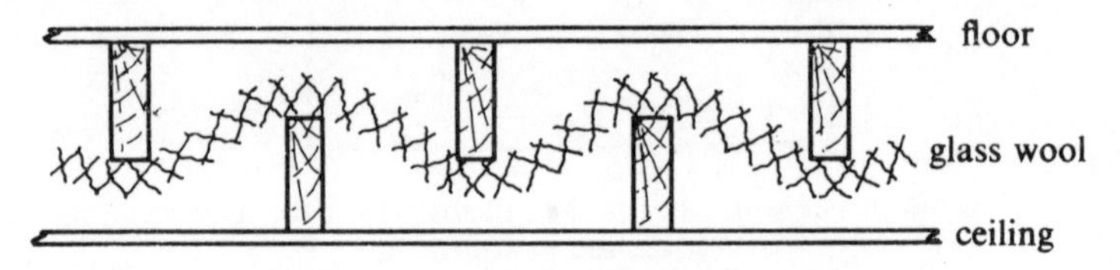


Fig. 16-5 Sound insulation between floors

direct structure path between floor and ceiling. At the same time air-borne noise between the two may be reduced by a layer of glass wool, which, incidentally, also provides thermal insulation.

An ordinary 9-in brick wall absorbs about 2% of the incident sound and reflects about 98%, with negligible transmission through the wall. However, an external sound source may set the entire wall into vibration, so that it acts as a giant sounding board, transmitting sound into the room; this is called *diaphragm action* of the wall. Such sound transmission is reduced by a double wall. The inner wall may simply be the lining of the room put in to increase absorption and adjust the reverberation time (Fig. 16-1). To be effective this inner wall should be only loosely joined to the main wall.

## 16-7 Conclusion

As we have shown, the acoustic design of a good concert hall or theatre is far from simple. Even after the most careful design the resulting hall may not be acoustically satisfactory. The result must always be a compromise between different observers' ideas as to what the acoustic requirements should be, quite apart from the different requirements for speech and the various forms of music. The acoustic requirements must also be tempered by aesthetic considerations.

A modern hall is normally designed to contain as large an audience as possible for its size. However, an audience is a very highly absorbing body, resulting in a short reverberation time. This often means that even if the least-absorbing materials are used for the walls and ceiling the time is still too short for good acoustics. Moreover, highly reflecting walls lead to echoes and the need for more absorbing surfaces. One solution is to increase the volume of the hall but, although this may increase the reverberation time, this solution is self-defeating, since the larger hall will inevitably be used to hold a larger audience.

When first built, the Royal Festival Hall, London, had a reverberation time of about 1·5 s whereas a time of about 2 s was required. The short time gave great clarity of sound, but the sound lacked the richness of the longer time. The system used to tune the Royal Festival Hall involves the use of cavity resonators. By varying the volume of the cavity, by adjusting the distance of the base to the neck, a cavity can be tuned to resonate at a particular frequency. Each cavity contains a microphone, which is, in turn, connected to an amplifier and a loudspeaker some distance away. Thus a particular frequency is selected by the cavity resonator, and then that frequency is amplified electronically. By adjusting the gain of the amplifier, the reverberation time for that particular frequency may be adjusted to any required value. For example, if the gain of the amplifier exactly balances the losses due to absorption, the reverberation continues indefinitely. It was found that a range of cavity resonators with their associated microphones, amplifiers, and loudspeakers to cover the frequencies 3 Hz to 1,000 Hz at 3 Hz intervals would be adequate. This large number of cavities, which for the low frequencies are quite large in volume, are accommodated in the ceiling, with the loudspeakers spread about the hall and even in the light fittings, so as to achieve a uniform sound level throughout the hall. Such a large number of duplicated electronic systems is a very expensive solution, but it does mean that a hall can be tuned as desired and that reverberation time may be adjusted to suit the size of the audience present and to cover the whole range of requirements for speech and music, from a string quartet to a full orchestra with organ.

## PROBLEMS

**16-1** What is the importance of reverberation time in architectural acoustics? How may it be adjusted to its optimum value?

**16-2** Discuss the importance of the size and shape of a hall, and its decorative finish, to the acoustics of the hall.

**16-3** A source of sound in front of a flight of stone steps, of tread depth 9 in, produces a musical echo from the steps of frequency 720 Hz. Calculate the velocity of sound in air under these conditions. [1,080 ft/s]

**16-4** A bare chamber with a surface area of 54 m$^2$ has a reverberation time of 3 s. When 18 m$^2$ of wall area are lined with tiles of absorption coefficient 0·8 the reverberation time is halved. What is the mean absorption coefficient of the untiled surfaces? [0·2]

**16-5** A hall has a volume of 80,000 ft$^3$, with a plaster ceiling of area 4,000 ft$^2$. The clear floor is of wood block, totalling 1,000 ft$^2$. The walls are covered with 4,000 ft$^2$ of wood panelling and 1,500 ft$^2$ of heavy-weight curtaining. If the hall seats 800 people on upholstered chairs, calculate the reverberation time (a) when the hall is empty, (b) when it is full, and (c) when it is half full. Values for the absorptions are given in Table 16-1. [(a) 1·45 s; (b) 0·79 s; (c) 1·02 s]

# 17 Ultrasonics

## 17-1 Introduction

The ear has an upper frequency limit beyond which it cannot detect sound waves. Vibrations having frequencies above this limit, about 20 kHz, are called *ultrasonic vibrations.* The upper frequency limit of ultrasonic vibrations is in turn set by the generator, at about $5 \times 10^5$ kHz. Taking the velocity of sound in air as about 330 m/s, this is equivalent to a wavelength range of 1·6 cm to $0{\cdot}6 \times 10^{-4}$ cm. For liquids, in which the velocity of sound is of the order of 1,400 m/s, the equivalent wavelength range is 7 cm to $2{\cdot}8 \times 10^{-4}$ cm, and for solids, in which the velocity is of the order of 4,000 m/s, it is 20 cm to $8 \times 10^{-4}$ cm. This means that the shortest wavelength ultrasonic vibration obtained is of a similar order of magnitude to that of light waves (that is, 0·4 to $0{\cdot}8 \times 10^{-4}$ cm). The ultrasonic vibration is, however, a longitudinal wave involving actual vibrations of the particles of the carrier medium and is thus fundamentally different from light waves, although as their wavelengths are similar they behave in many ways the same. Since the ultrasonic vibrations are high frequency sound waves, they may be produced in just the same way as ordinary sound waves, except that the source must vibrate at a correspondingly high frequency. They may be produced by a number of means of which the best is a transducer, which converts electrical energy into mechanical energy. A high frequency electric current, produced electronically, makes the transducer vibrate and act as a source for the ultrasonic wave. The most common types of transducers are made from magnetostrictive, piezoelectric, or electrostrictive materials.

## MAGNETOSTRICTIVE TRANSDUCERS

## 17-2 Magnetostriction effect

The *magnetostriction effect* is the change in dimensions produced in a ferromagnetic body when placed in a magnetic field. As the field is varied, all the dimensions change. The Joule effect is the most important for the generation of ultrasonic waves; this involves a change in length along the axis of the applied magnetic field. Transverse changes as well as overall volume changes also occur, but are small compared with the longitudinal change and may be neglected in this context.

At the atomic level, a magnetic field is produced by electrons orbiting about the nucleus and, due to the random orientations of the atoms within the solid, these effects cancel out in most materials. However, in certain substances, termed ferromagnetics, the atomic magnetic fields are similarly orientated within small volumes. In these small volumes, or *domains*, the atomic magnetic fields line up parallel with each other. Generally, although these domains are separately magnetized to saturation, the solid as a whole is unmagnetized due to the random orientation of the domains so that they cancel out each other's effects. If the fields within the

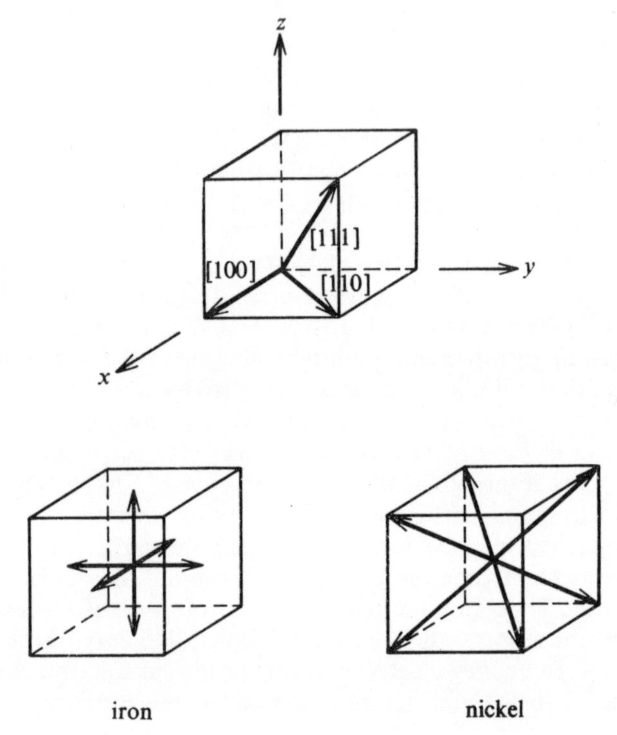


Fig. 17-1 Directions of easy magnetization

domains are aligned by means of an external magnetic field, the solid as a whole becomes a magnet. Each domain has directions of easy magnetization, depending on the crystal structure, and the field within each domain is in one of these directions. For example, iron and nickel each have six such directions (Fig. 17-1). In a polycrystalline material, individual crystals are orientated randomly and thus only a small number are magnetized with their directions of easy magnetization approximately parallel to an externally applied field. As the external field strength is increased, those domain fields not magnetized parallel to the external field are forced to rotate to line up. Thus, those domains whose direction of magnetization is parallel to the external field grow at the expense of the other domains until the body becomes one large domain—a magnet. This rotation of the unaligned domain fields causes the body to expand or contract externally.

If a rod of a ferromagnetic material is placed in a magnetic field parallel to its length, its length changes. This change of length is independent of the sign of the field and may be an increase or decrease in length depending on the composition of the rod. The magnitude of the change depends on the previous history of the rod material, its state of initial magnetization, and the temperature. For practicable field strengths the change in length is very small, with a relative elongation $\delta l/l$ of the order of $10^{-4}$ to $10^{-6}$, where $l$ is the length of the rod (Fig. 17-2). Nickel and annealed cobalt show a regular decrease in length with increasing field. These metals

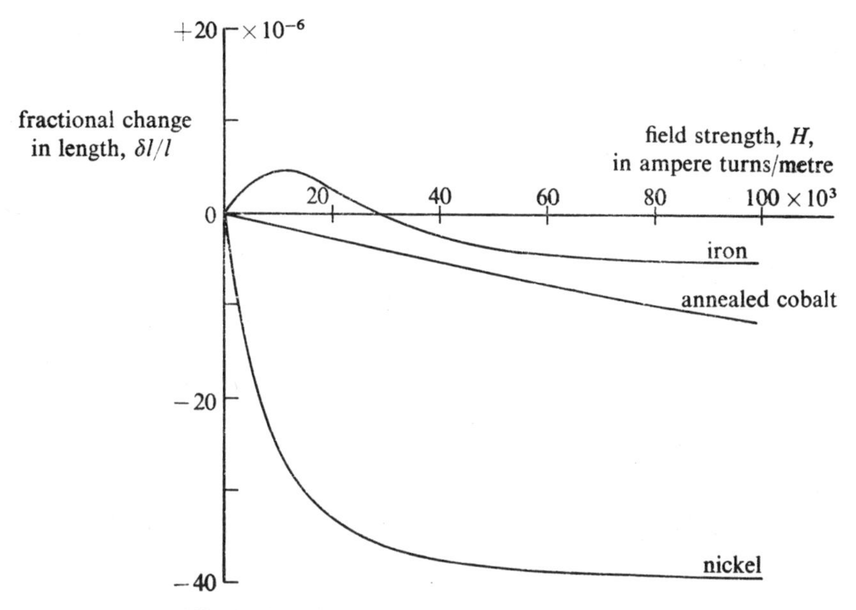


Fig. 17-2 Change of length by magnetostriction

contract when they are magnetized parallel to any of their principal crystallographic directions. Iron shows a more complicated behaviour: if it is magnetized along any of its easy directions—that is, the [100] directions parallel to the cube edges—it extends; in the [111] directions—that is, the most difficult directions of magnetization lying parallel to the diagonals of the cube—it contracts. In the [110] directions—that is, parallel to the face diagonals, it first increases in length and then at high magnetizing fields contracts. (See the resultant curve in Fig. 17-2, for a polycrystalline body with randomly orientated crystallites.)

## 17-3 Magnetostrictive transducers

When a nickel rod is placed in an alternating magnetic field with the direction of the field parallel to the long axis of the rod, the rod shortens periodically. If the rod is initially unmagnetized, it shortens periodically with twice the frequency of the alternating field, since it changes irrespective of the sign of the field. If the rod, or transducer, is now polarized —that is, given a premagnetization of magnitude greater than the exciting field—it shortens periodically with the same frequency as the alternating field.

Since the change in length of the transducer is normally very small, the frequency of the alternating field is made the same as the natural elastic period of vibration of the transducer; this makes the transducer resonate. The amplitude of the oscillation is then maximum and movement of the end of the transducer generates sound waves of the same frequency. The mechanical amplitude of the transducer at resonance is limited by the fatigue properties of the metal. For this reason it may be necessary to limit the amount of energy fed into the transducer. The natural frequency of vibration, $f$, of a rod of length $l$ is given by

$$f = \frac{n}{2l}\sqrt{\frac{E}{\rho}} \tag{17-1}$$

where $E$ is the modulus of elasticity, in this case Young's modulus of the rod material, and $\rho$ is its density (Section 14-5). $n$ is the order of the harmonic and equals 1, 2, 3, .... The fundamental vibration is formed when $n = 1$; since this is the strongest vibration, transducers in the form of rods are normally gripped at the centre, which then becomes a node for the fundamental and so excludes harmonics. For example, the Young's modulus of nickel is $21{\cdot}0 \times 10^{10}$ N/m$^2$ and its density is $8{\cdot}8 \times 10^3$ kg/m$^3$. Thus, the natural frequency of vibration of a nickel rod 10 cm long is 24·4 kHz. The highest frequency obtainable with a usable amount of power is about 60 kHz, which corresponds to a nickel rod about 4 cm long. If a higher frequency is required it is necessary to use harmonics. These may be obtained by a suitably designed, grooved rod transducer, clamped at the nodes of the particular harmonic so that it vibrates in that particular mode.

Solid rod transducers have the disadvantage that eddy currents are set up which tend to oppose the driving magnetic field, causing a loss of energy and heating of the transducer. These eddy current losses may be reduced by using tubular transducers or, better still, transducers made from laminations. If nickel laminations are used, they should first be heated in air to about 800°C and then allowed to cool slowly. This forms a superficial film of hard green carbonate, which acts as an insulator between the laminations. Laminations made of other materials are interleaved with mica or potted in a resin cement. If they are not potted, the laminations are bolted together. In all cases, however, it is most important that careful annealing be carried out so that maximum output may be obtained.

Magnetostriction effects decrease with temperature, disappearing completely at the Curie point. For example, nickel has a Curie point of 360°C, and above about 100°C its efficiency as a transducer is already beginning to fall off. It is, therefore, necessary that adequate cooling be arranged. Permandur (an alloy—Co 49%, Fe 49%, V 2%) has the greatest thermal stability, being usable up to about 800°C.

## 17-4 Magnetostrictive materials

The best ferromagnetic magnetostrictive materials, besides pure nickel, are nickel alloys, such as Invar (Ni 36%, Fe 64%) and Monel Metal (Ni 68%, Cu 28%, Fe, Si, Mn, C), cobalt alloys, such as Permandur, and certain *ferrites*, especially nickel–zinc ferrites. A ferrite is a compound of the type $MeO.Fe_2O_3$ where Me stands for any metal. Ferrite transducers are usually made up as mixed crystals of simple cubic ferrites in a ceramic base. They are easily moulded into convenient shapes and then sintered together. Their high electrical resistance reduces eddy current losses to a negligible amount, making ferrite transducers especially suitable for use at high frequencies. They also have the advantage over other magnetostriction materials that their elastic modulus is less dependent on temperature. However, the best ferrites, although the most efficient transducers, have lower mechanical strength than other magnetostriction materials; ferrites are somewhat brittle and, if the acoustic power obtained is not limited, they are liable to shatter.

## 17-5 Power supplies

The alternating magnetic field necessary to drive a magnetostrictive transducer is provided by means of a coil (Fig. 17-3) fed with an alternating current derived from an oscillator, followed by a preamplifier and

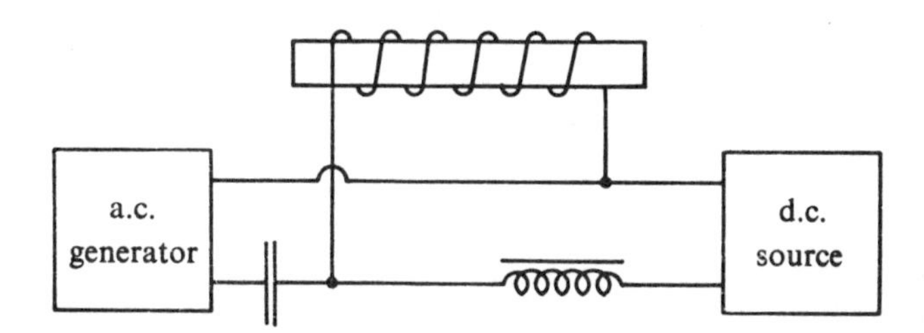


Fig. 17-3 Power supply for magnetostrictive transducer

power amplifier. The frequency stability depends only on the stability of the oscillator.

With an initially unmagnetized rod, the magnetostrictive deformation varies at twice the frequency of the excitation current. If the rod is suitably polarized by means of a permanent magnet or by a constant d.c. potential applied to the magnetizing coil, the deformation has the same frequency as the excitation current. Since the two power supplies are connected to the same coil, a blocking capacitor and choke are required to stop the two power supplies shorting each other. The oscillator frequency is adjusted to be the same as the natural frequency of vibration of the rod transducer.

## PIEZOELECTRIC TRANSDUCERS

### 17-6 Piezoelectric effect

When certain crystals are subjected to compression or extension forces, electric charges appear on opposite sides of the crystal. This is the *direct piezoelectric effect*. This effect, like the magnetostriction effect, is reversible, that is, under the action of an electric field the crystal becomes deformed. This, the *inverse piezoelectric effect*, is used as a source of ultrasonic sound waves.

For a crystal to have piezoelectric properties, it must have a crystalline structure including one or more polar axes. A polar axis is a crystallographic axis such that when the crystal is rotated through 180° about this

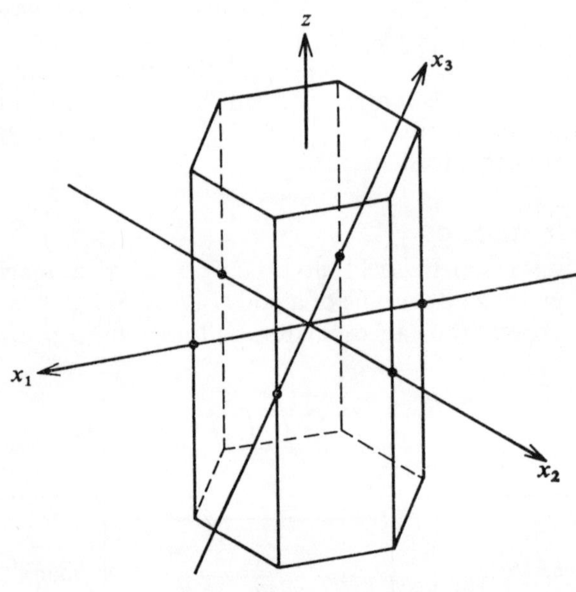


Fig. 17-4

axis, the atoms in the crystal do not coincide with their configuration prior to the rotation. For example, quartz ($SiO_2$) is crystallographically trigonal and has three polar axes, $x_1$, $x_2$, and $x_3$ (Fig. 17-4). If the crystal is rotated through 180° about any of these axes, the new positions of the silicon and oxygen atoms do not coincide with the previous positions of the silicon and oxygen atoms. The $z$-axis is not a polar axis, as a 180° rotation reproduces the atomic configuration and it is in fact the optic axis.

When the crystal is deformed, the maximum charges appear at the ends of a polar axis. The reason for the appearance of a charge may be explained in a somewhat simplified manner as follows. When a mechanical force is applied in the direction of a polar axis, the crystal lattice is deformed and,

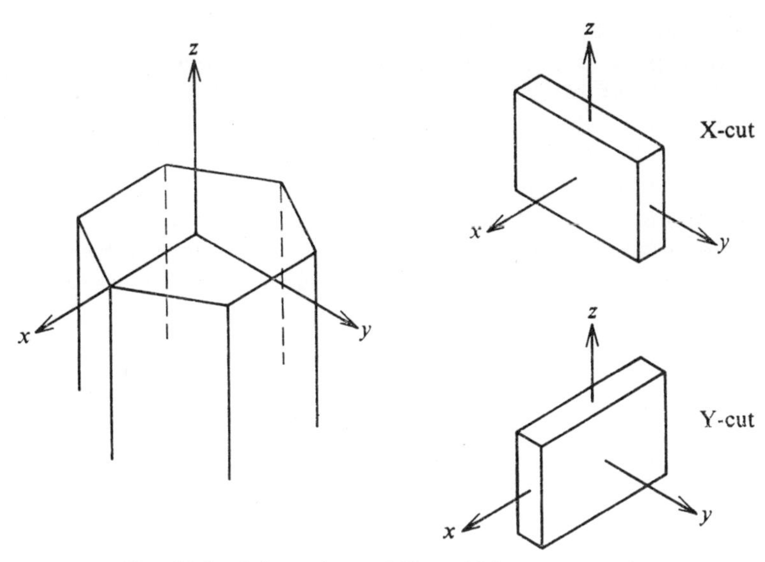


Fig. 17-5 Orientations of X- and Y-cut quartz plates

since the crystal is anisotropic, the relative spacings of the ions in the lattice are changed. This means that, if there are electric charges on the ions, they no longer cancel out and free charges must appear. Conversely, if there is an electric field across the crystal, the crystal lattice is deformed.

To make an efficient transducer it is necessary to cut from the crystal a slab of material with flat parallel sides so that a uniform electric field may be applied across it. It is also necessary that the maximum deformation is produced with the electric field applied. To obtain this, the slab is cut from the crystal with one of two main possible orientations, depending on its proposed use, although other orientations are sometimes chosen for special purposes, such as shearing distortion. The two main possible orientations are shown in Fig. 17-5: (a) the normal to the electrode faces is parallel to a polar axis of the crystal, termed an X-cut plate; and (b) the normal to the electrode faces is parallel to the $y$-axis of the crystal, termed a Y-cut plate. In an X-cut plate, the deformation is called longitudinal because it is in the direction of the applied electric field. In a Y-cut plate the deformation is transverse—that is, it is in a direction perpendicular to the applied field (Fig. 17-6).

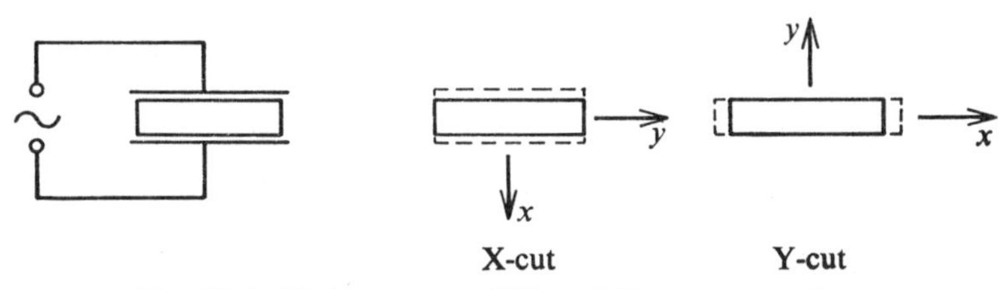


Fig. 17-6 Deformation of X- and Y-cut quartz plates

## 17-7 Piezoelectric transducers

If the crystal plate is put in an alternating field—such as, for example, between the plates of a parallel plate capacitor connected across an a.c. source—the crystal expands over one half cycle of the a.c. field but contracts over the other half cycle. Thus, the crystal plate is forced to vibrate with the same frequency as the a.c. field. If this field is vibrating at ultrasonic frequencies, then the crystal plate acts as a source of ultrasonic sound waves. Unfortunately, the deformation produced for a practical electric field is very small. For an X-cut plate transducer, which is the usual cut for producing longitudinal waves, the change in thickness is given by

$$x = Vd \tag{17-2}$$

$x$ is the deformation or change in thickness in the $x$-direction, measured in metres. $V$ is the total volts applied across the plate, causing the deformation. $d$, the piezoelectric modulus, is a constant relating the applied potential with the deformation produced in the direction specified. For X-cut quartz, $d = -2{\cdot}3 \times 10^{-12}$ m/V. Thus, 100 V applied across the plate produces a change of thickness of $2{\cdot}3 \times 10^{-10}$ m, or 2·3 Å (angstrom) units, which is independent of the thickness of the plate.

To obtain sufficient acoustic power for practical purposes, the amplitude of vibration is increased by making the thickness of the plate such that its natural mechanical frequency is the same as the frequency of the applied electric field. If the slab is considered as having infinite area, its natural frequency of vibration is given by

$$f = \frac{1}{2t}\sqrt{\frac{E}{\rho}} \tag{17-3}$$

$t$ is the thickness, expressed in metres. $\rho$ is the density of the slab. $E$ is the modulus of elasticity measured in the direction of the displacement. For an X-cut plate of quartz, the modulus of elasticity measured in the $x$-direction—that is, in the direction of the deformation—is equal to $86{\cdot}2 \times 10^{9}$ N/m². The density of quartz is $2{\cdot}65 \times 10^{3}$ kg/m³, giving as the natural frequency of vibration

$$f = \frac{2{,}850}{t}\ \text{Hz} \tag{17-4}$$

where $t$, the thickness of the plate, is measured in metres.

Quartz vibrating in air has a very sharp resonance peak and therefore to obtain resonance the frequency of the applied field must be very accurately adjusted to correspond with the natural frequency of the plate. It is for this reason that quartz crystals are used as standards of frequency and for stabilizers of oscillating circuits and transmitters.

A crystal plate used as a transducer is coated on two sides with a thin metal layer to act as electrodes. The metals used are usually gold, silver, or aluminium deposited by vacuum evaporation or sputtering, with a wire soldered onto each side. For use in liquids the transducer may be directly immersed, providing the liquid is electrically non-conducting. To make a conducting liquid vibrate ultrasonically, the transducer is immersed in an oil bath and the liquid, in a beaker, stood in the oil bath. The ultrasonic wave then travels through the oil to any immersed object. Alternatively, the transducer may be clamped to the outside of the bath.

For solids, the transducer may be applied directly to the side of the object, with some liquid, usually oil, acting as a couplant to transmit the vibrations from the transducer (Fig. 17-7). As well as the main vibrations in the $x$-direction, an X-cut plate produces weaker vibrations in the $y$- and $z$-directions which cause weak waves to travel along the surface of the solid. These weak waves may produce troublesome interference with the effects produced by the main longitudinal wave.

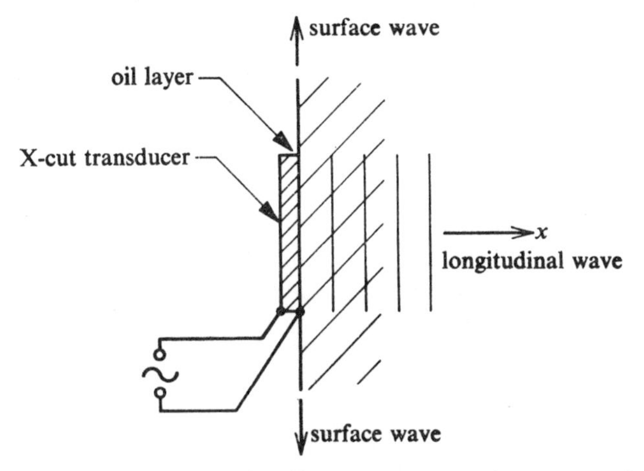


Fig. 17-7 Coupling of an X-cut quartz transducer to a solid

## 17-8 Piezoelectric materials

Quartz has proved to be the most generally suitable piezoelectric transducer for the following reasons:

1. It is hard and therefore may be well polished. A polished surface may easily be cleaned; a clean surface is essential for a deposited metal film to be adherent. Also, quartz can be worked very accurately to a parallel slab of known thickness, which is not easily changed in subsequent use by wear.
2. It is transparent. This allows the crystallographic axes to be determined easily by using polarized light, and means that the parallelism between the surfaces can be checked by interference.
3. It is not porous and therefore can easily be coated with metal in a vacuum chamber.
4. It is resistant to chemical attack and is suitable for immersion in most liquids.
5. It has a low coefficient of thermal expansion and therefore its natural frequency of vibration changes only slightly with temperature.

For example, a change in temperature of 1degC causes the natural frequency of an X-cut quartz plate to change by about 1 Hz in 70 kHz.

6. It is a very good insulator, having a specific resistance of $1{\cdot}2 \times 10^{12}$ ohm-metre, enabling a high potential difference to be applied across the electrodes.

Quartz has one main disadvantage, however; it is brittle. Thus, a very high frequency transducer, being correspondingly very thin—for example, a 50-MHz plate is only 0·057 mm thick—is very delicate. It may be broken not only in mechanical handling but also by too high an applied potential. Thus, for frequencies above about 15 MHz it is usual to use the harmonics produced in plates of lower natural frequencies. Although this is acoustically less efficient, the loss is partially offset by the possible increase in the magnitude of the applied potential across the plate.

A number of other piezoelectric crystals are used, but generally only for special purposes where they have an advantage over quartz. For example, Rochelle salt ($KNaC_4H_4O_6 . 4H_2O$), which has a high piezoelectric activity but is hygroscopic and melts at 56°C, and ADP (ammonium dihydrophosphate $NH_4H_2PO_4$), which has the disadvantage of becoming electrically conductive at temperatures above about 100°C, are both commonly used as underwater transducers. Both are water soluble and must therefore be mounted in oil-filled cells and also, due to their temperature characteristics, little power can be obtained from them without overheating. They are therefore normally used only for pulse operation.

### 17-9 Power supplies

The most suitable high frequency generators for driving the transducers are conventional valve oscillators, which are easily constructed to drive the transducers at any required frequency and intensity. Some oscillators are constructed to give a variable frequency that may be tuned to the resonant frequency of the crystal plate; others are constructed to operate at one fixed frequency, regulated by a second vibrating quartz crystal tuned to that resonant frequency. To the oscillator, the transducer appears as a capacitor.

## ELECTROSTRICTIVE TRANSDUCERS

### 17-10 Electrostrictive effect

All dielectrics exhibit an electrostrictive effect but for most the effect is negligible. However, in a few dielectrics, termed *ferroelectrics*, it is quite large. The effect is analogous to magnetostriction but, instead of a magnetic field in a given direction, an electric field is applied which produces a distortion in the dielectric proportional to the *square* of the applied field strength. Like the magnetostriction effect, the distortion is independent of the sense of the applied field. Thus the frequency of vibration is equal to twice the frequency of the alternating applied electric field. A vibration, having the same frequency as that of the applied frequency and in the same sense, may be obtained if the dielectric is first polarized as with the magnetostrictive transducers. The dielectric is first heated to a temperature above the Curie point (which is about 120°C for barium titanate); at this temperature, it loses its electrostriction properties. If a d.c. electric field is now applied in the direction in which the dielectric is subsequently required to vibrate and the dielectric is allowed to cool slowly, it behaves as a piezoelectric crystal. An alternating electric field, small in comparison with the original polarizing field and applied in the same direction, causes the dielectric to vibrate at the same frequency and with a magnitude substantially *proportional to* the applied voltage. In other words, it is now behaving exactly as a piezoelectric transducer, and for this reason polarized ferroelectric transducers are commonly referred to as being piezoelectric. The large d.c. polarizing field has the effect of approximately lining up the domains, while the smaller a.c. field has the effect, not of reversing the domains, but of lining them up to a greater or lesser extent as it adds to or subtracts from the polarizing field.

As for a piezoelectric crystal, the change in thickness is given by $x = Vd$, where $x$ is the deformation or change in thickness in metres, measured in the direction of the polarizing field; $V$ is the total volts causing the field in the direction of the deformation; and $d$ is the piezoelectric modulus, 60–190 $\times$ $10^{-12}$ m/V for barium titanate. Thus, 100 V applied across a barium titanate transducer produces a deformation of about 100 $\times$ $10^{-10}$ m —that is, about 100 Å—which is about 50 times greater than that produced in quartz by the same voltage.

As with the transducers previously described, the thickness of the ferroelectric is adjusted to give resonance with the exciting field. Barium titanate has a density of 5·6 $\times$ $10^3$ kg/m$^3$ and a Young's modulus of about 100 $\times$ $10^9$ N/m$^2$, giving the natural frequency of vibration for a parallel plate

$$f = \frac{2{,}110}{t} \text{ Hz} \tag{17-5}$$

where $t$ is the thickness in metres.

## 17-11 Electrostrictive transducers and materials

Barium titanate ($BaTiO_3$) has the disadvantage that its low Curie temperature of about 120°C and its low-temperature Curie point of $-5$°C limits its working temperatures to between these values. Other compounds have been devised—for example, replacing 5% by weight of the barium titanate with calcium titanate ($CaTiO_3$) extends the working range downwards, by shifting the low-temperature Curie point to $-40$°C. The further addition of about 10% lead titanate ($PbTiO_3$) further extends the range, from about $-80$°C to an upper Curie point of about 130°C. Mixtures of lead titanate and lead zirconate ($PbZrO_3$) have also been made which raise the upper Curie point to above 300°C.

To form these materials into transducers, a mixture of the ground ingredients is moulded to shape and fired in a furnace at about 1,200 to 1,400°C, to produce a hard ceramic. After cooling, electrodes are attached by painting on silver paint and refiring at some lower temperature. These ceramics are electrostrictive materials and must be polarized by heating in a d.c. field to above the Curie temperature and allowing to cool slowly to develop their piezoelectric properties. The addition of lead titanate to the normal barium titanate has the effect of causing greater retention of the remnant polarization.

Ceramic transducers have a number of advantages over quartz:

1. A ceramic material may easily be formed into any shape. For example, there are tubular transducers and concave transducers which have focusing properties, shapes which are impossible to produce from single crystals.
2. Being polycrystalline, they have no crystallographic axis. The direction of vibration is produced by the d.c. polarizing field, after the shape is formed.
3. The characteristics of the transducer can be controlled by a suitable choice of the component mixture.

These transducers are, therefore, versatile and cheap to manufacture as compared with quartz. Together with their high piezoelectric modulus and high dielectric constants (about 1,200 for barium titanate ceramic), which eases the design requirements of the power supply, these factors have caused them to replace quartz completely for many purposes. They do have some disadvantages, however—for example, their high internal heat dissipation together with their comparatively low Curie temperature limits their power output without adequate cooling, and the manufacturing process means that they are not exactly reproducible.

The power supplies for these ceramic transducers are very similar to those required for the natural piezoelectric crystals, except that the design is somewhat simplified by the lower voltage requirements due to the high piezoelectric modulus, and the consequently reduced problems of insulation. Also, the frequency need not be maintained as precisely as is necessary for quartz.

## INDUSTRIAL USES OF ULTRASONICS

Ultrasonics have many and varied uses, but only those of special interest to engineers are given here.

### 17-12 Ultrasonic flaw detectors

X-ray and gamma-ray methods are used for the inspection of castings and metal objects. However, they have the disadvantage that both sides of the test piece must be accessible, which is not always practicable, and the object must not be too thick otherwise the rays are unable to penetrate.

Moreover, the equipment is bulky and there is always the possibility of a safety hazard.

Ultrasonic methods are quicker and have none of these disadvantages. An electronic pulse fed into the transducer in contact with the test piece causes an ultrasonic pulse to be propagated into the test piece. To ensure good mechanical contact, it is necessary to have a layer of liquid, usually oil or soapy water, between the transducer and the test piece. The ultrasonic wave travels on until it meets a discontinuity, which may be the opposite side or any flaws—for example, cavities or inclusions. The discontinuity reflects the pulses back so that they induce electric pulses by

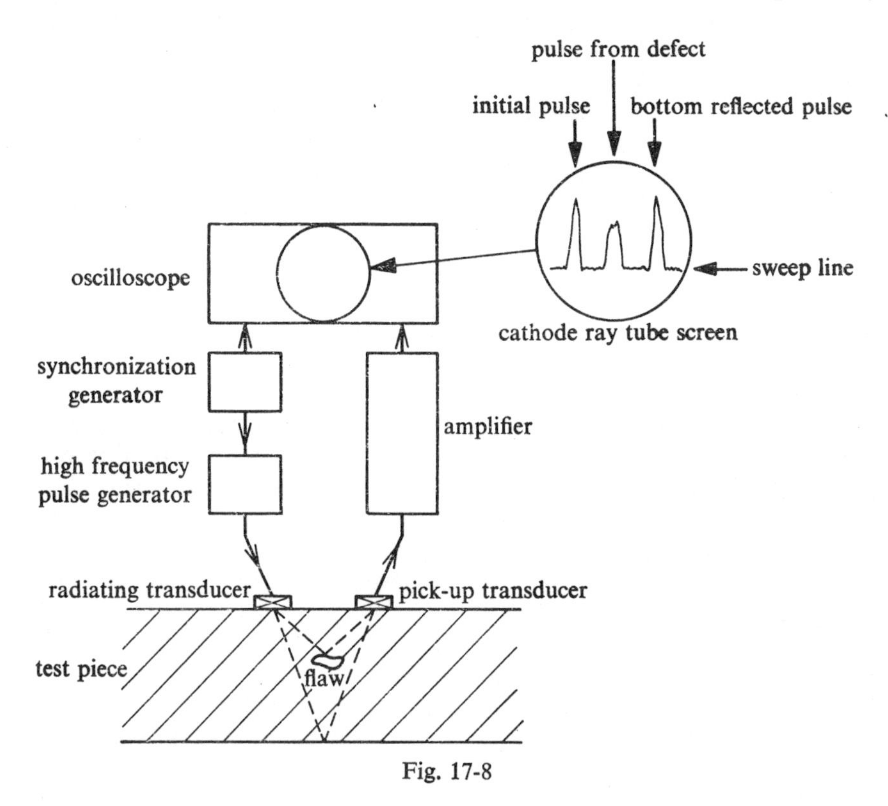


Fig. 17-8

mechanically distorting the transducer, which may be the original one or one close by. The time taken for the pulses to return is a measure of the distance of the discontinuities from the transducer. The pulses are displayed on an oscilloscope screen (Fig. 17-8), where the relative positions of the pulse from the defect and the pulses from the top and bottom indicate the position of the defect within the test piece. The form of the reflected pulse indicates the size and nature of the defect. By moving the probes along the surface, a picture of the internal state of the specimen may be built up.

The method is quite straightforward for test pieces having parallel sides as in the Fig. 17-8 and with a defect lying roughly parallel to the surfaces. Detection is more difficult where the defect is normal to the surfaces or close to one of them, or if two defects are overlapping in different levels. Although the method is basically simple, it in fact requires considerable skill and experience on the part of the operator to distinguish pulses due to true defects from those caused by stray internal reflections, especially if the test piece has a complicated shape.

For defects not parallel to the surface and for welds, it is usually better to use an angled probe. This is a transducer mounted on a wedge-shaped block of perspex, which is then applied to the test piece with a layer of liquid in the usual way.

For high resolution, the ultrasonic wavelength should be of the same order as, or smaller than, the dimensions of the defect under investigation and therefore high frequencies are required. However, with too high a frequency and a specimen that has natural porosity, for example, there can be considerable scattering which results in 'noise' on the oscilloscope screen and masks any genuine flaws.

When the surface is very rough, it is difficult to get good contact between the transducer and the specimen, even with a liquid couplant, and there may be a considerable loss of energy. Similar difficulties may be experienced if the surface is covered with scale, because its non-uniform surface may lead to multiple reflections and interference effects.

## 17-13 Thickness gauges

The flaw detection equipment can also be used for thickness gauging. If the sweep time of the oscilloscope is known and the separation between the pulses caused by reflections from the top and bottom surfaces is measured, the thickness of the test piece may be calculated from a knowledge of the velocity of sound in the solid.

Better results are obtained, however, with a variable delay line. This usually consists of a nickel rod (Fig. 17-9) in which an electric pulse applied to the input coil causes a magnetostrictive pulse in the nickel. The ultrasonic wave formed travels along the nickel rod and in turn induces an electric signal in the output coil. In practice, pulses are sent simultaneously into the sample and the delay line and the two pulses, from the other end of the delay line and from the reflection at the other side of the specimen, are displayed on the oscilloscope screen. The length

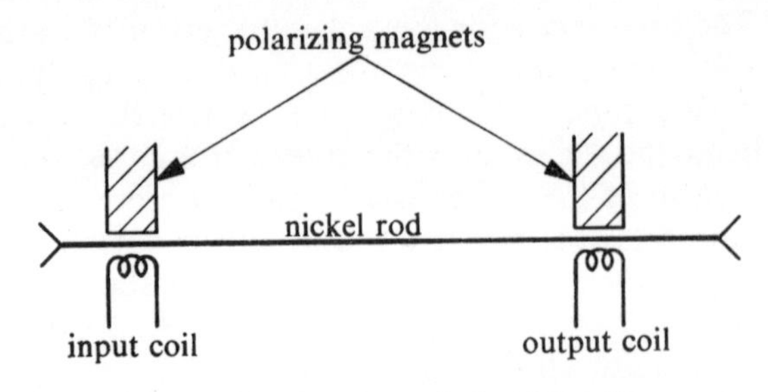


Fig. 17-9 Magnetostrictive delay line

of the delay line is then adjusted by moving the position of the output coil by means of a micrometer device until the two displayed pulses coincide. Then the thickness of the specimen is gauged from the previously calibrated length of the delay line. Unwanted reflections occurring from the ends of the delay line are usually damped out by covering the ends with grease.

Ultrasonic methods of gauging thickness have the advantage over other methods in that access to only one side of the specimen is required. The method is especially suitable, for example, for gauging the wall thickness of tubes, the thickness of hulls of ships from the inside, and even for the thickness of fat on bodies of living animals, as in the selection of pigs for bacon.

## 17-14 Ultrasonic cleaning

Articles to be cleaned are placed in a tank of liquid—for example, trichlorethylene for degreasing—with an ultrasonic transducer mounted in the bottom. This imparts high accelerations within the cleaning fluid which, coupled with the cavitation produced, causes an eroding action, and the emulsification of oily matter, producing a very effective cleaning action. Greater cavitation is produced at lower frequencies but the action is then localized to the region immediately adjacent to the transducer. On the other hand, high frequencies give better directional properties which are better for cleaning within recesses and cavities. Thus a combination of the effects is required, which is possible by using focusing barium titanate transducers.

## 17-15 Ultrasonic drilling

A suitable tool bit is fixed to the end of a transducer, which imparts an oscillatory motion to the tool, enabling it to chip its way through the work piece. The action is most effective if the article to be drilled or cut is hard and brittle. The cutting speed is increased if the cutting face is fed with an abrasive paste. The tool should be tough rather than hard to enable the abrasive particles to be embedded in the end. It is the abrasive that does the cutting, the tool merely acts as a guide. Usually, magnetostrictive transducers are used but, since the amplitude is small, 'velocity transformers' are used with them. These consist of stubs fitted to the transducer face, with a tapered section analogous to the acoustic horn used with loudspeakers, and with the cutting tool fixed to the small area end (Fig. 17-10). This increases the amplitude of oscillation of the tool. Various shapes of velocity transformer are possible, from exponential to stepped, but in all cases it is necessary that its natural frequency of vibration should be equal to or a whole number multiple of the natural frequency of the transducer. Usually the length is chosen to be equal to half the wavelength of the sound radiated by the transducer. In use, only light pressure is required, otherwise the motion of the tool is damped out. The cutting abrasive should be harder than the article being cut, although diamond may be cut using diamond dust.

Since the cutting action is oscillatory, not rotary, the holes cut need not be round and in fact holes with very complicated shapes may be drilled. Curved holes may be cut with a curved tool stub. Threads may be cut in carbides by first cutting a hole with the same diameter as the tapping drill and then using a tool in the form of a threaded rod, with the work piece being raised in a spiral motion to the cutting tool. Ultrasonic tools are used to make dies from sintered carbide materials and in the drilling of synthetic rubies for bearings.

The cutting action is best at low frequencies, since higher amplitudes may be obtained at the drill tip, but for operator comfort it is usual to operate at a frequency just above the audible limit, about 20–25 kHz.

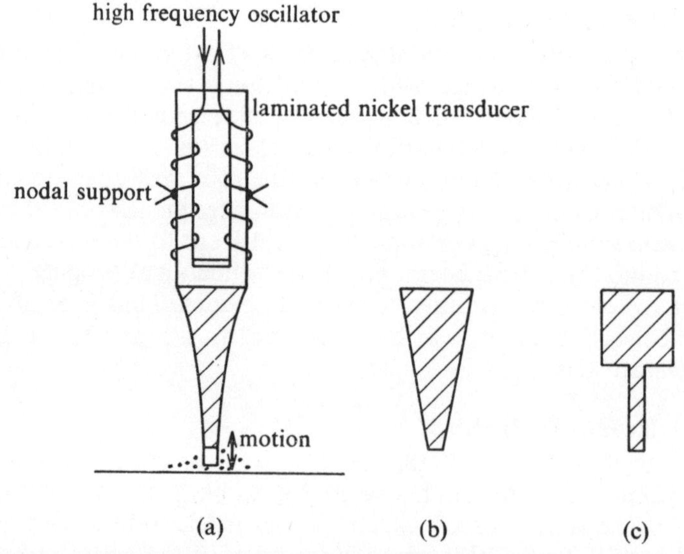


Fig. 17-10 (a) Ultrasonic drill with exponential velocity transformer, and (b) conical and (c) stepped velocity transformers

## 17-16 Other uses

Aluminium can be soldered without difficulty using an ultrasonic soldering iron. This consists of a stub, heated by a conventional thermal element, fitted to a magnetostrictive transducer. The soldering iron melts solder onto the aluminium, the ultrasonic action then causes cavitation in the molten solder which breaks up the aluminium oxide layer, and the solder can then wet the clean metal.

Thermoplastic components may be heat welded together in a simple manner. One plastic component is firmly clamped to the transducer and then brought up to the other component. The first component is then made to vibrate ultrasonically so that the contact surfaces rub together; the surface temperature rises due to friction and the surfaces melt and weld. The joint formed is neat and clean, and without the surplus glue smears involved with the conventional method. The temperature of the contact surface rises about 150 degC per second.

Many other uses have been found for ultrasonic waves; for example, the agglomeration of suspended particles in gases and liquids, which then precipitate due to their increased size; the emulsification of immiscible liquids; the dispersion of solid matter in liquids; the hardening of metals by the production of small grain sizes; and the settling of liquid concrete.

## PROBLEMS

**17-1** Write brief notes on (a) magnetostrictive, (b) piezoelectric, and (c) electrostrictive transducers and contrast their uses.

**17-2** A 3-mm thick quartz disc vibrates in its fundamental thickness mode with a frequency of $10^6$ Hz. If the density of quartz is $2{\cdot}65 \times 10^3$ kg/m$^3$, what is the modulus of elasticity for thickness changes in this disc? [$95{\cdot}4 \times 10^9$ N/m$^2$]

**17-3** Give an account of the use of ultrasonics in industry.

# Bibliography

*Light*

Ditchburn, R. W. *Light*. Blackie, London, 1963.

Jenkins, Francis A. and Harvey E. White. *Fundamentals of optics*, 3rd edn. McGraw-Hill, London, 1957.

Kingslake, R. *Applied optics and optical engineering* (3 vols). Academic Press, London, 1965.

Longhurst, R. S. *Geometrical and physical optics*. Longmans, Green, London, 1963.

Welford, W. T. *Geometrical optics*. North Holland Publishing Co., Amsterdam, 1962.

*Sound*

Blitz, J. *Elements of acoustics*. Butterworths, London, 1964.

Blitz, J. *Fundamentals of ultrasonics*. Butterworths, London, 1963.

Richardson, E. G. and E. Meyer (Eds). *Technical aspects of sound* (3 vols). Elsevier Publishing Co., Amsterdam, 1953–62.

Stephens, R. W. B. and A. E. Bate. *Acoustics and vibrational physics*, 2nd edn. Edward Arnold, London, 1965.

Wood, A. *Acoustics*. Blackie & Sons, Glasgow, 1963.

Wood, A. B. *A textbook of sound*. Bell & Sons, London, 1960.

# Index